Robert Stawell Ball

Elements of Astronomy

Robert Stawell Ball

Elements of Astronomy

ISBN/EAN: 9783337276621

Printed in Europe, USA, Canada, Australia, Japan

Cover: Foto ©berggeist007 / pixelio.de

More available books at **www.hansebooks.com**

ELEMENTS

OF

ASTRONOMY

BY

SIR ROBERT STAWELL BALL, LL.D., F.R.S.

ANDREWS PROFESSOR OF ASTRONOMY IN THE UNIVERSITY
OF DUBLIN : ROYAL ASTRONOMER OF IRELAND

NEW EDITION

LONDON

LONGMANS, GREEN, AND CO.

AND NEW YORK : 15 EAST 16th STREET

1889

PREFACE.

———◦◦◦———

THE READER of this volume is expected to possess such
knowledge of Mathematics as may be gained by studying
the Elements of Euclid and Algebra, together with a rudi-
mentary acquaintance with the geometry of Planes and
Spheres.

While the book is mainly intended for beginners, yet it
is believed that more advanced students may find portions
of it useful. This will perhaps be especially the case in the
last chapter, where a somewhat extended account is given
of the important constants of Astronomy, with references to
the original sources of information.

Many of the illustrations have been taken from Delaunay's 'Cours Elémentaire d'Astronomie,' and a few from
Secchi's 'Le Soleil,' from Guillemin's 'Le Ciel,' and other
sources. In the last chapter extensive use has been made
of Houzeau's 'Répertoire des Constantes Astronomiques.'

ROBERT S. BALL.

DUNSINK : *June* 4, 1880.

CONTENTS.

———◦◦◦———

CHAPTER I.

ON THE INSTRUMENTS USED IN ASTRONOMICAL OBSERVATIONS.

CHAPTER II.

THE EARTH.

CHAPTER III.

THE DIURNAL MOTION OF THE HEAVENS.

CHAPTER IV.

THE SUN.

CHAPTER VII.

THE PLANETS.

CHAPTER VIII.

COMETS AND METEORS.

CHAPTER IX.

UNIVERSAL GRAVITATION.

CHAPTER X.

STARS AND NEBULÆ.

CHAPTER XI.

THE STRUCTURE OF THE SUN.

CHAPTER XII.

ASTRONOMICAL CONSTANTS.

ASTRONOMY.

——◦◦◦——

CHAPTER I.

ON THE INSTRUMENTS USED IN ASTRONOMICAL OBSERVATIONS

§ 1. *Introduction.*—The sun, the moon, and the other objects which shine in the heavens on a clear night constitute what are known as the *celestial bodies*. To study the celestial bodies is to pursue the branch of science which is called *astronomy*. This includes all that we can learn of the motions of the celestial bodies, of their actual dimensions and distances, and of their physical constitution.

§ 2. *Astronomical Instruments.*—The instruments which are employed in making astronomical observations may be conveniently divided into three classes. Instruments of the first class are for the purpose of *extending our powers of vision*, so as to enable us to use our eyes to greater advantage than would be possible without such assistance. By instruments of this class we are enabled to view the features of the celestial bodies as if they were much nearer to us than they really are. We can thus become acquainted to a certain extent with the shapes and appearance of many of the celestial bodies which would otherwise be very imperfectly known. By the help of these instruments we are also enabled to perceive large numbers of celestial bodies which

B

are not sufficiently conspicuous to be detected by the unaided eye. The instruments of this class are termed *telescopes*.

We shall subsequently see that the position of a celestial body on the surface of the heavens is suitably expressed by certain angular measurements, and therefore the means by which these measurements can be made with accuracy must be explained. The various appliances which are used in making such *measurements of angles* form the second class of instruments.

The celestial bodies are in constant motion, real or apparent. In order to study these motions it is not only necessary to ascertain the different positions which the bodies assume, but it is also essential to have the means of *measuring with accuracy the intervals of time* which elapse during which the celestial body is passing from one position to the positions which it subsequently occupies. We must therefore be provided with instruments for the purpose of measuring time. Clocks or chronometers, therefore, constitute the third class of instruments.

§ 3. *Angular Magnitude.*—It will first be necessary for us to consider briefly the circumstances of ordinary vision with the unaided eye. We shall then be able to comprehend what is the nature of the assistance which is afforded when a telescope is interposed between the eye and a distant object.

FIG. 1.

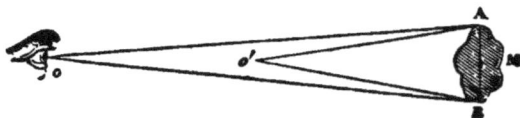

Let A B (Fig. 1) be an object viewed by an eye placed at the position *o*. Then the line A B which connects two points of the object is seen to subtend an angle A *o* B at the point *o*. If the eye be moved to the point *o'*, which is at one-half the distance from M, the line A B will subtend an

angle A o' B. It is obvious that the angle A o' B is greater
than the angle A o B; and if the line A B be very small in
comparison with its distance from o, the angle A o' B will be
approximately double the angle A o B. It thus appears that
when the distance of an object is halved the angle under
which the object is seen is doubled. In the same way it
can be shown that if the distance of the object be reduced
to one-third, the angle which it subtends will be increased
threefold, or more generally that if the distance be diminished in any ratio the angle subtended at the eye will be
increased in the same ratio. It will be convenient to speak
of the angles A o B, A o' B as the apparent magnitudes of the
line A B, and we are thus led to the following law :—

*The apparent magnitude of one of the dimensions of an
object varies inversely as the distance of the object from the eye.*

Since the areas of similar figures are proportional to the
squares of their linear dimensions, it is obvious that the apparent area of an object is proportional to the square of the
apparent magnitude of one of its dimensions, and hence we
are led to the result that—

*The apparent area of a distant object varies inversely as
the square of the distance of the object from the eye.*

. § 4. *Brightness of a Distant Object.*—A luminous object,
whether it shines by its own light or by the light from some
other body reflected
from it, radiates light
in all different direc-
tions. Let us consider
a certain point m of the
body M (Fig. 2), and en-
quire how the apparent
brightness of this point
depends upon the dis-
tance at which the eye
is placed. From the
point m rays of light diverge in all directions, and of these

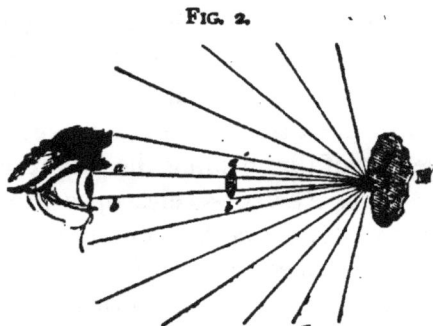

FIG. 2.

rays a certain number enter the pupil *a b* of the eye. The rays that enter the pupil will all be contained in the cone of which *m* is the vertex, while the base of the cone is the area of the pupil. If, however, the eye be moved to a distance which is one-half its former distance, so that the pupil occupies the position *a' b'*, then the rays which enter the eye will be those contained in the cone, which has the pupil at *a' b'* for its base and the point *m* as its vertex. It is obvious that the angle *a m b*, which the pupil subtends at *m*, is half the angle *a' m b'*, which is subtended when the pupil moves into the new position. It therefore follows that the apparent area of the pupil *a' b'* subtended at the point *m* is four times as large as it is when the pupil has the position *a b*. The quantity of light received by the pupil from a given point of the body must therefore vary inversely as the square of the distance. It does not, however, follow that the brightness of the object increases as the eye approaches thereto ; for although the quantity of light received from a small spot of the body increases as the eye approaches the body, yet the apparent area of the spot increases in the same ratio. The quantity of light received is, therefore, always proportional to the apparent illuminated area, so that the intrinsic brightness of the object is always the same whatever be the distance from which it is viewed.

Some modification of this statement would be required when, as in the case of distant terrestrial objects, a great thickness of air has to be traversed by the rays. In this case the absorption of the light by the atmosphere has the effect of diminishing the brightness of a distant object.

§ 5. *Lenses.*—Telescopes are usually made by inserting into a tube a certain number of specially shaped pieces of glass, termed *lenses*. It will be necessary for us to describe some of the different forms of lenses and their properties, in order to understand the combination of lenses which forms a telescope.

If a converging lens D (Fig. 3) receive a pencil of rays diverging from a point A, the rays will, after refraction, converge towards a certain fixed point *a*. A converging lens

FIG. 3.

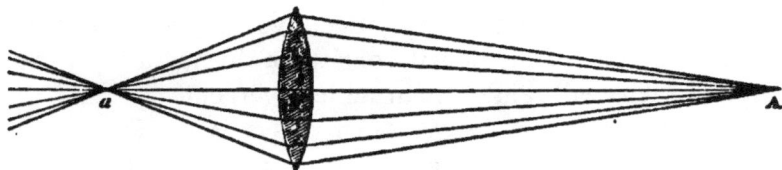

will act in the same way upon a pencil of rays which diverge from a point A, not situated exactly in the axis of symmetry of the lens, and will come to a point of intersection *a*. Let

FIG. 4.

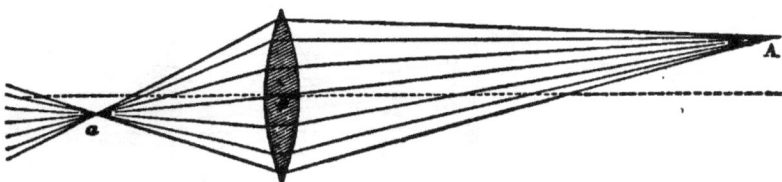

us suppose that the point A recedes farther and farther from the lens, then the point *a* will draw in closer and closer towards the lens. When the point A has withdrawn to such

FIG. 5.

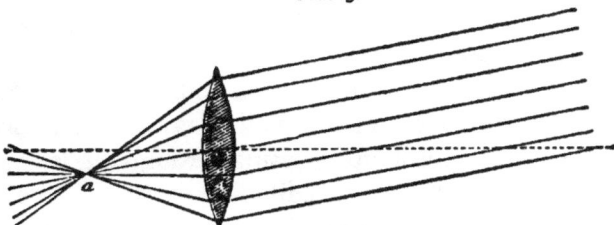

a distance that the rays which diverge from it have become sensibly parallel, the point *a* will have moved in to a certain distance *o a*, which is called the *focal length* of the lens (Fig. 5).

The influence of diverging lenses (Fig. 6) upon a pencil of light is opposite to that of converging lenses. If a convergent

FIG. 6.

pencil of rays falls upon a diverging lens it renders them less convergent, or even parallel or divergent. If the pencil consists of parallel rays, the diverging lens renders them divergent; if the pencil consists of divergent rays, the diverging lens will render them more divergent.

§ 6. *Formation of Images.*—An object A B (Fig. 7) is placed in front of a converging lens, at a distance from that lens which is greater than its focal length.

This object is supposed either to be self-luminous or to be illuminated by the rays from some source of light. From each point of the object a pencil of rays will diverge; this

FIG. 7.

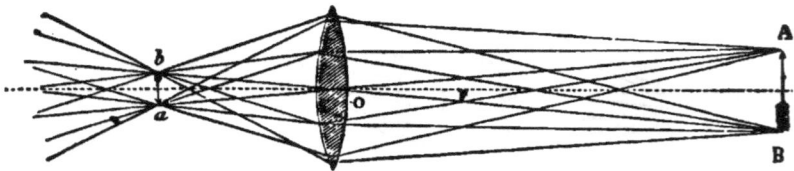

pencil will fall upon the converging lens, and after passing through the lens the rays will converge to meet at a point on the other side of the lens. Thus from the point A of the image a pencil of rays will diverge, which, after passing through the lens, converge to the point *a*. In a similar manner the pencil of rays proceeding from the point B will converge to the point *b*.

It therefore appears that each point of the original object will be the source of a pencil of rays, which converge to a corresponding point after passing through the lens. The figure produced in this way is called an *image*. If the object A B be at an exceedingly great distance, then the image will be formed at the principal focus of the lens.

When we wish to observe the details of a small object it is usual to employ a single lens for the purpose, which is held between the object and the eye. In this way the eye can be brought much closer to the object than would be possible for distinct vision without the assistance which the lens affords. The structure of the eye is such that distinct vision is only possible when the divergence of the rays which enter the pupil does not exceed a certain amount. If the eye be brought too close to an object for distinct vision, then the divergence of the rays from that object is too large. By placing a simple converging lens between the eye and the object the divergence of the rays is diminished, and thus distinct vision is rendered possible.

This effect of a lens can be illustrated by a figure. Let A B be a small object which is to be examined, and let it be

FIG. 8.

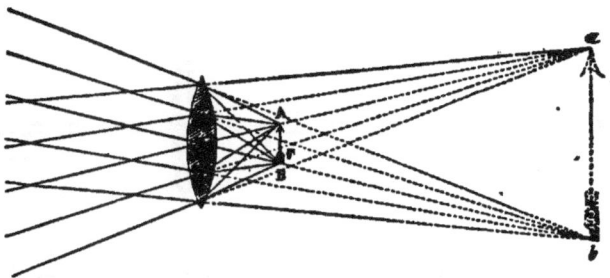

placed *between* the lens and its focus F, the eye of course being situated at the other side of the lens. The rays which diverge from the point A will, after passing through the lens, have their divergency diminished; they will therefore appear as if they diverged from a point *a* more remote from the lens, and situated upon the line joining the centre of the lens to the point A. In a similar way the rays from the point B will appear to come from the point *b*, and thus the whole object A B will appear of the same dimensions as its image *a b*. The distance of the image from the eye will depend upon the distance of the object from the lens, and the latter may be adjusted so that the image shall be situated at the distance of distinct vision. The eye is thus able to see distinctly the details of the image, because the image is larger than the object would appear if it were placed at the distance of distinct vision and viewed without the aid of the lens.

§ 7. *The Telescope.*—The telescope in its most simple form consists of a combination of two lenses, one of which forms an image of the distant object while the other magnifies that image.

Let L be a lens, called the object lens, which receives rays coming from a distant object. An image *a b* is formed of this object where *o* A and *o* B are directed from the centre of the lens to the extremities of the distant object. The image *a b* is then magnified by the lens into the image *a' b'.* The lens L' is termed the eye piece. The distance at which

FIG. 9.

the image *a b* is situated from the object lens depends upon the distance of the object itself from the object glass. When the latter is so great that the rays from a point may practically be considered to form a parallel beam, then the distance from *a b* to the object glass is equal to the focal length of the object glass. This is of course the case when the telescope is used for astronomical observations. The distance at which the lens L' is placed from the image *a b* depends upon the eye of the observer, because the image *a' b'* must be placed at the distance of distinct vision, which is different in different persons.

We can now ascertain the effect which the telescope produces in magnifying a distant object. The angle which the object subtends at the object glass is equal to A *o* B.

Since the length of the telescope may be regarded as quite inappreciable when compared with the distance of the object, it follows that A *o* B is also equal to the angle which the object would subtend if viewed by the unaided eye. But when we view the object through the telescope we see, instead of the object, the image *a' b'*, and this image subtends an angle *b o' a*. It therefore appears that the telescope augments the angle at which an object is seen in the ratio of the angle *b o' a* to the angle *b o a*. Since these angles are small we shall be approximately correct in assuming that the angle *b o a* bears to the angle *b o' a* the same ratio as the distance from *b a* to the eye lens bears to the distance from *b a* to the object lens. The distance from *b a* to the eye lens is very nearly the same thing as the focal length of the eye lens, because the rays, on emerging from the eye lens, have to be nearly parallel in order to have them in a suitable form for distinct vision. We have also seen that the distance from *b a* to the object lens is equal to the focal length of the object lens, and hence we deduce the important result which is thus stated:—

The magnifying power of a telescope is equal to the ratio of the focal length of the object glass to the focal length of the eye piece.

An important practical consequence follows from this law. We are enabled, by simply changing the eye piece, to augment indefinitely the magnifying power of the telescope. The advantage of increasing the magnifying power beyond a certain point is, however, more than neutralised by the fact that the brilliancy of the object is diminished.

Let us next consider the effect of the telescope upon the brilliancy with which the object is seen. From each point of the object a pencil of rays diverge, which may be considered to be parallel. Thus the entire surface of the object glass receives rays from each point of the distant object. These rays, after passing through the object glass, converge to a point *m* in the image *a b* (Fig. 10). Diverging there, they fall upon the eye lens, and thence into the eye of the observer. If the eye lens had not been used, the eye of the observer would have to be placed at the distance of distinct vision, about 8 or 10 inches from *m*. The beam would

FIG. 10.

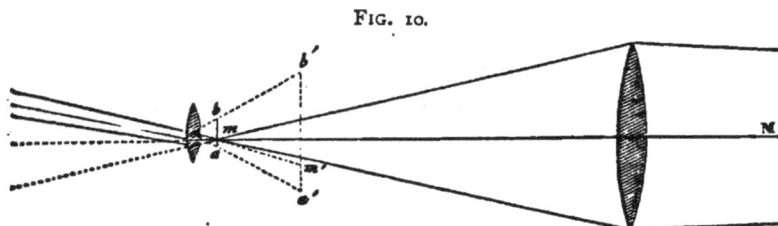

therefore have attained a considerable degree of divergence, and only a portion of the rays would have been able to have entered the pupil. A portion, therefore, of the light grasped by the object glass would have been wasted. By the intervention of the eye piece the eye is enabled to be brought much closer to the point whence rays are diverging, and thus the beam is intercepted by the eye while the section of the diverging cone is still smaller than the pupil. In this way the eye piece enables us to avail ourselves of all the light grasped by the object glass.

When all the light enters the pupil, then the quantity of light which illuminates the point *m* of the image is equal to the quantity of light from that point which fell upon the object glass. It is therefore proportional to the area of the object glass. It should be observed, however, that in what has been said we have overlooked the loss of light which occurs in consequence of its absorption by the glass both in the object glass and the eye piece, and also the loss which takes place by reflection at the different surfaces ; the effect of these losses is to render the quantity of light received by the eye from a point of the object somewhat less than it would have been had the object glass and the eye piece been perfectly transparent.

We have, then, seen that the magnifying power of a telescope depends entirely upon the ratio of the focal lengths of the object glass and the eye piece, while the quantity of light transmitted from each point of the object to the eye depends upon the aperture of the object glass.

It will now be obvious that the quantity of light with which the image in a telescope is illuminated depends upon the *size* of the object glass. If, therefore, the magnifying power of the telescope be increased while the aperture of the object glass remains the same, the intrinsic brilliancy of the image is diminished, for the same quantity of light has now to be spread over a larger area.

The same considerations will also explain to us how it is that a telescope can render visible an object which without such aid would be invisible. Take, for example, a telescopic star. The rays from such a star, falling upon the pupil, fail to produce an impression upon the retina, because the aperture of the pupil is so small that a sufficiency of rays to produce the required impression cannot enter the eye. When the aid of a telescope is called in, all the rays which fall upon the surface of the object glass are so modified that they can enter the eye. In fact, if we may use the illustration, the telescope acts as a sort of funnel, and pours a large quantity of the rays in through the small aperture of the pupil. Omitting the sources of loss already referred to, we may say that the quantity of light which enters the eye with the aid of the telescope bears to the quantity which enters without that aid the same proportion which the area of the object glass bears to the area of the pupil.

§ 8. *Achromatic Object Glasses.*—A beam of ordinary white light consists of a number of rays of different coloured lights mingled together. When a beam of white light is refracted through a prism, the different coloured lights are separated, and what are called prismatic colours are observed. It will be found that the bluish rays are more bent than the red, while the greenish tints occupy an intermediate position. When a beam of white light falls upon a converging lens, the

blue rays being more bent than the red, the focal length of
the lens for the blue rays is somewhat shorter than it is for
the red. The image of an object which emitted only blue
rays would therefore be formed somewhat nearer to the
object glass than the image of an object which emitted
only red rays. The light from the different celestial bodies
consists generally of a mixture of several different colours,
and it follows that a number of images will be produced at
the foci corresponding to the different colours. It is true that
these foci are all near together, but the effect is still sufficient
to render the image of the object, as seen through the
object glass, somewhat indistinct and fringed with colours.

It was, however, discovered by Dolland in 1758 that the
difficulty arising from this cause could be obviated. He
found that by a suitable combination of two lenses an
object glass could be constructed, in which the focus of the
red rays and that of the blue rays could be made to coincide.
The *achromatic* object glass, as this compound lens is termed,
is shown in Fig. 11. One of these lenses is convergent, and is
made of what is called *crown glass* ;
the other is divergent, and is
formed of *flint glass*. There is a
remarkable difference between the
action of flint glass and crown
glass upon light. If we take a convergent lens of crown glass
and a convergent lens of flint glass, and if these lenses be such
that each of them has the same focal length for the red rays,
then the focal length of the blue rays will be shorter for
the flint lens than for the crown. In fact, the effect of the
flint glass in separating the rays of a compound beam is
more marked than that of the crown. If we take a diverging lens formed of flint glass, and unite it to a converging
lens formed of crown glass, then when a beam of light pours
through the compound lens the red rays are bent by the
converging lens and unbent by the diverging lens ; the blue
rays are also bent by the converging lens and unbent by

FIG. 11.

the diverging lens. But the lenses can be so apportioned that the total effect will be to bend the red rays and the blue rays by the same amount, and therefore to concentrate them all at one focus. The effect of the union of the two lenses is thus to produce what is equivalent to a single lens, which acts equally upon the blue rays and the red rays.

When the blue rays and the red rays have thus been brought to coincide, the intermediate rays of orange, yellow,

FIG. 12.

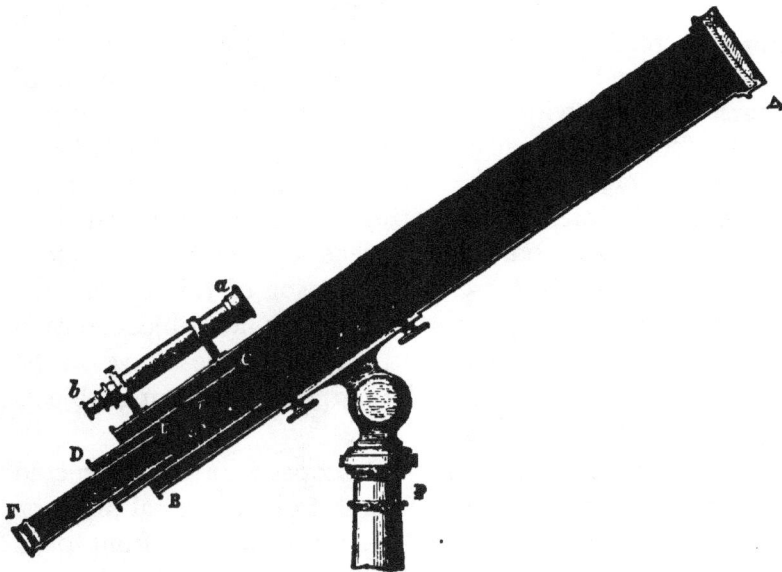

and green may practically be considered to have been brought to the same point. The construction of the achromatic object glass so that the various corrections shall be made with nicety is a most delicate mechanical operation. Until recently good achromatic object glasses above 6 or 8 inches in diameter were very rare, but now object glasses of 24 inches in aperture and upwards have been successfully accomplished. A section of an achromatic telescope is shown in Fig. 12. The achromatic object glass A is fixed at the extremity of a tube, at right angles to its axis. At F is the eye piece, which consists of two lenses fitted into a tube which slides in and out of a tube D C by the aid of a rack and pinion *r.* The eye piece sometimes consists of only a single lens, and this arrangement is undoubtedly preferable when the object under examination is situated near the centre of the field of view. There is, however, so much distortion produced by a single eye lens in those objects which lie near the margin of the field of view that the compound eye piece, consisting of two lenses, is generally preferred. By its aid the same magnifying power is retained while the distortion is greatly diminished. On the other hand, the introduction of a second lens is of course attended with some sacrifice of light.

In focussing the telescope the eye piece is drawn in or out by the rack until the object is seen distinctly. If the telescope be intended for terrestrial objects it is usual to replace the astronomical eye piece by another eye piece which contains four lenses. The reason of this is that the astronomical telescope always exhibits an object turned upside down, and the second pair of lenses have to be added to the eye piece for the purpose of showing objects in their ordinary position.

§ 9. *Reflecting Telescope.*—Telescopes are also constructed by means of the *reflection* of light from spherical mirrors. Suppose a luminous object A B is placed in front of a spherical mirror M (Fig. 13). The rays from the point A form a diverging beam, which, after reflection from the surface of the mirror, form, in accordance with the laws of reflection, a converging beam, which comes to a focus at the point *a.* In a similar way the rays diverging from B are brought to a focus at the point *b.* Thus an image of the entire object A B will be formed at the position *a b.*

There are various different forms in which the principle of the mirror can be employed in the construction of a reflecting telescope. The most simple arrangement, and probably the best, is what is known as the Newtonian telescope (Fig. 14). The rays of light from a distant object A B fall upon a concave mirror M, and tend to form an image of the object at the point *a b*. Before, however, the rays reach the position *a b*, they are intercepted by a small diagonal

FIG. 13.

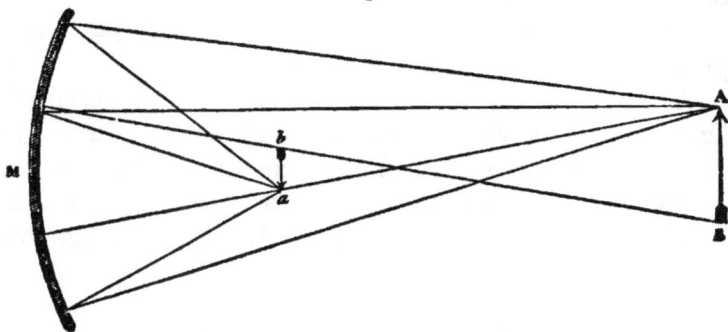

mirror N, which directs the rays to the side of the tube, where they form the image *a' b'*. This image is then viewed by the observer at the side of the telescope with the aid of the eye piece O.

The magnifying power of a Newtonian reflecting telescope is equal to the ratio of the focal length of the large mirror to

FIG. 14.

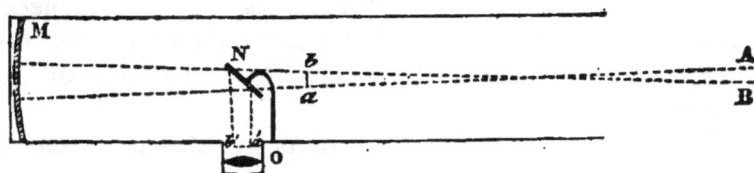

the focal length of the eye piece. The light-grasping power is proportional to the area of the large mirror.

The construction of a large mirror is an exceedingly delicate and difficult operation ; the size is, however, not so limited as that of achromatic object glasses. The largest telescope which has yet been constructed is of the Newtonian form. This magnificent instrument was constructed by the late Earl of Rosse at Parsonstown, and the large mirror is six feet in diameter. The amount of concavity in the mirror is designedly exaggerated in the figure. In the case of the great mirror at Parsonstown the depression at the centre is only about half an inch, and the focal length is about sixty feet. The mirror, in fact, differs but little from a small portion of a sphere of which the radius is about 120 feet.

Another form of reflecting telescope is represented in Fig. 15, and is called the Gregorian telescope. In this case the large mirror M is pierced by a circular aperture, and the small mirror N is also concave, and is situated in the axis of the tube. The rays from the distant object, after reflection at

FIG. 15.

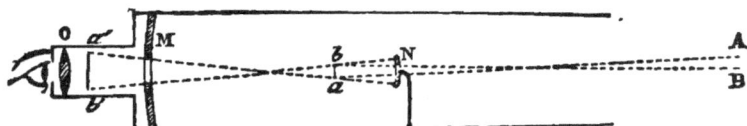

the surface of the large mirror, form an image of the object at *a b*. The rays diverging from this image fall upon the small mirror at N. After reflection from the small mirror they are returned through the aperture in the large mirror, and form another image at *a' b'*. They are then viewed by the eye piece placed at o. In the Cassegrain telescope the small mirror is convex, and is placed at the other side of the focus of the large mirror.

The mirrors were formerly composed of what is called speculum metal, an alloy consisting of two parts of copper and one of tin. Of late years, however, they are usually constructed of silvered glass.

§ **10.** *The Measurement of Angles.*—If a right angle be divided into ninety equal parts, each one of the parts thus obtained is termed a *degree*. If a degree be subdivided into sixty equal parts, each one of these parts is termed a *minute* and if a minute be subdivided into sixty equal parts, each of these parts is termed a *second.*

An angle is, therefore, to be expressed in degrees, minutes, and seconds, and, if necessary, decimal parts of one second. For brevity certain symbols are used : thus 49° 13′ 11″·4 signifies 49 degrees, 13 minutes, 11 seconds, and four-tenths of one second.

We shall now explain what is meant by a graduated circle. Let the circumference of a circle A D B (Fig. 16) be divided into 360 parts of equal length. The division

FIG. 16.

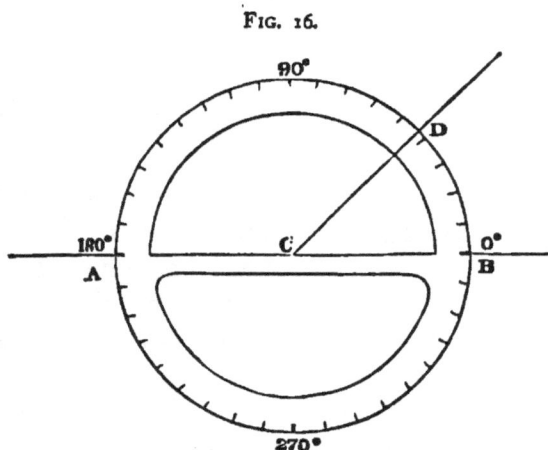

lines separating these parts are denoted by 0°, 1°, 2°, up to 359°. It is usual in small circles to engrave upon the circle only those figures which are appropriate to every tenth division. The actual numbers found on the circle are, therefore, 0°, 10°, 20°, &c. There is, however, no difficulty in finding at a glance the number appropriate to any intermediate division. To facilitate this operation the divisions 5°, 15°, 25°, which are situated half-way between each of the numbered divisions, are sometimes marked with a larger line, so that they can be instantly recognised.

The interval between two consecutive divisions on a

circle is often, for convenience, termed a degree. The reader must, however, carefully remember that the word *degree means an angle and not an arc.* With this caution, no confusion will arise from the occasional use of the word to denote the small arc of the circle instead of the angle which this arc subtends at the centre.

For the more refined purposes of science the subdivision of the circle must be carried much further than the division into degrees. The extent of the subdivision of each arc of one degree into smaller arcs depends upon the particular purpose for which the graduated circle is intended.

The most familiar instance of a graduated circle is the ordinary drawing instrument termed a *protractor.* The protractor is employed for drawing angles of a specified size. For example, suppose that from a point c in a straight line A B (Fig. 16) it is required to draw a line C D so that the angle B C D shall be equal to 43°. The centre of the protractor is to be placed at c, and the division marked 0° upon the protractor is to be placed upon the line A B. Then a dot is to be placed upon the paper at the division 43°, and a line C D drawn through the dot from the point c is the line which is required.

The extent of the subdivisions is limited by the size of the instrument. Thus in a protractor six inches in diameter the length of the arc of one degree is about the twentieth of an inch. If much further subdivision be attempted, the divisions are so close together that they cannot be conveniently read without a magnifier. A protractor of this size is, therefore, usually only divided to 30-minute spaces

For astronomical instruments the graduated circles are generally subdivided to a greater extent than 30-minute spaces. In the instrument known as the meridian circle, the divisions of the circles are engraved on silver, and two consecutive divisions are only five minutes, or in some instruments only two minutes, apart. In the latter case the entire circumference contains $30 \times 360 = 10800$ divisions.

The circles in this case are about three feet in diameter, and the divisions are read by microscopes.

Mechanical ingenuity has, however, obviated the necessity for carrying the subdivisions of the circumference of a graduated circle to an excessive degree of minuteness. By the graduated circles attached to large astromonical instruments we are now able to 'read off' angles to the tenth part of a second. If this were to be effected by divisions alone there would have to be 12,960,000 distinct marks upon the circumference, and this is clearly impossible with circles of moderate dimensions. We shall presently explain the contrivances by which this vast extension of the accuracy with which angles can be measured has been obtained.

§ 11. *Circular Measure.*—There is another mode by which the magnitude of angles may be expressed, which, though unsuited for the graduation of astronomical instruments, is still of the greatest importance in many astronomical calculations. If we measure upon the circumference of a circle an arc of which the length is equal to the radius of the circle, and if we draw straight lines from the two extremities of this arc to the centre of the circle, the joining lines include a certain definite angle, which is termed the *unit of circular measure.* This is sometimes called the *radian.* It can easily be shown that the unit of circular measure is independent of the length of the radius of the circle. Whether the radius of the circle be an inch or a yard, the number of degrees, minutes, and seconds in the unit of circular measure is a constant.

We can now see how the magnitude of any angle whatever may be expressed by the number of radians and fractional parts of one radian to which the given angle is equivalent. This number is called the *circular measure oj the angle.* It is often necessary to convert the expression for the magnitude of an angle in radians to the equivalent expression in degrees, minutes, and seconds. To accomplish this we first calculate the number of seconds in one radian.

Since the circumference of a circle is very nearly equal to 3·14159 times its diameter, the arc of a semicircle is very nearly equal to 3·14159 times the radius. The angle subtended at the centre of the circle by the arc of a semicircle is of course 180°. It therefore follows that 180° must be almost exactly equal to 3·14159 radians. By reducing 180 degrees to seconds, and dividing by 3·14159, it is found that one radian is very nearly equal to 206265 seconds.

It is frequently necessary in astronomical calculations to make use of the principle that when the length of an arc of a circle is very small in comparison with the radius, the length of the arc may generally be taken as the length of the chord.

To illustrate the application of this principle we shall state here a problem which very often occurs in astronomy.

FIG. 17.

A distant object, of which the apparent diameter is A B (Fig. 17), subtends a known angle at the eye placed at o. If the distance o A from the eye to the object be known, it is required to find the length A B. It generally happens in astronomical calculations that although the length A B may be exceedingly great, yet the ratio of the length A B to the distance o A is comparatively small. If a circle be described of which o is the centre and o A the radius, this circle will pass through the points A, B, and the length of the chord A B will practically be equal to the length of the arc. We may therefore compute the *arc* instead of the chord. The angles subtended at the centre of a circle by arcs upon its circumference are proportional to the lengths of those arcs; it therefore follows that the required distance A B must bear to the length o A the same ratio which the angle A O B bears to the angle subtended by an arc equal to o A—i.e. to an angle of one radian. If θ be the angle O A B expressed in seconds, we therefore have the relation

$$\text{A B} : \text{O A} :: \theta : 206265,$$

from which A B is determined.

§ 12. *Instruments for Measuring Angles.*—In practical astronomy it is frequently necessary to measure the angle which two distant points subtend at the eye. Take, for example, two stars, which may be regarded as almost mathematical points on the surface of the heavens; then it may be required to measure the angle which is formed by the two lines which may be conceived to be drawn from the eye of the observer to the two stars. In effecting this measurement it is first of all necessary to bring two radii of a graduated circle into coincidence with the two sides of the angle which is to be measured, and then to ascertain the number of degrees, minutes, and seconds between the two radii.

Before the invention of the telescope one of the methods of bringing the radius of the graduated circle into coincidence with the visual ray was by the operation known as 'sighting.'

At one end A of a bar A B (Fig. 18) a vertical slit is fixed, while at the other end B is a pin. The eye being

FIG. 18.

placed at the slit, the bar can be so placed that the object under observation appears exactly behind the pin. The bar containing the sights can turn about the centre of a graduated circle, of which a quadrant is represented (Fig. 19). When the sights have been properly directed to one of the objects to be observed, the position of the bar on the graduated circle is read off. The sights are then directed to the other object, and the position of the bar is again read off.   The difference between the two readings gives the angle between the two objects.

The great drawback to this contrivance is its want of accuracy, which renders it quite unfitted for astronomical observations in the present state of the science.   In fact, it is impossible to avoid errors, which may amount to some or even many minutes of arc, in measuring angles by such a contrivance.   Even the improved sights used by Tycho Brahe could

Fig. 19.

Fig. 20.

not distinguish less than a minute.   The sighting is therefore now replaced for astronomical purposes by the incomparably superior contrivance which the invention of the telescope has placed in our hands.

At the focus of the object glass of a telescope a diaphragm with a circular aperture is placed, and across this circular aperture two extremely fine lines, usually taken from a spider's web, are stretched (Fig. 20).   The intersection of these lines marks a certain point in the field of view, and the telescope is to be directed to the distant object, so that the object or a specified point thereof shall coincide with the intersection of the wires.

It remains to be shown what specific line in the telescope is to be actually regarded as the line of vision which is to replace the line joining the two sights in the comparatively rude apparatus already described. Let A represent (Fig. 21) the distant object, the rays from which fall

FIG. 21.

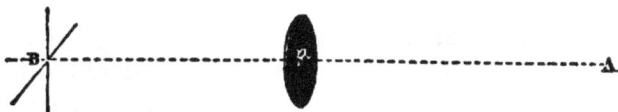

upon the object glass at *o* and are then brought to a focus at a point B, which coincides with the intersection of the two cross wires. The rays which fall upon the object glass must in general be deflected in passing through the glass, but there is *one* ray among those which diverge from A which the object glass does not deflect. If the line A B be drawn, it cuts the object glass at a point *o*. A ray passing along the direction A *o* cannot be deflected by the object glass, for otherwise it would not travel to the point B after passing through the object glass. This ray must pass through a certain point, which is called the optical centre of the object glass. The line joining the optical centre of the object glass to the intersection of the cross wires is actually defined in the telescope itself, without reference to external objects. This line is adopted as the line of sight of the telescope, and is known as the *optic axis*. It is necessary to observe that the *optic axis* need not necessarily coincide with the actual axis of the cylindrical tube of the telescope itself, nor is it either the line joining the centres of the object glass and the eye piece. The eye piece, in fact, may be displaced without altering the optic axis. If it be desired to effect a small change in the position of the optic axis of the telescope, this is effected by moving the diaphragm containing the cross wires in a direction transverse to the tube itself; the line joining the optic centre of the object glass to the intersection of the cross wires can thus be moved within certain limits.

We can now understand how the telescope, when provided with the cross wires, affords a much more exact instrument than the sights placed on a bar. In the latter case the slit of the one sight and the pin of the other have to be of considerable dimensions, in order to be readily seen, and consequently the sighting of the object can only be very coarsely accomplished. But in the telescope the optic centre of the object glass may be considered almost a mathematical point, while the intersection of two exceedingly fine lines is perhaps the most definite method we possess for indicating the position of a point. It follows that the optic axis of the telescope has the most extreme precision, for it is the line joining two definite points.

The lines furnished by the spider possess in a high degree the necessary qualities for forming the cross lines we have described. In the first place, they are exceedingly fine and uniform in thickness, and they are also sufficiently elastic to ensure that when once adjusted with the right degree of tension they shall remain constantly stretched.

The spiders' webs with which we are all so familiar are often used for this purpose, but the cocoons formed by some species of spiders are perhaps the most convenient source whence these lines can be derived.

At night it is necessary to illuminate the wires, as otherwise they could not be seen against the dark background of the sky. This is accomplished in two different ways. In the first of these methods the light from a lamp is thrown into the tube, so as to illuminate the entire field, and then the lines are seen as dark objects against the bright background. This is the most satisfactory method of rendering the wires visible whenever it is applicable. In the case, however, of very faint stars the illumination of the field will sometimes render them invisible. It then becomes necessary to resort to a different method of making the wires visible. By certain optical arrangements, which need not here be described, the light from a lamp is thrown *across* the field so that it illuminates the lines while the field still remains dark. In this case, therefore, bright lines are seen in a dark field, while in the other dark lines are on a bright field. Arrangements are provided by which either kind of illumination can be adopted, according to circumstances. The illuminated lines are, however, not so sharply seen as the dark lines, and therefore the latter are always preferred when the objects under examination are sufficiently brilliant to be visible in the illuminated field.

The telescope is to be attached to the divided circle in place of the sights which we have already described. The optic axis of the telescope could not conveniently be placed in the actual plane of the divided circle. It is, however, quite sufficient that it should be parallel thereto, nor is it even necessary that the optic axis should pass through a perpendicular to the plane of the graduated circle drawn through its centre.

Let us suppose that the angle between two distant points has to be measured. The plane of the graduated circle is first to be placed so that it passes through the two points; the telescope is to be turned so that the image of one of the points coincides with the intersection of the wires. The graduated circle is then to be 'read off' by means of a pointer, fixed quite independently of the telescope and the circle. The reading having been made, the telescope, bearing with it the graduated circle, to which it is rigidly attached, is to be turned round until the optic axis is directed towards the other point, the plane of the graduated circle remaining unaltered. The circle is again to be read off by the same fixed pointer. The difference between the two readings determines the angle through which the optic axis of the telescope must be turned in order to be moved from one object to the other. This angle is the angular distance of the two objects, which has thus been determined quite independently of the position of the telescope itself with respect to the graduated circle.

The degree of precision which can be attained in the measurement of angles by the aid of a telescope may be estimated by a consideration of the thickness of a spider's line. A line taken from a cocoon ordinarily used has a thickness of about 0·003 inch. In a telescope of about 20 feet focal length the thickness of a line of this size will subtend at the object glass an angle of about one-quarter of a second. This, therefore, indicates the degree of exactitude with which the telescope might be directed towards a distant point.

For many purposes it is found more accurate to have two parallel spider lines tolerably close together, and then to bring the image of a distant point so as to bisect the distance between the lines. As the eye can judge with considerable delicacy of the equality of the distances of the point from the two lines, this method has much to recommend it for certain kinds of observation.

§ 13. *Method of Reading an Angle.*—The reading microscope consists of an object glass at B (Fig. 22) and an

FIG. 22.

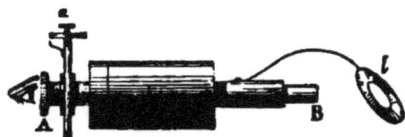

FIG. 23.

eye piece at A, connected by a tube. At the focus of the object glass a pair of cross lines are inserted, and the diaphragm which carries the cross lines is capable of being moved by a screw, which carries a graduated head at *a*. A portion of the graduated circle, which is rigidly attached to the telescope, is shown at C D (Fig. 23). The

microscope is adjusted so that a distinct image of a portion
of the graduated limb of the circle can be seen when the
eye is placed at A. The appear-
ance presented in the field of the
microscope is shown in Fig. 24.
It may be supposed that the circle
is so divided that the distance from
one division line to the next is
equal to 5 minutes. A few of
these division lines will be seen in
the field, and also the cross wires

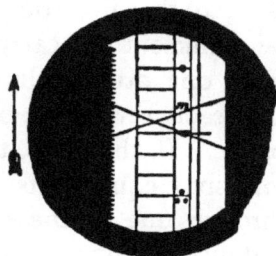

FIG. 24.

which are attached to the screw of the micrometer. As
the telescope, carrying with it the graduated circle, is
moved the division marks are seen to move through the
field of the microscope, while the cross lines of course
remain fixed. If the microscope screw be set so that the
divided head stands at zero, then the optic axis of the
microscope intersects the graduated limb of the circle at a
definite point, which is the reading of the circle. If this
optic axis actually passed through one of the division lines,
then this division line would be seen to pass through the
intersection of the cross lines, and the degrees and minutes
corresponding to that division line would be the reading of
the circle.

It will, however, generally happen that the optic axis of
the microscope will intersect the limb of the graduated circle
at a point lying between two adjacent division lines. In
this case the two division lines will be seen, as in Fig. 24, to be
lying one on each side of the point defined by the intersection of the cross wires. To accomplish the reading of the
circle it becomes necessary to determine the accurate reading,
which corresponds to the point in which the intersection of
the cross wires divides the distance between the two
adjacent divisions. Assuming that the divisions of the
graduated circle increase from the top of the figure to the
bottom, then the distance from the mark *m* to the intersec-

tion of the cross wires has to be ascertained. The distances between the division lines being 5 minutes, we shall suppose that the thread of the screw at the head of the microscope is such that five complete revolutions of the screw would be sufficient to carry the intersection of the cross wires from one division line to that next adjacent. It follows that each revolution of the screw corresponds to one minute of arc. If, therefore, the head of the screw be subdivided into sixty equal parts, it is plain that a rotation of the head through one of these parts will carry the cross lines over a distance of one second.

The reading is effected by moving the cross lines in the direction of the arrow until their intersection is brought to coincide with *m*. The number of entire revolutions will give the number of minutes, and the number of fractional parts the number of seconds, between the division mark *m* and the point of the graduated limb, which was intersected by the optic axis of the microscope when the cross wires were set at zero.

In order to illuminate the graduated limb at night, a mirror is placed between the object glass and the graduated limb. The light from a lamp falling upon the mirror is reflected upon the limb, while a hole in the centre of the mirror allows the limb to be seen by the microscope.

§ 14. *Error of Eccentricity.*—In large circles it is usual to have four microscopes, or sometimes more, placed symmetrically round the circumference, and the reading is made at each of the microscopes separately. By this means an important source of error is entirely eliminated. This must now be explained, and for simplicity we shall take an example.

A meridian circle was turned through a certain angle, and the value of that angle was measured by four microscopes placed round the circumference at angles of 90° apart. The value of the angle by

Microscope I. is 104° 56′ 47″·1
II.   104 56 51·3
III.   104 56 47·1
IV.   104 56 41·4

If everything had been perfectly right, then it is obvious that these four angles ought to be equal to each other. But various sources of error are present.

1. There are doubtless actual errors of *judgment* in making the coincidence of the cross wires with the division mark. These errors are, however, but small, and in a good instrument should not amount to a single second.

2. There are also errors of *workmanship* in the execution of the division marks upon the limb of the graduated circle, and also in the screws of the different microscopes. Still these errors are but small, and in the instrument with which the measurements now under discussion were made they certainly do not exceed a single second.

It is therefore plain that we must look to some other source for an explanation of the discrepancies between the four values which have been found.

The chief source of the error is what is called *eccentricity*, and it arises from a want of absolute coincidence between the centre of the graduated circle and the axis about which the telescope and the graduated circle revolve. The existence of an error from this cause is inevitable whenever a graduated circle is mounted upon an axis, but fortunately we can easily eliminate its effects, so that they shall not vitiate our results.

It must first be observed that what we *want to measure* is the angle A O B (Fig. 25), through which the telescope has been rotated. What we *actually do measure* is the number of divisions of the circle between A and B. It is, however, obvious that the arc of the circle (which is proportional to the number of divisions) will not be proportional to the angle at O, unless the point O, about which the telescope rotates, coincides with the centre of the graduations. If, however, we have two microscopes diametrically opposite, one of them

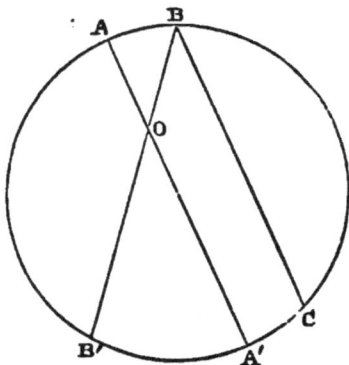

FIG. 25.

will indicate the arc A′ B′ and the other the arc A B. Draw B C parallel to A A′ ; then since C A′ is equal to A B, the sum of the two measured arcs is C B′ ; but the angle which C B′ subtends at the circumference is obviously equal to the angle at O. It follows that $\frac{1}{2}$ C B′ subtends an angle O at the centre, and therefore we have the following important result :—

The arithmetic mean of the readings of two diametrically opposite microscopes is independent of the error of eccentricity.

In the case at present under consideration the mean readings of the diametrically opposite Microscopes I. and III. is

$$104° \ 56' \ 47''\cdot1,$$

while the mean given by Microscopes II. and IV. is

$$104° \ 56' \ 46''\cdot4$$

The small discrepancy between these two results may fairly be attributed to some of the other sources of error to which we have referred. It will, therefore, be natural to take the mean of all four microscopes as the final result, which is accordingly

$$104° \ 56' \ 46''\cdot75.$$

§ 15. *The Measurement of Time.*—In many astronomical investigations it is necessary to have an accurate method of measuring time. This is best effected by the astronomical clock. The clock is really only an arrangement for counting the vibrations of a pendulum, while at the same time it sustains the motion of the pendulum. It will first be necessary to explain fully the principles upon which the motion of the pendulum depends.

If a weight A (Fig. 26) be attached to a cord A B which is suspended from a fixed point B, the contrivance forms what is known as a simple pendulum.

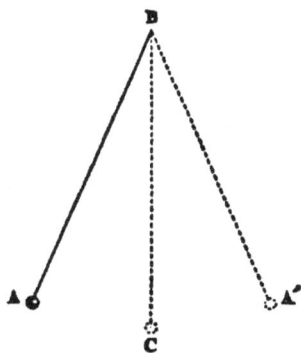

FIG. 26.

When at rest the cord will hang in the vertical position B C, with the weight directly under the point of support B. If the cord B C be drawn out of the vertical position into the position B A, and if the weight A be released, it will descend until the cord occupies the vertical position B C, after passing which the cord will ascend into the position B A'. If we could suppose that the resistance of the air were entirely removed, then it would be found that the weight would rise at A' to precisely the same vertical height that it had at A, or that the angle A B C is equal to the angle A' B C. After a momentary pause in the position B A', the weight will again descend, pass through the position C, and rise again to A. Again pausing, it will commence to descend, and again rise to A', and so on indefinitely. Owing, however, to the resistance of the air, and also, it should be added, to the rigidity of the cord at the point of suspension, the weight A will not on each occasion rise quite to the same height that it occupied before, and consequently the amplitude of the swing will gradually diminish until finally the weight comes to rest again in the position C. We shall, however, in what follows overlook the effect of the resistance of the air; for the motion of the pendulum can be indefinitely sustained, notwithstanding the resistance of the air.

A certain definite time is required for the weight to pass from the point A to the point A', and we shall first proceed to enquire into the circumstances by which the duration of that time—called the *time of vibration*—is to be determined. In the first place, it is to be observed that if the arc A A' form but a small portion of the entire circumference of the circle which could be described around the centre B with the radius B A, the time taken to move from A to A' is independent of the length of the arc. For example, if the length of the cord A B be one yard, and if the length of the arc A A' be one inch, then the time taken by the weight to move from A to A' is very nearly the same as it would have been had the arc A C A' been half an inch or had it been two inches, the length of the cord A B being the same in all three cases. It is no doubt true that there is a *minute* difference of time, that the short arc is described in a somewhat shorter time than the long arc ; but this difference bears such an exceedingly small ratio to the total time that for all practical purposes it may be omitted. We are thus enabled to state the following remarkable law :—

The time of vibration of a simple circular pendulum through a small arc is independent of the length of the arc.

This property of the pendulum is often known as its *isochronism.*

It is also easy to show that the time of vibration does not depend upon the *weight* of the pendulum. If the weight A be replaced by one which is double the amount or half the amount, the time of vibration is still found to be the same, and still found to be nearly independent of the length of the arc.

Nor is the case altered when the material of which the weight of the pendulum is composed is changed. A ball of lead, of iron, or of wood will vibrate precisely in the same time, whatever be its weight or whatever be the arc (supposed small) through which it swings, provided only that the length of the cord by which it is suspended remains unaltered.

We have thus ascertained that the time of vibration depends solely upon the length of the cord by which the weight is supported. It can be shown from theoretical considerations, and it can also be readily verified by actual experiment, that the following law expresses the true relation between the length of a pendulum and the time of its vibration.

The time of vibration of a simple circular pendulum is proportional to the square root of its length, and the length of the pendulum which at the latitude of London vibrates in one mean solar second is 39·139 *inches.*

It will thus be perceived that we have in the pendulum an admirable method of determining intervals of time. It is only necessary to count the number of vibrations made by a pendulum of known length, and the required interval of time is at once ascertained. For this purpose we must sustain the motion of the pendulum by occasionally giving it a suitable impulse; but, owing to the principle of isochronism, it is fortunately unnecessary to impose the condition that the arc of vibration remain strictly constant.

§ 16. *The Clock.*—A pendulum of the simple form which we have been describing would not be adapted for use in a clock. The pendulum in a clock consists of a heavy mass of metal attached to a rigid rod, which is suspended from a fixed point by a steel spring, which acts as a sort of *hinge* to permit the vibrations of the pendulum. It is, however, possible to apply the principles of the simple circular pendulum to the clock pendulum, only observing that the length of the circular pendulum, which would vibrate in precisely the same way as the clock pendulum, is somewhat less than that of the latter.

The first duty of the clock is to sustain the motion of the pendulum, which it is the constant effort of the resistance of the air to bring to rest. To attain this end the clock is supplied with a motive power, which gives the pendulum a small impulse in each vibration, and thus retains the arc of vibration approximately constant.

The motive power employed in the clocks which are used for astronomical purposes is generally afforded by a weight, which is periodically wound up. The cord attached to the weight is wound around a drum, and as the weight

FIG. 27.   FIG. 28.

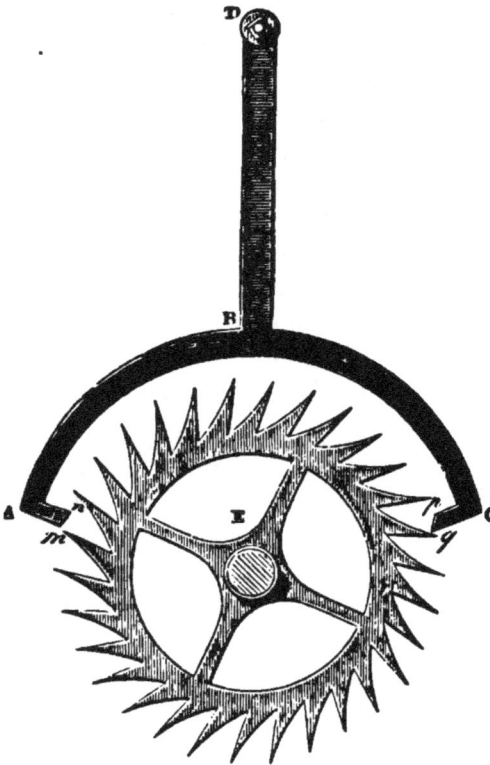
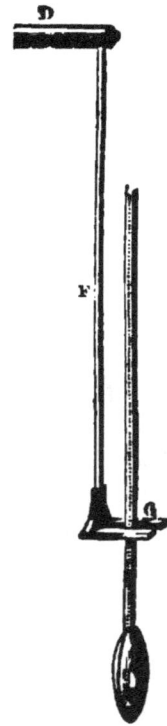

descends the drum is forced to revolve. On the axis of the drum is a toothed wheel, which sets in motion a series of toothed wheels, by which the hands are turned round, and the motion of the whole is controlled by the pendulum. This control is exercised through the medium of that important part of a clock which is known as the escapement, while the escapement also discharges the important duty of sustaining the motion of the pendulum.

The action of the escapement may be understood from Fig. 27. A toothed wheel E is connected with the wheelwork driven by the weight. Above this is suspended a piece A B C D which is able to oscillate about the centre D. From the axis D a rod F (Fig. 28) descends, which by means of a fork at G catches the rod of the pendulum. Thus as the pendulum oscillates so does the rod F, and hence also the piece A B C. In the course of the vibration the portion c *p* enters between a pair of consecutive teeth of the escapement wheel, and the tooth then falls upon the arc. In this condition the escapement wheel remains fixed until, in the return motion of the pendulum, the part c *p* withdraws from between the teeth sufficiently to allow the point of the tooth to glide down the inclined face *p q*, and the tooth *escapes*, as it is called. Immediately afterwards, however, the tooth at the other side drops upon the arc at *m*, and is there detained until the pendulum gains a position which will enable the point of the tooth to slide down the face *m n*. It will thus be seen that the revolution of the escapement wheel, and therefore of the whole train of wheels, is controlled by the motion of the pendulum. In the act of descending the inclined planes at *p q* and *m n* the point of the tooth imparts a slight impulse to the piece A B C ; this impulse is, through the medium of the fork G, transmitted to the pendulum, and so the oscillations of the pendulum are sustained.

When the clock is wound up a store of energy is imparted thereto, and this is doled out to the pendulum in a very small impulse which it receives at each vibration. The clock weight is of such a magnitude that it shall just be able to counterbalance the retarding forces when the pendulum has a proper amplitude of vibration. In all machines there is a certain amount of energy lost in setting the parts in motion and in overcoming friction and other resistances ; in clocks this represents the whole amount of the energy consumed, as there is no external work to be performed.

A good construction of the escapement wheel is essential to the correct performance of the clock. Although the pendulum must have its motion sustained by receiving an impulse at every vibration, yet it is desirable that the vibration of the pendulum should be hampered as little as possible by mechanical constraint. The isochronism of the pendulum, on which its utility as a timekeeper depends, is only a property of a pendulum which is swinging quite freely. Hence we must endeavour to approximate the clock pendulum as nearly as possible to a pendulum swinging quite freely. To effect this and at the same time to maintain the arc of vibration constant is the property of a good escapement.

The operations are so timed that the pendulum receives its impulse when at the middle of its stroke, and the pendulum is then unacted upon until it reaches a similar position in the next vibration. There is still a certain resisting force acting to retard the pendulum; this arises from the pressure of the teeth upon the circular surfaces, where some friction is unavoidable, however carefully the surfaces may be polished.

It is essential for the correct performance of a clock that the pendulum should vibrate at a proper rate, as a very small irregularity may produce an appreciable effect upon the clock. Thus suppose the pendulum vibrates in 1·001 second instead of in one second; the clock loses one-thousandth of a second at each beat, and, since there are 86,400 seconds in a day, it follows that the clock will lose about 86 seconds, or nearly a minute and a half, daily. The time of vibration depends upon the length of the pendulum, and therefore the rate of the clock will be constant, provided the length of the pendulum remain constant. To alter the rate of the clock the length of the pendulum must be altered; thus, for example: if the length of the pendulum be altered by one-tenth of an inch the clock will lose or gain nearly two minutes daily, according to whether the pendulum be lengthened or shortened. This explains the well-known practice of raising or lowering the bob of the pendulum when the clock is going too slow or too fast.

§ **17.** *The Compensating Pendulum.*—Let us suppose that the length of the pendulum has been properly adjusted, so that the clock keeps accurate time. It is necessary that the pendulum should not alter in length. But, as all bodies expand by heat, a pendulum which consists of a single rod to which the weight is attached must be longer in hot weather than it is in cold weather, and hence a clock will generally have a tendency to go faster in winter than in summer. For a pendulum with a steel rod the difference of temperature between summer and winter will cause a difference in the rate of five seconds daily, or about half a minute in a week. The amount of error thus introduced is of no great consequence in clocks which are intended for ordinary use, but in astronomical clocks, where seconds, or even portions of a second, are of the utmost consequence, inaccuracies of this magnitude would be quite inadmissible.

There are, it is true, substances—for example, slips of white deal—in which the rate of expansion is less than that of steel ; consequently the irregularities introduced by employing a pendulum whose rod is a slip of deal would be less than that of the steel pendulum we have mentioned ; but no substance is known which would not undergo greater variations than are admissible in the pendulum of an astronomical clock.

We must, therefore, devise some means by which the effect of changes of temperature in altering the time of vibration of a pendulum can be avoided. Various contrivances have been proposed for this purpose ; we shall describe one which is often adopted.

In Fig. 29 is shown what is known as the mercurial compensating pendulum. The rod *a*, by which the pendulum is suspended, is made of steel, and at its lower extremity two cylindrical glass jars are supported, which

are partly filled with mercury. The distance of the centre
of gravity of the mercury from the point of suspension
is very nearly equal to the length of the simple pendulum,

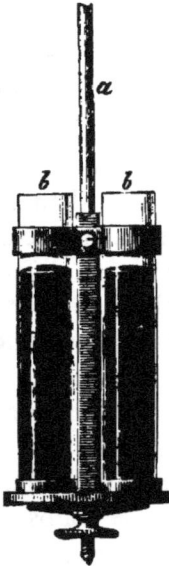

Fig. 29.

which would vibrate isochronously with
the compound arrangement here described.
Mercury and steel both expand when their
temperature is raised, but the rates at which
they expand are widely different, and it is
upon this difference in the *coefficients of
expansion* that the action of the compensating pendulum depends. For a given rise
of temperature the linear coefficient of expansion of mercury is several times greater
than that of steel. If the height of the
column of mercury in the two jars *b, b* has
a proper ratio to the length of the steel rod,
the compensation will be complete. For
suppose the temperature of the pendulum to
be raised, the steel rod would be lengthened,
and therefore the vessels of mercury would
be lowered. On the other hand, the surface
of the column of mercury in the jars would
be varied by the expansion of the murcury. The centre of
the column of mercury will be raised by half the amount
which its surface is raised. It can be arranged that the
centre of the column of mercury is raised by its own expansion as much as it is lowered by the expansion of the
steel. By this contrivance the time of oscillation of the
pendulum is rendered independent of the temperature.

The mercurial compensating pendulum possesses the
great advantage that it is easy to alter the quantity of mercury so as to adjust the compensation with precision.

CHAPTER II.

§ 18. *Form of the Earth.*—To an observer who is limited, as we are, to the surface of the earth, the contrast appears at the first sight to be very wide indeed between the appearance of the earth and the appearances presented by the sun and moon. The earth appears to be a flat plain, more or less diversified ; the sun and moon appear to be globular ; the earth appears to be at rest, while the sun and moon are apparently in constant motion ; and, lastly, the earth appears to have a bulk incomparably greater than that of either the sun or the moon.

If, however, we could change our point of view to a suitable position in space, we should form a more just conception of the relation of the earth to the sun and moon. We should then see that each of the three bodies was really spherical, that each of them was really in motion, and that the earth, though larger than the moon, was very much less than the sun.

It need hardly be said that it is impossible for observers on the earth to obtain such a bird's-eye view as we have here described. A balloon might indeed convey the observer to a point from which he would have a very extensive view, but it would be necessary to ascend to a height vastly beyond that to which any balloon could attain before the shape of the earth would be discerned as we discern the shape of the sun and the moon. Our knowledge of the figure of the earth is only to be attained by indirect means, the nature of which we shall now explain.

The most simple method of becoming actually convinced that the surface of the earth (or rather of the sea) is *not* an indefinitely extended plain is by taking a station on a high

cliff near the sea-side, from which an uninterrupted view of
the sea can be obtained. We shall also suppose that a
number of vessels are dotted over the surface of the sea at
different distances, and that the station which the observer
has chosen is *at a greater vertical height above the sea than
are the tops of the masts of any of the vessels.*

If the surface of the sea extended as an indefinite plane,
then this plane must be intersected by a line drawn from
the eye of the observer to the topmost point of the mast of
any one of the vessels. It is, therefore, obvious that the
entire of each of the vessels must be seen projected upon the
surface of the water. But this is obviously not the case.
The view which the observer has of the sea is shown in Fig.
30. The vessel in the foreground is no doubt entirely

FIG. 30.

projected on the surface of the sea, but the more distant
vessel is almost entirely projected against the sky, while the
most distant vessel of all is so far from being projected on
the surface of the water that the hull of the vessel is
rendered invisible by the interposition of a protuberant
mass of water, while the masts are seen projected against
the sky. We are, therefore, forced to the conclusion that
the sea is not a flat plain, but that it is a convex surface.

At whatever part of the earth the observations which we
have just described be made, it is invariably found that the
surface of the ocean is convex. The degree of the curvature
may be estimated by the distance at which the hull of the
vessel becomes invisible. The observer being stationed at
a certain height, and the hull of the vessel having a certain
size, the greater the curvature the less is the distance at

which the hull is invisible. Tested in this way, the curvature of the ocean appears nearly uniform over its entire extent. The surface which has a curvature which is uniform at all points must be a sphere. We are hence forced to the conclusion that the surface of the sea is approximately spherical.

It is somewhat more difficult to perceive that the general surface of the land is also approximately spherical. The irregularities on the surface of the earth, arising from hills and mountains and valleys, appear at first sight to preclude the possibility of making any general statement with reference to the figure of the earth. It is, however, to be observed that these irregularities on the surface are of quite trivial extent in comparison with the vast bulk of the earth itself.

It has been found that the figure of the earth, though very approximately a sphere, differs therefrom to an appreciable extent. It appears that the true figure of the earth is nearly that of the surface produced by the revolution of an ellipse A D B C (Fig. 31) about its minor axis A B. The eccentricity of the ellipse is designedly much exaggerated in the figure. According to the most recent determinations by Colonel A. R. Clarke, C.B., we have the following dimensions of the ellipsoid of revolution which most closely approximates to the figure of the earth :—Let a, b be the equatorial and polar semi-diameters respectively, then

FIG. 31.

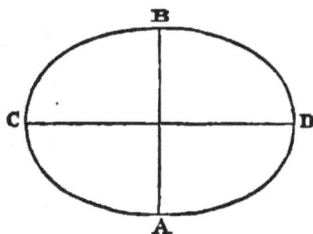

$$a = 6378207 \text{ metres} = 20926062 \text{ feet,}$$
$$b = 6356584 \text{ metres} = 20855121 \text{ feet;}$$

while for the ellipticity

$$\varepsilon = \frac{a-b}{a} = \frac{1}{294\cdot98}.[1]$$

---

[1] For certain purposes, and perhaps on the whole, the actual figure of the earth can, it is believed, be represented with still greater accuracy

§ **19.** *The Earth an Isolated Body.*—It having been ascertained that the earth has an approximately spherical figure, the next important point is to show that the earth is free in space, and not supported by or attached to any other object. This is clear from the circumstance that though the surface of the earth has been traversed in nearly every direction, no trace has been found of any such support or connection, and we must therefore admit that the earth is really an isolated object. At first it may seem difficult to believe that this is really the case. It might be thought that, as the earth is not supported, it ought to *fall* somewhere or other. This will be subsequently explained.

§ **20.** *The Atmosphere.*—At the exterior of the surface of the earth, and entirely enveloping the earth, is what is known as the atmosphere. At whatever height we ascend upon the mountains, we still find that air is present. We know that there is a limit to the atmosphere ; but the height is far greater than any to which we can attain, though we can ascertain its amount to some degree of approximation.

It can easily be shown that air is a substance which has weight. A glass balloon filled with air weighs more than it does when all the air has been removed from it. Air is eminently compressible, and the air near the surface of the earth, which has to bear the weight of all the superincumbent air, is much more compressed than the air which is

as an ellipsoid with three unequal axes. Captain Clarke's most recent determination gives for the lengths of the semiaxes—

$$a = 20926629 \text{ feet} = 6378390 \text{ metres,}$$
$$b = 20925105 \text{ feet} = 6377920 \text{ metres,}$$
$$c = 20854477 \text{ feet} = 6356390 \text{ metres.}$$

The ellipticities of the two principal meridians are

$$\frac{1}{289 \cdot 5} \text{ and } \frac{1}{295 \cdot 8}.$$

The longitude of the greatest axis is 8° 15′ W.

higher up. It follows that the density of the air is a maximum at the surface of the earth, and that it diminishes as
we ascend until the confines of the atmosphere are reached,
where the density is zero. It is, therefore, plain that the real
height of the atmosphere must be much greater than it
would have been were the atmosphere homogeneous throughout.

We shall subsequently allude to the phenomena of *shooting
stars,* and we shall show by what observations their height
is ascertained. It is known that these objects only become
visible when they enter our atmosphere, and therefore we
have the means of ascertaining a minor limit to the height
of the atmosphere. It is found that shooting stars are sometimes seen at a height of more than two hundred miles, and
therefore the atmosphere must
extend to at least that height.
There can be little doubt that
for a very great depth at the
upper surface of the atmosphere the density is exceedingly small.

FIG. 32.

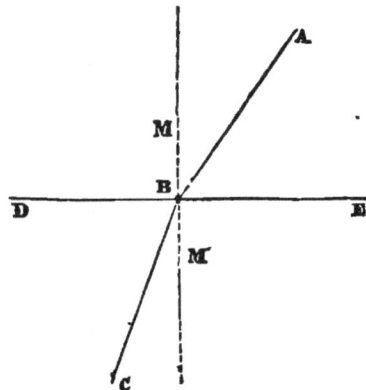

§ 21. *Atmospheric Refraction.*—In viewing a celestial
body we have always to look
completely through the atmospheric shell by which the
earth is surrounded. This
produces an apparent dis-
placement of the celestial body by what is called *refraction.*
Let D E (Fig. 32) be the surface of any transparent medium,
and let a ray A B fall upon the surface D E at B. Then
when the ray enters the medium it undergoes a certain
deflection, insomuch that it pursues the path B C instead
of the original path. The angle A B M is called the angle of
incidence, and the angle C B M′ is the angle of refraction
where M M′ is perpendicular to the surface D E. The rays A B

and c ʀ both lie in a plane which contains the normal, and the sine of the angle of incidence is in a constant ratio to the sine of the angle of reflection.

We can now explain the effect which the atmosphere has on the refraction of the light from a celestial body— suppose a star. We may for this purpose assume the earth to be a spherical body whose centre is at o (Fig. 33). The atmosphere may then be considered to consist of a number of concentric spherical shells, each of uniform density, but the density of each shell being greater than that of the shell immediately out-side it. If these shells be regarded as indefinitely thin, this supposition may accurately repre-sent the atmosphere of our earth.

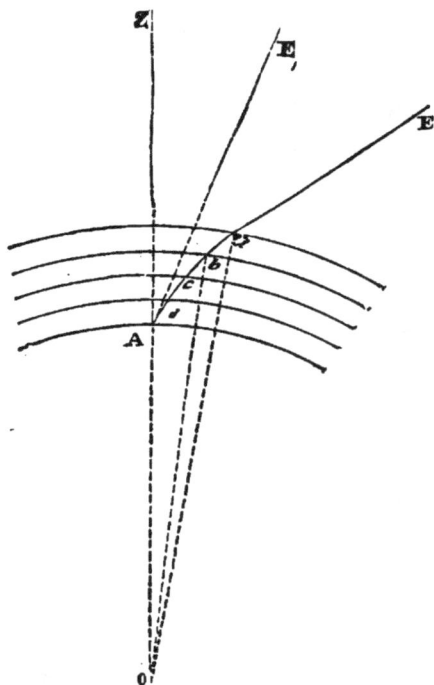

FIG. 33.

A ray of light, after passing through space in the direction E *a*, falls upon the first layer of atmosphere at *a* ; after entering the layer the ray moves along the di-rection *a b*, which is less inclined to o *a* than was the original line E *a*.

Having traversed the first layer, the ray enters the second layer ; it is again deflected into the line *b c.* The same process goes on as the ray enters each successive layer, until finally it reaches the eye at the point A on the surface of the earth. The actual direction of the ray at the moment it entered the eye is A E′. This, then, will be the direction in which the

star is seen, and therefore the real position of the star would
be judged erroneously.

The line o A points upwards to the zenith of the
observer at A. The angle subtended at the eye by the
zenith point and the star is the zenith distance of the star.
The apparent zenith distance of the star is represented by
the angle z A E'. The real zenith distance of the star is the
angle between A z and a line through A parallel to *a* E.
Thus the effect of refraction is always to diminish the zenith
distance of a star, or to make the star appear nearer to
the zenith than it would be were refraction absent. The
effect of refraction on a star actually at the zenith is zero,
while the effect is a maximum on a star situated on the
horizon.

To compute with accuracy the precise amount of deviation which a ray of light experiences in traversing the entire
thickness of our atmosphere, it would appear to be necessary
to know the precise density of each of the shells of which
we have supposed the atmosphere to be constituted. Our
knowledge of the pressure and temperature in the upper
regions of the atmosphere is, however, so imperfect that it
is impossible to form any accurate law which connects the
alteration of density in the atmosphere with the increase of
height. It is, however, fortunate that the refraction can be
computed in a much more simple manner whenever the
zenith distance is comparatively small.

The thickness of the atmosphere is, even on the largest
estimation, only a very small fraction of the radius of the
earth. We may therefore, without appreciable error, consider
the surface of the earth to be a plane, and consequently on
this assumption the successive shells of different densities
will be horizontal layers. When this assumption is made,
the problem of computing the refractions becomes greatly
simplified. It is, under these circumstances, quite unnecessary to know what the densities of the successive layers may
be, or to what height the atmosphere extends. The refrac-

tion will, in fact, be the same as if the entire atmosphere were of absolutely uniform density throughout, that density being the same as is indicated by the barometer and thermometer at the surface of the earth at the moment when the observation is made.

This very remarkable result can be demonstrated with facility. Let M, M', M'', M''' represent media of different

FIG. 34.

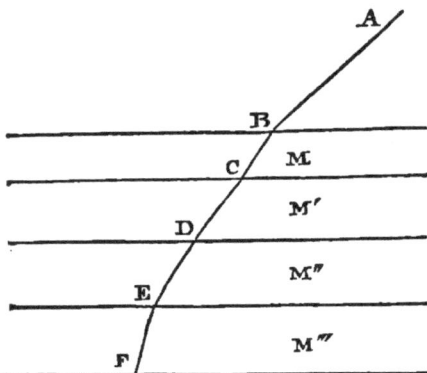

FIG. 35.

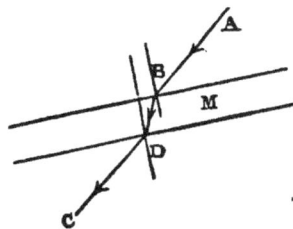

degrees of refrangibility, through which a ray of light originally moving in the direction A B is to be refracted. It is assumed that the *bounding surfaces of these media are all parallel planes.* After refraction in the medium M the ray travels in the direction B C, then it is refracted by the medium M' into the direction C D, and so on till in the last medium the ray has the direction E F. It is to be proved that the direction E F is the same as it would have been had the ray A B fallen directly upon the medium M''' without the intervention of M, M', and M''. It should, however, be observed that we are here only concerned with the *direction* of the ray ; the absolute situation, no doubt, is to a certain extent altered by the superincumbent media.

This will be made clear by Fig. 35. A B is a ray of light which impinges upon a single medium M bounded by two parallel planes. After refraction at the first surface the ray traverses the direction B D, then falls upon the second surface and emerges in the direction C D. Since the surfaces are parallel, it is clear that the ray B D makes the same angle with the normal at B which it does with the normal at D; it therefore follows that after emerging from the medium at D the ray C D must be inclined to the normal at D at the same angle at which A B was inclined to the normal at B, and that consequently the ray A B is parallel to the ray C D. Hence the effect of the transmission of a ray of light through a medium bounded by parallel planes is merely to change the situation of the ray while leaving its direction unaltered.

This principle can be applied to the explanation of the property illustrated in Fig. 34. After the ray has traversed M we may suppose it emerges parallel to its original direction, and then immediately impinges upon M' in the same way; after passing M' it emerges still parallel to A B and falls upon M''; thence emerging parallel to its original direction, the ray falls upon M'''. It is, therefore, obvious that the refraction produced by the medium M''' is the same as if the ray A B had directly impinged thereon.

It will be observed that the entire argument depends upon the parallelism of the planes bounding the surfaces of the media. The actual bounding surfaces in the atmosphere being curved surfaces closely approaching to spheres, the property we have just proved is only approximately true. It has, however, been found that for all zenith distances not exceeding 75° the calculations made upon this assumption are sufficiently accurate for most purposes.

When the zenith distance exceeds 75° the amount of the refraction depends upon the constitution of the different layers of the atmosphere. It is also subject to considerable changes in accordance with the variations in the temperature and pressure of the atmosphere.

The following table gives the amount of the refraction at different zenith distances from 0° to 90°, the height of the barometer being 30 inches and the temperature being 50° :—

Apparent Zenith Distance	Refraction	Apparent Zenith Distance	Refraction	Apparent Zenith Distance	Refraction
°	"		"	°	' "
0	0·0	35	40·8	70	2 38·8
5	5·1	40	48·9	75	3 34·3
10	10·3	45	58·2	80	5 19·8
15	15·6	50	1 9·3	85	9 54·8
20	21·2	55	1 23·4	87	14 28·1
25	27·2	60	1 40·6	89	24 21·2
30	33·6	65	2 4·3	90	33 46·3

Thus, for example, the apparent zenith distance of an object being 60°, the real zenith distance is found by augmenting the apparent zenith distance by 1′ 40″·6, and therefore the true zenith distance is 60° 1′ 40″·6.

At moderate zenith distances we have the following law :—

The amount of refraction is proportional to the tangent of the zenith distance.

From the table of refractions it appears that the refraction at the zenith distance 45° is 58″·2, whence we must have in general the expression

$$58''·2 \tan z$$

for the value of the refraction at the distance *z*. It is, however, obvious that this law is incorrect near the horizon. Under these circumstances the tangent of *z* approaches to infinity, while the actual value of the refraction at the horizon is only 33′ 46″·3.

CHAPTER III.

THE DIURNAL MOTION OF THE HEAVENS.

§ 22. *The Celestial Bodies.*—On a clear night the surface of the heavens glitters with innumerable points of light, commonly called stars. The objects which are popularly known by this term comprise two different classes. The great majority belong to what are called *fixed stars*, while a few are known as *planets*.

The first feature of the *fixed stars* to which we must direct attention is their apparent fixity in the heavens with respect to each other. If we form figures by lines, joining the different stars together, then it is found that these figures remain unaltered even while centuries pass away. In fact, if we could regard the surface of the heavens as a vault of solid material, the stars appear as if they were rigidly stuck on the interior of this vault.

The *planets*, on the other hand, do not remain fixed ; they are, as their name implies, wanderers, and they move about on the surface of the heavens among the stars. The word *planet* is usually restricted to a very special group of objects. There are other celestial bodies which are also wanderers. Among these we may mention the moon, the satellites of the planets, and comets.

At a superficial glance a planet resembles an ordinary fixed star so closely that it is difficult to realise how funda-mental and important is the distinction which separates them. To test whether an object is really a planet or a fixed star the position of the object should be carefully noted with respect to the bright stars in its vicinity. After some time the object should be again examined and the place compared with the adjacent fixed stars. If it be found that the place has altered, then the object is a planet. There

E

is, however, another method of discriminating between a star and a planet, which has the advantage of enabling the decision to be pronounced immediately without waiting until the planet, shall disclose its real character by movement. If a telescope be directed to a fixed star the object appears to be merely a point of light. No increase in the power of the telescope will enable us to see the dimensions of the small point from which the light comes, although the brilliancy is increased with each increase of the power of the telescope. The case is, however, widely different when the telescope is directed to a planet. It is then seen that the planet really has a circular disk of dimensions which are quite appreciable. This test can be applied with facility to all the more conspicuous planets. In the case of planets which are extremely minute or extremely distant this test cannot be applied easily. In fact, there are numbers of planets so minute that the telescope cannot show them to be different from the stars in their vicinity. In these cases the motion test has to be applied.

§ **23.** *The Celestial Sphere.*—A sphere is a surface such that every point upon it is equidistant from one point in the interior which is called the centre. If a plane be drawn through the centre of the sphere it cuts the sphere in a circle which is called a *great circle*. A plane which cuts the sphere, but which does not pass through the centre, has also a circle for the line along which it intersects the sphere ; this is called a *small circle*. The radius of a great circle is of course equal to the radius of the sphere. The radius of a small circle may be of any length less than the radius of the sphere. We may suppose that a sphere is produced by the revolution of a circle about its diameter, and the radius of the sphere is then equal to the radius of the circle from which it has been produced.

Let o be the centre of a sphere, and A B any two points on its surface. Then a plane through the three points o A B cuts the sphere in a great circle. This may, for simplicity, be termed the great circle A B. The length of the arc of the great circle connecting two given points on a sphere of known radius is most conveniently measured by the angle which the arc subtends at the centre.

If we imagine the angle of a pair of compasses to be placed at the eye while each leg of the compasses is directed towards a particular star, the angle between the legs of the compasses is said to be the angular distance between the two stars. By an instrument founded on this principle, it is possible to measure the angular distance between two stars with great accuracy, and from such measurements a *celestial globe* can be constructed. Two points, A and B, are first to be marked on the surface of the globe, so that the angle which A and B subtend at the centre of the globe is equal to the angle subtended at the eye by the two stars to which A and B correspond. The angular distance of a third star S is to be measured from both A and B, and the star S is to be marked on the globe, so that the two arcs S A and S B shall subtend at the centre of the globe the angles which have been observed. In this way all the principal stars on the surface of the heavens may be accurately depicted upon the surface of a globe.

We shall now introduce a convention which is very useful. The stars are, no doubt, at very varied distances from the earth, but, nevertheless, we have seen that the appearance of the heavens can be adequately represented on a globe where all the stars are at the same distance from the centre. Let us suppose a colossal globe to be described with the earth at its centre and an enormously great radius. Then if the stars were all bright points stuck on the interior surface of this globe, the appearance of the heavens would be the same as we actually find it. This imaginary globe we call the *celestial sphere.*

§ 24. *The Fixed Stars.*—It may be observed that the *planets* which can be conspicuously seen with the unaided eye are only five in number (viz. Mercury, Venus, Mars,

Jupiter, Saturn). Uranus can be seen like a very faint star, and one or two of the remaining planets have occasionally been detected by exceptionally sharp vision. It is thus evident that of the vast multitude of celestial objects visible to the unaided eye every clear night, by far the greater number are fixed stars.

The first feature connected with the stars to which we shall direct attention is their very different degrees of brightness. Astronomers have thus been led to classify the stars accordingly. About twenty of the brightest stars in the heavens are said to be of the first magnitude. Among these we may mention a few which are particularly conspicuous in northern latitudes. The brightest star in the whole heavens is Sirius. Then come Vega, Capella, Aldebaran, Rigel, Arcturus, Spica, and Betelgeuze.

Next in order to the stars of the first magnitude come those of the second magnitude. Of these we may mention, as examples which must be familiar to many, the four brightest stars in the constellation sometimes known as Ursa Major, the Great Bear (Fig. 36).

The stars immediately below those of the second magnitude in brightness are called stars of the third magnitude; next come the fourth, and so on down to the very smallest stars that can be seen in the most powerful telescopes. Argelander has computed the number of stars of each of the different magnitudes. The result of these investigations for the first nine magnitudes are here given :—

1st	.	.	20	4th	.	.	425	7th	.	.	13,000
2nd	.	.	65	5th	.	.	1,100	8th	.	.	40,000
3rd	.	.	190	6th	.	.	3,200	9th	.	.	142,000

From this it will be seen that the fainter stars are much more numerous than the brighter ones. The number of stars of each magnitude is greater than the number in any preceding magnitude, and less than in any following magnitude. Thus, though there are but 20 stars of the first magnitude, there are 142,000 stars of the ninth magnitude.

Of the total number of stars only a comparatively small number are visible to the unaided eye. Stars of the fifth magnitude are faint, those of the sixth very faint, and it requires very good vision to perceive stars of the seventh magnitude without the assistance of a telescope. The number of stars which can be seen with the unaided eye in England may be estimated at about 3,000.

It is hardly possible to form any very accurate conception of the numbers of the stars of each magnitude when that magnitude is expressed by a larger number than nine. This arises partly from the prodigious numbers of the stars, and partly from some uncertainty with which the estimation of the magnitudes of very small stars are attended. Argelander has published a most valuable catalogue of the stars in the northern hemisphere. This catalogue is accompanied by a series of maps on which these stars are depicted. All stars of the first nine magnitudes are included in this catalogue, as well as a very large number of stars which are between the ninth and tenth magnitudes. The total number of stars contained in the catalogue and on the maps amounts to 324,188.

Maps have been formed of isolated portions of the heavens which include stars much smaller than those here referred to. Some of these maps contain stars of the eleventh and twelfth magnitudes, if not actually smaller. The enormous numbers of the very small stars render the formation of such maps exceedingly laborious. The total number of stars visible in powerful telescopes doubtless amounts to many (perhaps very many) millions.

The prodigious multitudes of minute stars with which the heavens are strewn is well illustrated by the nature of what is commonly known as the *Milky Way*. The milky way is an irregular band of faint luminosity which encircles the whole heavens, and may be seen on any dark and clear night in the absence of the moon. The telescope shows that this faint luminosity really arises from myriads of minute stars, which, though individually so faint as to be invisible to the naked eye, yet by their countless numbers produce the appearance with which doubtless everyone is familiar.

§ 25. *Constellations.*—For the purpose of marking out different regions in the heavens, modern astronomers have retained the method which was principally due to the poetical imagination of the ancients. The whole surface of the celestial sphere is, on this method, supposed to be covered by imaginary human figures, and representations of other objects. The stars on each of these figures are by some grotesque conception supposed to point out the form of the object. In this way different regions of the heavens are known by special names, and the stars on each of these regions are collectively termed a *constellation.* We thus have, for example, the constellation of Orion, Ursa Major, Ursa Minor, Leo, Lyra, &c. &c.

This arrangement provides a very convenient method of indicating the stars. It is for this purpose only necessary to mention the name of the constellation to which the star belongs, and to append a letter or number by which the different stars of each constellation may be distinguished. Bayer published in the year 1603 a series of maps of the stars, in which the letters of the Greek alphabet were attributed to the principal stars of each constellation. To the brightest star of the constellation the letter α was assigned, the next brightest is denoted by β, the next by γ, and so on throughout the alphabet. When all the Greek letters have been exhausted, the remaining stars in the constellation are usually denoted by Latin letters and numbers.

It is exceedingly desirable that the learner should make himself acquainted with the principal constellations. To facilitate him in doing so, we shall give a few outline charts, in which the brighter stars in several of the constellations will be shown.

There is no difficulty in recognising at once the constellation of Ursa Major, of which the seven principal stars

FIG. 36.

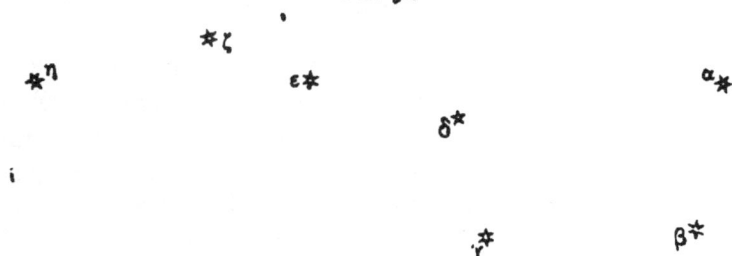

are shown in Fig. 36. This constellation is perhaps the most conspicuous object in the northern skies, and in these latitudes it never sets. Ursa Major can thus be seen every clear night in the year, and in April it is near the zenith at 11 P.M. When once this group of stars has been recognised, many of the other important stars and constellations can be determined with facility.

We shall first point out how the position of the Pole Star is to be ascertained by the help of Ursa Major. If we imagine a line drawn through the stars β and α in Ursa

FIG. 37.

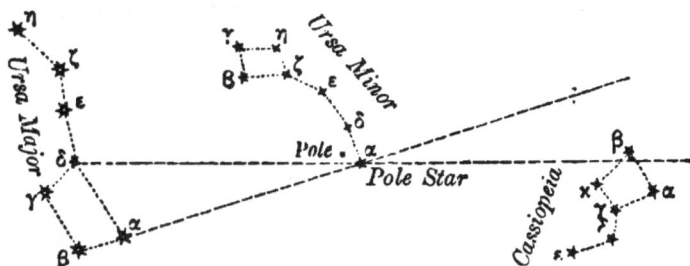

Major and then continued on from α to a distance which is about five times the distance from β to α, the extremity of this line will be found to be close to a bright star. This bright

star is called the Pole Star. The Pole Star is also the star
a of the constellation *Ursa Minor*, Fig. 37. This group
contains seven principal stars, of which β and γ, in addition
of course to the Pole Star, are the most conspicuous. The
point which is known as *the pole* of the heavens, lies exceedingly near to the Pole Star (Fig. 37), whence the latter has
received its name.

If we join δ in Ursa Major to the Pole Star, and produce
the joining line nearly as far on the other side of the Pole
Star, we come to a remarkable group of stars, forming the
constellation *Cassiopeia*. Thus Ursa Major and Cassiopeia
are so situated, that the Pole Star lies almost exactly half-
way between them. These are the constellations which
should first be made familiar to the eye and memory of the
student, and then he can proceed to the others now to be
described.

The great *Square of Pegasus* is a remarkable group.
Unlike Cassiopeia and Ursa Major, the constellation now
under consideration rises and sets daily. It cannot be seen

FIG. 38.

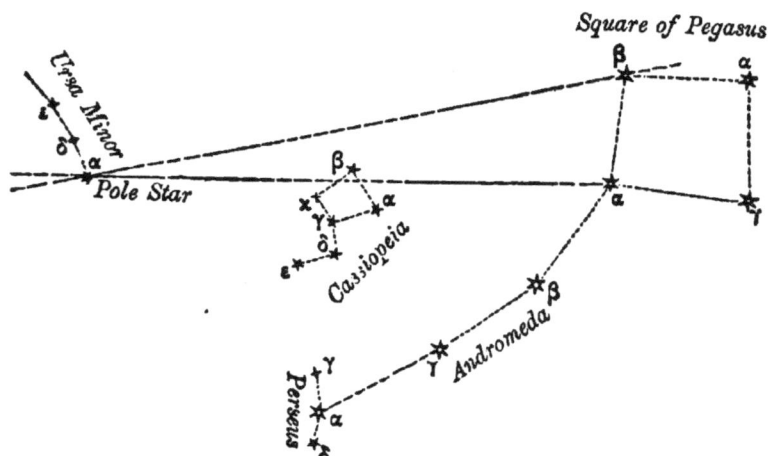

conveniently during the spring and summer, but during
autumn and winter the four stars forming the great Square

of Pegasus form a remarkable object every evening. These stars are of the second magnitude. The square may be determined by drawing two lines through α and δ Ursæ Majoris to the Pole Star. These lines continued on beyond Cassiopeia to about the same distance as Cassiopeia is from the Pole Star, include between them the great square of Pegasus (Fig. 38).

Three of the stars in the square, viz. α, β, γ, belong to the constellation Pegasus, but the fourth star, which is also marked α, belongs to the constellation *Andromeda*, of which it is the brightest object. Two other stars, β and γ, belonging to Andromeda can also be readily determined from the circumstance that they are near the corner of the square, where α Andromedæ is situated (Fig. 38). The line joining β and γ in Andromeda, produced away from Pegasus, points to the brightest star α in the constellation *Perseus*.

The seven stars, viz. α Persei, α, β, γ Andromedæ, and α, β, γ Pegasi, form a very remarkable group of stars, and once they have been recognised they will aid in the indication of several other constellations.

α Persei lies between two other stars γ and δ belonging to the same constellation. These stars form an arc (Fig. 39) which, produced on a little distance, includes the bright star *Capella*, belonging to the constellation *Auriga*. On the convex side of this arc lies the remarkable variable star β Persei, more usually known as *Algol*. If the arc formed by γ, α, δ Persei be prolonged so as to be concave towards Algol, it points out the stars ε, ζ Persei ; and continued on further in the same direction, we come upon the very remarkable

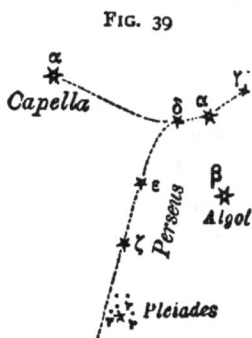

FIG. 39

group of small stars close together which are called the *Pleiades*.

Joining the Pole Star to Capella, and producing the join-

ing line beyond Capella, the splendid constellation of *Orion*
is arrived at (Fig. 40). The brightest star of this constellation
is α Orionis, often called *Betelgeuze,* which is remarkable for
its ruddy hue. The Belt of Orion is formed by the three stars
δ, ε, ζ. The belt is surrounded by the quadrilateral formed
by the four stars α, β, γ, κ. Of the seven stars α and β are

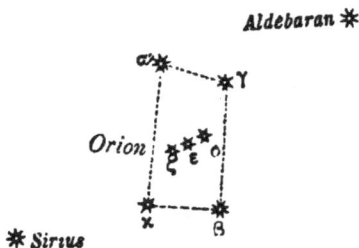

FIG. 40.

FIG. 41.

of the first magnitude, while the five
remaining stars are of the second magnitude.

The three stars in the Belt of Orion
point downwards to a most conspicuous
star called *Sirius.* This is by far the
most brilliant of all the stars. In conjunction with a few inconspicuous
stars it forms the constellation of *Canis
Major.*

If the line of the Belt of Orion be
prolonged in the other direction, it
points out the star *Aldebaran.* This
is the brightest star in the constellation
of *Taurus.*

The line joining the stars β and δ in Ursa Major, and
produced on sufficiently far, points out the two stars of the
second magnitude termed Castor and Pollux in the constellation Gemini (Fig. 41). This same line continued a little
farther passes near the star Procyon, of the first magnitude,
in the constellation of *Canis Minor* (Fig. 41).

The line joining the two stars α, β in Ursa Major, prolonged on beyond β to a distance five times as far as from α to β, points out the constellation of *Leo* (Fig. 42). The stars in this constellation have a remarkable form. The four principal stars—α, β, γ, δ —are shown in the figure. α, otherwise called *Regulus*, is of the first magnitude. The three remaining stars of the quadrilateral are of the second magnitude.

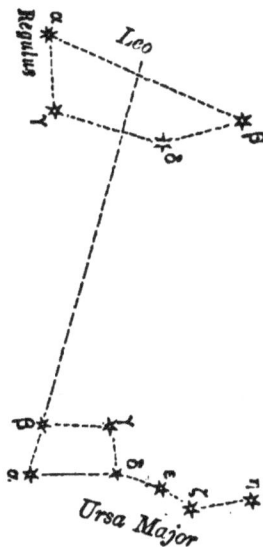

Fig. 42.

.   The tail of Ursa Major, when prolonged, points to the brilliant star of the first magnitude called *Arcturus*, which is the principal object in the constellation *Bootes* (Fig. 43). The stars β, γ, δ, ε in the same constellation are also shown in the figure.

Close to Bootes, and in a direction which may be found by following the line of the stars β, δ, ε, ζ in Ursa Major, is the constellation *Corona Borealis*. This consists of a number of stars arranged nearly in a semicircle, the largest of them being of the second magnitude.

Fig. 43.

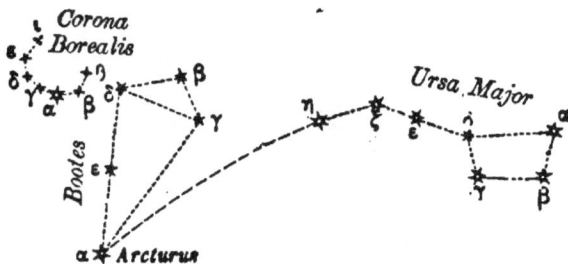

If a line be drawn through α and γ in Ursa Major, and if this line be curved so as to present its convexity towards

Arcturus, as in Fig. 44, it points to a brilliant star termed
Spica, which is the brightest star in the constellation *Virgo*.

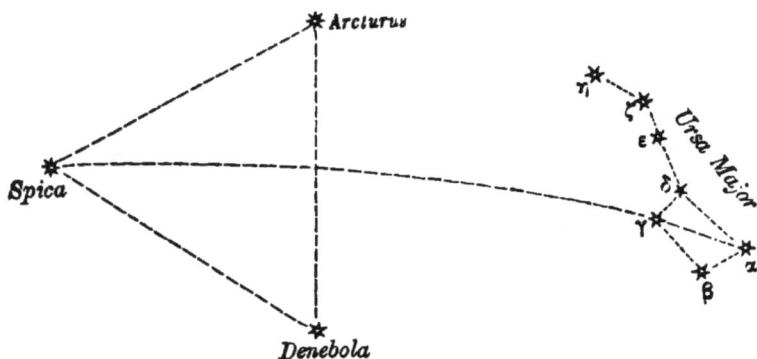

FIG. 44.

The star β Leonis, otherwise called Denebola, which is situated at the tail of Leo, forms one vertex of a very striking

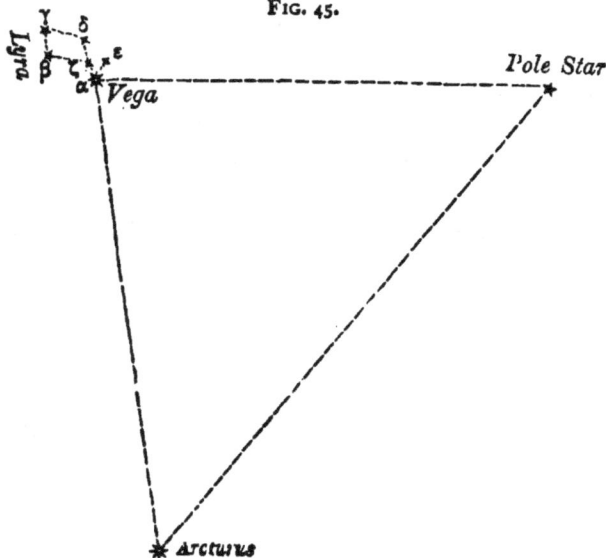

FIG. 45.

equilateral triangle of which Arcturus and Spica are the
two remaining vertices.

One of the most brilliant stars in the northern sky is the bright star in the constellation *Lyra*, which is known generally by the term *Vega*. Vega may readily be recognised as the vertex of a nearly right-angled triangle constructed on the line joining the Pole Star to Arcturus. Four other stars in the constellation Lyra form a parallelogram which may be noticed. These stars are β, γ, δ, ζ.

The constellation *Cygnus* will be recognised as lying between Vega and the great Square of Pegasus. There are five principal stars in Cygnus which form a remarkable configuration (Fig. 46).

FIG. 46.

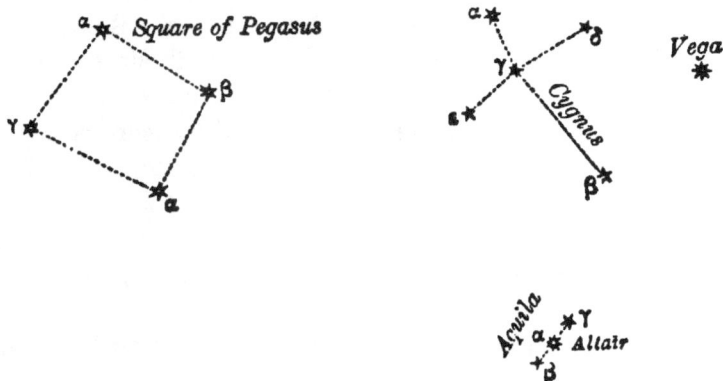

The last *constellation* which will be noticed is *Aquila*. This will be easily determined, because a line drawn from Vega across β Cygni passes a little above the group of three stars which form the most conspicuous part of Aquila (Fig. 46). The star a Aquilæ, otherwise called *Altair*, is of the first magnitude.

§ 26. *Diurnal Motion of the Heavens.*—After the learner has become familiar with the appearance and names of the leading constellations, he is recommended to notice the apparent movements which they perform on the surface of the heavens.

The constellation Ursa Major is to be looked at early in

the evening, and its position with regard to the trees, or houses, or other terrestrial objects is to be noted. If the observation be renewed a few hours later, a very remarkable change will be perceived. The *relative* positions of the stars have not indeed altered. All the angular distances between the several pairs of stars in the constellation are the same on the two occasions. Nor have the stars changed their positions with reference to the stars in the adjacent constellations, but the whole system of stars has moved bodily.

Observation will also show that the position of the Pole Star does not appreciably change its place either with respect to the stars or with respect to the terrestrial objects by which its position may be indicated. At different hours of the night, or at different seasons of the year, the Pole Star will constantly be seen in the northern sky at about the same elevation above the horizon. We shall hereafter have to explain that with more careful methods of measurement this statement would not be found strictly accurate. The Pole Star *does* really change its position, though the amount of that change is not sufficient to be readily appreciable in the coarse, naked-eye observations which we are at present discussing.

A marked contrast is thus perceptible between the fixity of the Pole Star and the large movements which are made by the stars in Ursa Major. It is, however, to be noticed that the two stars α and β (Fig. 37), which point towards the Pole Star, continue to point to the Pole Star, notwithstanding the great movements which they undergo in common with all the other stars of the constellation. The idea is thus suggested that the stars of Ursa Major really move as if they were all fastened together by invisible rods, and as if each of the stars was also fastened by an invisible rod to the Pole Star, about which the whole system is free to turn. If the observations be renewed at intervals through the night, and then with the help of a telescope through the following day, it would be seen that Ursa Major, after ascending from the east, passes over the observer's head, then down towards the west, under the Pole Star in the north, and round again to the east, and the observer would find that in about twenty-four hours the constellation had returned to its original position.

This movement of the constellations by which each of them moves (or appears to move) round the heavens in about twenty-four hours is termed the *Diurnal Motion.* It will be instructive to trace the same series of movements in some other celestial objects. Take, for example, the remarkable group of small stars known as the Pleiades (Fig. 39). This beautiful group is visible at night throughout the greater part of the year, but it need not be looked for from the middle of April to the middle of June. Winter is the best season for observing it. In November this little group may be detected in the east shortly after sunset. It will then gradually rise until about midnight, when it reaches its greatest height. After passing the highest point the group begins to descend, and gradually gets lower and lower, until it disappears in the west. There is, however, this very remarkable difference between the motion of the Pleiades and that of Ursa Major. The latter could be followed (with a telescope) through a complete revolution (at least in our latitude), but this is not the case with the Pleiades, for they actually disappear in the west, and after some hours reappear again in the east. If, however, the time be noted which elapses between two consecutive returns of the Pleiades to the same position, the interval will be found to be equal to the time of revolution of Ursa Major.

§ **27.** *The Equatorial Telescope.*—To study the apparent diurnal motion of the heavenly bodies with the accuracy which its great importance demands, we call in the aid of the astronomical instrument called the *Equatorial Telescope.*

The equatorial in its essential features consists of an astronomical telescope attached at its centre o to an axis A B, called the *polar axis.* The telescope is capable of being turned round the axis passing through o, while the polar axis is capable of being turned round the two pivots at A and B. By the combination of these two motions it is possible to direct the telescope towards any required point. To render this instrument serviceable for astronomical purposes, the polar axis must be carefully adjusted in a very special direction. The axes of the two pivots at A and B

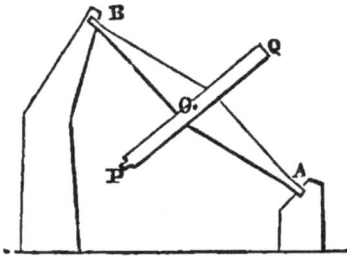

FIG. 47.

being supposed to form parts of the same straight line, the direction of this straight line should point towards a particular point on the surface of the heavens in the immediate vicinity of the Pole Star. This point is the *Pole*. It will be subsequently explained how this adjustment of the polar axis of the equatorial is to be made, but for our present purpose we may assume that the adjustment is perfect.

The equatorial telescope may be employed in watching the apparent diurnal movement of a star by the following method. Point the telescope to the star shortly after the star has made its appearance in the east. Then *clamp* the telescope so that it can no longer turn around the axis through o, but leaving the polar axis carrying the telescope along with it quite free to turn round the pivots A, B. It will then be found that by the simple operation of turning the polar axis round at the proper speed, the star can be kept continually in the field of view, although the magnitude of the angle B O Q made by the telescope with the polar axis remains unaltered.

To appreciate the full significance of the lesson which the observations with this instrument teaches, let the telescope be directed to any other star. For this purpose it must be unclamped, and after the angle B O Q has been suitably altered, so as to enable the star to be seen in the

telescope, the instrument is again clamped, so as to preserve the angle B O Q from alteration. It is again found
that the movement of the star can be followed by simply
rotating the polar axis, which carries with it the telescope.

If the polar axis of the equatorial be not directed exactly
to the correct point of the heavens, it will not be found that
the diurnal motion of a star can be followed by simply
rotating the polar axis without altering the angle B O Q.
It is to be particularly observed that when the polar axis
is set correctly for one star it is set correctly for all stars.
The Pole Star itself is no exception to this law. If the
instrument be directed to the Pole Star, we shall now see
what the coarser methods of observing failed to indicate
—namely, that the Pole Star is itself in motion. In this case
the telescope, when adjusted on the Pole Star, is inclined at a
very small angle to the polar axis. The angle B O Q is, in
fact, only about 1° 21′.

The other extremity of the polar axis may be supposed
to be prolonged downwards through the earth, and it will
then point towards a point in the southern heavens which is
called the South Pole. It happens, unfortunately for astronomers in the Southern hemisphere, that there is no bright
star situated so conveniently near to the South Pole as the
Pole Star is to the North Pole.
These two poles of the celestial
sphere are of the very utmost importance in astronomy.

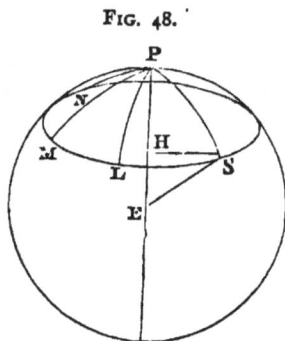

Fig. 48.

The polar axis of an equatorial
having been correctly adjusted, we
shall suppose that its direction is
E P (Fig. 48) ; and if P be the point
in which the direction of the polar
axis intersects the celestial sphere,
then P is the pole of the heavens.
Let s denote the position of a star on the celestial sphere
which is being observed, then the axis of the telescope

F

has the direction E S. We have seen that the diurnal motion of the star can be followed by rotating the telescope about the polar axis without at the same time making any alteration in the angle at which the telescope is inclined to the polar axis. Let us, then, suppose that the star, in consequence of the diurnal motion, assumes the various positions L, M, N, &c. When the star has arrived at L, the telescope must have moved into such a position that its axis is directed along the line E L. It follows that the angle between the polar axis and the telescope must be equal to the angle P E L. But this angle has remained unchanged notwithstanding the diurnal motion, and therefore we see that the angle P E S is equal to the angle P E L. The arcs P S and P L on the surface of the celestial sphere subtend at the centre the angles P E S and P E L respectively, and as these two angles are equal the two arcs must also be equal. In the same way it is shown that the arcs P M, P N, &c., are each equal to P S.

We have therefore ascertained that the diurnal movement of a star on the celestial sphere is always subject to the condition that the length of the arc drawn from the pole to the star is a constant. From s let fall a perpendicular S H upon the polar axis P E. Then, since the length of E S, the angle P E S, and the right angle at H remain unaltered during the motion of the star, it follows that the triangle S E H remains unaltered, and therefore also the lengths of the sides E H and S H. The line S H will therefore be always contained in the plane drawn through the point H perpendicular to the axis E P. The path of the star will always be in this plane, and, as the path is also on the celestial sphere, it follows that the actual path apparently described by the star will be the intersection of a plane and a sphere, and will therefore be a small circle of the sphere. The same process may be applied in the case of other stars, and therefore we are led to the following very important result :—

The apparent diurnal motions of the stars are performed in small circles of the celestial sphere, and all these small circles lie in a system of parallel planes.

The next question to be considered is the *rate* at which the apparent motion of each star in its small circle is performed. While the star moves from s to L (Fig. 48) the polar axis of the telescope must be rotated through an angle equal to that between the planes P S E and P L E. This angle is equal to that between the two arcs of the celestial sphere P S and P L. Thus while the star moves from s to L the polar axis of the telescope must be turned through an angle s P L. By means of a graduated circle, called the hour circle, of which the plane is perpendicular to the direction of the polar axis, it is easy to measure the angle through which the polar axis has been rotated when the telescope is turned from the position E s to the position E L. The observations are, then, to be made in the following manner :—When the star is at s, and when the telescope is pointed to it, note the position of the polar axis as indicated by the hour circle, and also the time as shown by a clock, and repeat the observation when the star arrives at other points L, M, N of its path. From the readings of the hour circle the angles s P L, L P M, M P N, &c., become known, while from the recorded clock times we have the times of moving from s to L, from L to M, and from M to N, &c. It is, then, found that the *time intervals are proportional to the angles through which the polar axis has been rotated.* The distance through which the star appears to move is therefore proportional to the time, and hence we have the following important law of the diurnal motion :—

The velocity of a star in the small circle which it appears to describe in the diurnal motion is uniform.

It follows from this law that when the equatorial telescope is to be moved so as to keep the star continually in the field of view, the angular velocity with which the polar axis revolves must be uniform. To secure this condition a clockwork arrangement is usually attached to the telescope, which carries the polar axis (and of course the telescope attached thereto) with the correct velocity.

So far we have only been considering the movement of a single star in its small circle. We have now to make a comparison between the movements of different stars in their appropriate small circles. This can also be effected by the equatorial, in the following manner :—Point the telescope to a star, and note the time when the star passes the centre of the field ; then, without altering the position of the telescope or the polar axis, observe the time when the star, after having performed a complete revolution of the heavens, returns again to the centre of the field. This time will be found to be equal to

$$23^h \ 56^m \ 4^s.$$

Turn the telescope to any other star, and make the same observation. It will be found that for the second star the time will be precisely the same. This conclusion may be extended to other stars, and thus we are led to the following important result :—

The time occupied by a star in performing its diurnal path around the heavens, is the same for all stars, and is equal to $23^h \ 56^m \ 4^s$ *of mean solar time.*

§ **28.** *Circles of the Celestial Sphere.*—The study of the apparent diurnal motion of the heavens is of such funda-mental importance that we shall proceed to illustrate it in a somewhat different manner. It will first be necessary to explain some terms which are of frequent use.

An observer, situated on a ship out of sight of land, ob-serves that the celestial sphere is bounded by a circle, below which he cannot see. This circle, when regarded as one of the great circles of the celestial sphere, is the *horizon.*

If a weight be suspended by a thread from a fixed point, then when the weight is at rest the thread is said to be *vertical.* That point of the heavens to which the thread points, and which it would appear to reach if it could be prolonged indefinitely upwards, is the *zenith*, while if the direction of the thread be prolonged downwards through the earth it will point to that part of the celestial sphere which is termed the *nadir*.

A straight line which is perpendicular to a vertical line is called a *horizontal line*, and all the straight lines which can be drawn perpendicular to a vertical line through any one point in it lie in one and the same plane, which is called a *horizontal plane*.

If the face of an observer in the northern hemisphere be directed towards that part of the heavens where the sun is at noon, the part of the heavens in front of him is termed the *south*, that behind him is the *north*, while the *east* is on his left hand and the *west* upon his right.

The great circle of the celestial sphere which passes through the poles, and also through the zenith, is called the *meridian*. The meridian may also be defined as the great circle which passes through the north and south points of the horizon, and also through the zenith.

Fig. 49.

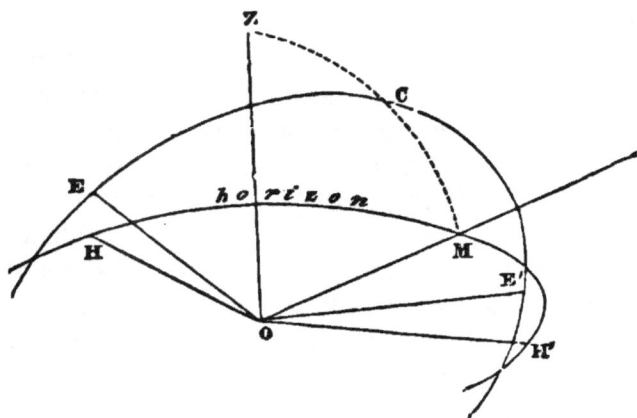

Let the circle H M H′ (Fig. 49) represent the horizon, and let z be the zenith, the position of the observer being at o.

If M be the south point of the horizon, then the great circle z M is a portion of the *meridian.* Let E C E' represent the apparent path of a star. Let H be the point on the horizon vertically beneath the star in the position E. This point will be found by drawing a great circle from z to E, and producing it on to cut the horizon at H. The angle E O H is termed the *altitude* of the star at E. It is observed that the altitude of the star increases until the star arrives at the meridian z M, which it crosses at the point C. When the star has this position its altitude is the arc C M, or the angle which that arc subtends at O. The altitude at this point is a maximum, and the star is said to *culminate.* After passing the meridian the star moves along the arc C E', and begins to descend towards the horizon ; when it reaches the point E' its altitude is only E' O H', and this altitude gradually diminishes until the star crosses the horizon, or *sets,* as it is called. In the case here represented only a portion of the path of the star is visible, the remainder lying below the plane of the horizon. It is important to observe that the path of the star is symmetrical with respect to the meridian. It cuts the meridian at right angles, and at equal distances on either side of the meridian the altitudes are equal.

§ 29. *Circumpolar Stars.*—In this case we have supposed the star is placed at a considerable distance from the pole ; let us now take the case of a star situated comparatively near the pole. For this purpose the observer must turn his face towards the north (Fig. 50). P is the pole, z is the zenith, and the great circle z P is the meridian which cuts the horizon at the northern point. Let E denote the position of the star, which is in the act of crossing the meridian by the diurnal motion. Then in the course of a little less than twelve hours the star will have moved round to the position E' below the pole. We have already seen that the diurnal motion is performed subject to the condition that the distance from the pole to the star measured along the celestial sphere remains constant ; we must therefore have the arc P E equal to the arc P E'. The star, when at the position E, is said to be at its *upper culmination*, while at E' the star is at its *lower culmination*.

FIG. 50.

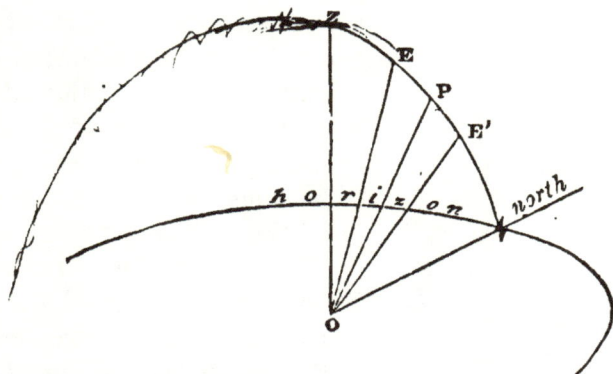

A star which is situated at a distance from the pole which is less than the altitude of the pole above the horizon will be visible both at upper and lower culmination. Such a star will never either rise or set, and is said to be a circumpolar star. If, however, the distance from the star to the pole be exactly equal to the altitude of the pole, then the star at its lower culmination will just graze the horizon. If the polar distance of the star exceed the altitude of the pole, then the star will pass below the horizon, and its lower culmination will not be visible.

It should be mentioned that the phenomena we have described are to a certain small extent modified by the influence of atmospheric refraction. For example, the arc P E (Fig. 50) is not precisely equal to the arc P E', because the refraction has acted unequally upon the position of the star in both cases. We can, however, allow for the amount of refraction, and thus we can ascertain what the distances would be if we could see them without the modification which the atmosphere produces. When these allowances have been made it is found that the laws enunciated are accurately fulfilled.

§ 30. *The Globe.*—We are now in a position to form a distinct idea of the varied series of phenomena which the diurnal rotation brings before us.

Let us take a globe A (Fig. 51), which is movable about an axis P Q. The axis P Q is supported by a ring M M, which is sustained by another ring H H. We may conceive the axis P Q to be adjusted so that it is parallel to the line joining the two poles of the heavens or to the polar axis of the equatorial (Fig. 47). The plane H H is to be made horizontal. Then the globe may be taken to represent the celestial sphere, of which P, Q are the two poles, the ring H H indicates the horizon, while M M, which passes through the zenith and the pole P, is the meridian. On this globe we may suppose that the principal stars of the celestial sphere are marked in their proper relative places, and thus we have a miniature representation of the celestial sphere.

FIG. 51.

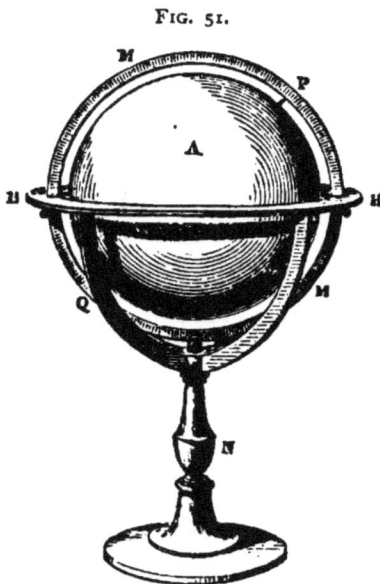

The globe is supposed to be free to turn around the polar axis P Q, and we may assume that by suitable clock-work such a motion is given to the globe as will turn it round its axis in 23^h 56^m 4^s, while the velocity with which the motion is effected is uniform. It is easy to show that under these circumstances all the phenomena of the diurnal motion will be faithfully represented.

A star will, in consequence of this motion, gradually

ascend above the circle H H and thus *rise*; it will ascend
farther and farther until it reaches the meridian M M,
or *culminates*; after culmination it will gradually descend
again to H H, or *set*. So also a circumpolar star will be
seen to pass the meridian M M at its upper culmination, and
then in $11^h 58^m 2^s$ afterwards to cross the meridian at
the lower culmination.

It will further be seen that the laws of the diurnal
motion will be preserved in this miniature representation
of the phenomenon. In the first place, the distance from
the pole to the star measured along the surface of the sphere
remains unaltered; the velocity of each star in its path is
also obviously uniform, while the time of a complete revolution is the same for each star as the time of revolution of
the sphere—i.e. $23^h 56^m 4^s$.

We are, therefore, led to the important conclusion that
the apparent diurnal motion of the heavens is performed in
the same manner as if all the stars were actually stuck on
the surface of a hollow sphere which revolved uniformly
about an axis passing through its two poles in $23^h 56^m 4^s$.

The period of revolution of the celestial sphere is practically a constant quantity. It is, therefore, very natural to
adopt this period as the unit of time, and it is known as the
sidereal day. The sidereal day is subdivided into twenty-
four hours, each hour into sixty minutes, and each minute
into sixty seconds. An interval of time expressed in these
fractional parts of a sidereal day is termed *sidereal time*.
The difference, amounting to nearly four minutes, by which
the sidereal day falls short of the ordinary day of civil
reckoning will be subsequently explained.

The polar axis of an equatorial telescope prolonged both
ways to the celestial sphere cannot be distinguished from
the axis about which the celestial sphere rotates. This
is true wherever the equatorial be situated. It would,
therefore, appear that the axis of the celestial sphere may
be considered to pass through any point on the surface of

the earth. It is, however, obvious that there is really only one axis about which the celestial sphere is turning. The only way by which' these discrepancies can be reconciled is by supposing that the dimensions of the earth are exceedingly small in comparison with the distance at which the stars are situated. · The axis must actually occupy some definite position in the earth ; but the earth is, comparatively speaking, so small that the observer must always be so near to the axis that the observed phenomena are the same as they would be were he actually situated on the axis. The polar axes of the equatorials all over the earth are absolutely parallel to each other, and they all intersect the celestial sphere in two small regions which for all practical purposes are undistinguishable from the north and south poles of the celestial sphere.

§ 31. *Rotation of the Earth.*—The apparent diurnal motion of the heavens might be no doubt explained by the hypothesis that the celestial bodies were all attached to the interior surface of a colossal globe of which the earth was the centre, and that this globe revolved around one of its diameters once every sidereal day. There is, however, another method of explaining the diurnal motion, which demands our careful attention. The earth itself is, as we have seen, an isolated body in the universe, and is attached to no other object. What is there, then, to prevent the earth being actually in motion? It may be said that we, as dwellers on the earth, do not feel the motion ; but this is no argument. If you are seated in the cabin of a canal boat and have no opportunity of looking out, you cannot tell whether you are in motion or not. If you can look out and see the trees and other objects on the bank apparently moving, you will at once refer the motion to yourself and acknowledge the trees are at rest. In this way if the earth were moving equably it would be impossible for us to detect that motion except by comparing our positions with the positions of external objects. It is, therefore, reasonable for us to ask whether we can obtain from external objects any information of the state of the earth as to rest or motion.

A solid body like the earth may have either of two different descriptions of movement, or, more generally, it may possess both forms of movement combined together. The body may have a movement of translation, by which all points of the body are at any moment moving in parallel lines, or it may have a movement of rotation about an axis, or it may have, and generally has, the movements of translation and of rotation combined. Setting aside for the present the question of translation, let us enquire whether the earth possesses the features that we might expect if it were actually rotating about an axis. The celestial bodies are the only external objects by the observation of which we are to see whether the earth has any movement of rotation or not. If the earth be rotating in the opposite direction about the axis of the sphere, and with the uniform velocity which would complete the revolution in a sidereal day, then the apparent diurnal motion would be completely explained. We have, therefore, two solutions of the problem of the apparent diurnal motion. We may suppose that the celestial sphere is revolving around the earth from east to west, while the earth is at rest ; or we may suppose that the celestial sphere is at rest, and that the earth is revolving from west to east, and thus produces the *apparent* motion.

Which of these two solutions are we to adopt? We shall see hereafter that many of the celestial bodies are vastly larger than the earth, that they are situated at very great distances from the earth, and that some of these distances are very much greater than others. It therefore seems much more reasonable to suppose that the earth, which is a comparatively small body, should be in a condition of rotation rather than that the vast fabric of the universe should all be moving round the earth once every day. Astronomers, therefore, now universally admit that the true explanation of the apparent diurnal motion of the heavens is to be found in the fact that the earth revolves on its axis once every sidereal day from west to east.

§ 32. *Shape of the Earth connected with its Rotation.*—A remarkable confirmation of this conclusion is presented by the shape of the earth itself. Conceive a straight line drawn from the centre of the earth towards the north pole of the celestial sphere. This straight line will cut the surface of the earth in a point which is called the north pole of the earth. By means of the surveying operations which have determined the figure of the earth we are enabled to ascertain the point on the earth's surface which is the extremity of the shorter axis of the ellipse by the rotation of which the figure of the earth can be produced. It will be noticed that the apparent diurnal motion has nothing whatever to do with the surveying operations, so that it is exceedingly remarkable to find that the north pole of the earth is close to, if not actually identical with, the extremity of the shorter axis of the ellipse. Thus we see that the axis about which the earth actually rotates coincides with the shortest diameter of the earth. In this we have another very remarkable proof of the reality of the earth's rotation. It is generally believed that at some very remote epoch the earth was in a fluid or a semifluid condition. At the time that this was so the effect of the centrifugal force would make the earth bulge out at the equator and flatten it down at the poles, and thus impart to it the shape of an ellipsoid of revolution. The shortest axis of this ellipsoid would coincide with the axis of rotation. We are thus led to the belief that the observed coincidence between the axis of the apparent diurnal rotation of the celestial sphere and the shortest axis of the earth is a proof that the apparent diurnal motion of the heavens is really due to the rotation of the earth on its axis.

§ 33. *Definition of Terms.*—To enable us to define the positions of the various celestial bodies on the surface of the celestial sphere it is necessary to imagine certain circles traced upon its surface, with reference to which the positions of the bodies may be specified. There is considerable convenience in choosing these circles, so as to be symmetrical with reference to the diameter of the celestial sphere about which the apparent diurnal motion is performed.

Let the point c be the centre of the celestial sphere (Fig. 52), and let the diameter P Q be the axis of the apparent diurnal rotation. The earth may be regarded as a particle of exceedingly small magnitude situated at the centre c. If through the point c a plane be drawn which is perpendicular to the polar axis P Q, this plane will cut the surface of the celestial sphere in a great circle E E, which is called the *celestial equator.* The celestial equator divides the celestial sphere into two equal portions, in which the poles P and Q

FIG. 52.

occupy two symmetrical positions. The hemisphere which contains the north pole is called the *northern hemisphere,* while that which contains the south pole is termed the *southern hemisphere.*

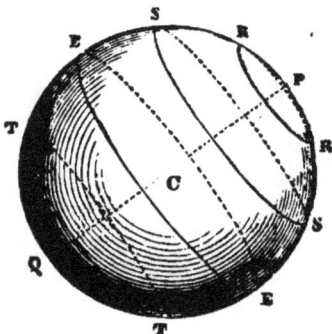

If the celestial sphere be cut by a plane s s which is parallel to the plane of the equator, the section of the celestial sphere is termed a *parallel.* The circles T T and R R are also *parallels.* The effect of the diurnal motion is to make each star move around the celestial sphere in the parallel which is appropriate to it. A plane passing through the axis P Q cuts the celestial sphere in a circle, called a *declination circle.*

§ 34. *Right Ascensions and Declinations.*—By the aid of the various parallels and declination circles drawn upon the surface of the celestial sphere we are enabled to specify the position of a star on that sphere with the greatest facility. It must be understood that we are not now speaking of the absolute position of the body in space ; for this it would be necessary to know the actual distance of the body from the earth, of which in the great majority of cases we are entirely ignorant. What we now refer to is merely the *direction* in space where the body is situated.

Let A be a star on the celestial sphere. Then we are to draw a great circle from the pole through A, and this great

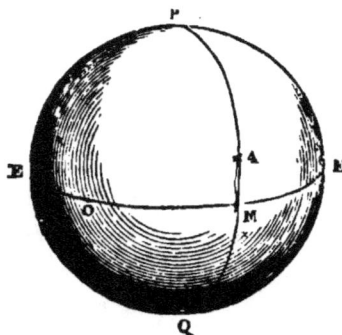

FIG. 53.

circle cuts the equator at the point M. Taking an arbitrary point O upon the equator from which to make our measurements, it is clear that the declination circle P A Q will be completely defined if we know the length of the arc O M. If the distance of A from M be also known, then we have a complete method of specifying the position of the point A.

Instead of the arcs O M and A M we may speak of the angles which those arcs subtend at the centre of the celestial sphere, and thus we may define the position of the star A by two angles as well as by their corresponding arcs.

The arc O M, which is the distance between the declination circle passing through A and the standard point O, is called the *right ascension* of the star A ; while the arc A M, which is the arc of the great circle drawn through the star perpendicular to the equator, is termed the *declination*.

The position of a star is thus completely specified when its right ascension and declination are known. It is only necessary to mark off a distance O M upon the equator equal to the given right ascension, and take off the distance M A equal to the given declination. In this way the position of the star can be expressed in a manner which is quite unambiguous. If the right ascension of the star only were given, then all we would know as to its position is that it is somewhere on the declination circle passing through A. If the declination of the star only were given, then all that would be known is that the star lies somewhere on the 'parallel' passing through A.

The point o, from which the right ascensions of the stars are measured, may be chosen arbitrarily. For example, we might agree that o should be defined to be the point in which a declination circle, passing through a certain conspicuous star, such as Vega or Sirius, cuts the equator. Astronomers have, however, adopted a different method, and the point o is defined by the motion of the sun, in a manner which will be explained subsequently.

Right ascensions are measured from west to east, and increase from 0° to 360°. Declinations are measured above and below the equator, and are taken as positive when the star lies between the equator and the north pole, and negative when the star lies between the equator and the south pole.

§ **35.** *The Transit Instrument.*—For the determination of the right ascensions of the celestial bodies use is made of the *transit instrument*, the principle of which will now be described.

A telescope A B (Fig. 54) is fixed to an axis X Y, the direction of which is at right angles to the line of collimation of the telescope. The shapes of the axis and the telescope are so designed as to secure as much rigidity as possible. At the extremities of the axis are two cylindrical pivots X, Y, which turn in suitable bearings sup-

FIG. 54.

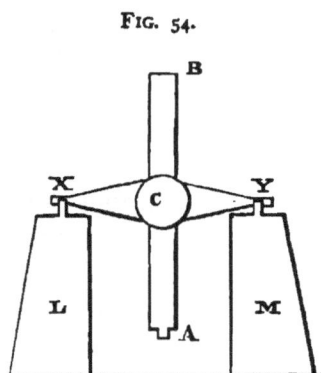

ported on solid masonry piers L, M. The transit instrument is capable of turning around these pivots, but it can have no other motion. As the axis of A B is at right angles to the line of pivots X Y, the movement of the line A B must obviously be limited to the plane perpendicular to X Y.

In the focus of the object glass of the telescope are stretched a number of parallel spider-lines A B, &c. (Fig. 55).

FIG. 55.

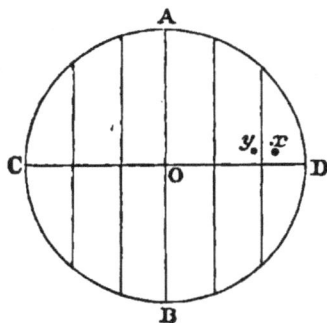

These lines are placed perpendicular to the axis X Y, and therefore parallel to the plane in which the movements of the axis of the telescope are confined. These lines are at equal distances apart, and their number depends upon the particular kind of observations which are made. It is sometimes as low as five, and sometimes as great as twenty-five. The vertical system of spider lines is also crossed at right angles by one or more horizontal spider-lines C D.

When the telescope is pointed to a star the image of the star is formed in the same plane which contains the system of spider lines. Consequently, when the eye piece has been adjusted, the star and the spider lines will be clearly seen together. If the telescope be at rest the apparent diurnal motion of the celestial sphere carries the star across the spider lines one after the other; and the instrument must be adjusted so that the path of each star across the field will be parallel to the horizontal wire C D, and therefore parallel to the axis about which the telescope revolves.

We have now to explain how the transit instrument is to be adjusted and placed in its proper position. If the point O, which is the intersection of the horizontal line with the central vertical line, be joined by an imaginary line to the centre of the object glass, this line is the axis of collimation.

The pivots at x and y, about which the instrument rotates, are presumed to be perfectly cylindrical, and the diameters of these cylinders are presumed to be equal and their axes to be collinear. The straight line which contains the axes of the two pivots is the axis around which the telescope revolves.

The first adjustment of the transit instrument which we shall consider is purely instrumental. It consists in fixing the telescope to the axis in such a manner that the line of collimation shall be exactly perpendicular to the axis of revolution. The astronomical instrument maker can, no doubt, effect this adjustment to a high degree of approximation so far as mere mechanical measurement is concerned. The excessive delicacy of astronomical observations renders errors visible which would entirely elude a less subtle method of detecting them, and it is therefore necessary to point out how we can avoid the effects of the errors into which inaccuracies in the construction of the instrument would lead us. This is attained not so much by endeavouring to correct these errors as by ascertaining their amount and then allowing for the effects which they produce on the various observations. For the present, however, we shall merely point out how the errors can be detected.

§ 36. *Error of Collimation.*—The amount by which the angle between the line of collimation and the axis of revolution exceeds or falls short of a right angle is termed the *error of collimation.*

Let A B (Fig. 56) be the axis of revolution and x y be the line of collimation. (We have purposely greatly exaggerated the error of collimation in the figure.) Suppose that the telescope is pointed to a fixed object, suitably placed at a considerable distance, and that a certain point P of the object

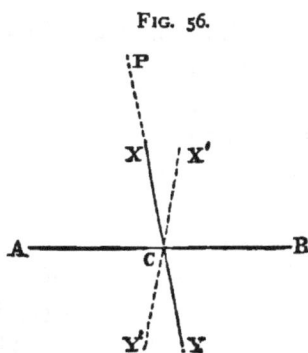

Fig. 56.

is noted, of which the image in the field of view is coincident with the point o in the field of view (Fig. 55). Let the telescope be lifted from its bearings and replaced with the pivots reversed—i.e. the pivot which was previously at the west is now at the east, and *vice versâ*—and let the telescope be again directed to the distant mark. If the axis of collimation be at right angles to the axis of revolution, then it is plain that the line of collimation would still point in the direction which it had before the reversal. But if the two axes are not at right angles, then the line of collimation, instead of having the direction x y, will have the direction x' y'. This will be made manifest at once from the circumstance that the point of the distant body, which had its image at o before the reversal, will not have its image at that point after the reversal. This points out the method of correcting the error of collimation. It is to be remembered that the line of collimation is found by joining o to the centre of the object glass. The system of spider lines are set in a frame, and this frame is attached to the tube of the telescope by adjusting screws. If the frame be moved by these screws, then the point o is moved, and thus the line of collimation is moved. To get rid of the error of collimation it is therefore necessary to move the frame containing the system of wires until the point on the distant mark whose image coincides with the point o is the same after the reversal of the instrument as it was before.

When this adjustment has been made the line of collimation moves in a plane, and this plane cuts the celestial sphere in a *great circle.* It is the object of the subsequent adjustments to arrange that this great circle shall be *coincident with the meridian.*

§ 37. *Error of Level.*—The second adjustment which we shall describe consists in placing the axis of revolution in a horizontal plane. This is accomplished by means of the spirit level A A (Fig. 57), which is suspended between the arms B, B. These arms are provided with hooks at the extremities, by means of which the level can be suspended from the pivots in the manner shown in Fig. 58. The

FIG. 58.

FIG. 57.

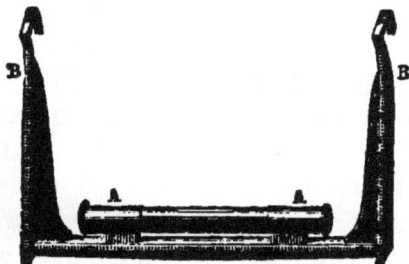

bubble in the level assumes a certain position, which can be read off by means of a graduated scale. The level is then to be *reversed*, so that the hook which previously hung from one pivot now hangs from the other, and after allowing a few moments for the bubble to come to rest its position is to be read off again. If the two pivots be accurately horizontal, then the position of the bubble will be the same after the reversal as it was before ; but if there be the slightest departure of the bubble from its original position the axis of revolution is not horizontal. This can, so far as its grosser portion is concerned, be corrected by raising or lowering one of the bearings of the pivots until the bubble retains the position after reversal which it had before.

We may here mention that by the help of the level we can also scrutinise another detail in which the instrument may be more or less imperfect. We have assumed that the diameters of the two cylinders which constitute the pivots are equal, but this may not be (indeed, generally is not) strictly true. Their equality may be tested in the following way :—Adjust the pivots so that they appear horizontal ; then reverse the axis of the telescope, so as to interchange the pivots in the bearings, just as was done in the case of determining the collimation. If the pivots be equal the level will show that they are horizontal after the reversal; but if the two pivots, being horizontal before the reversal, are found not to be horizontal after the reversal, then the pivots are not of the same size. By these observations the actual difference (if any) between the diameters of the two pivots can be ascertained, and the effect of this difference upon the observations can be calculated and allowed for.

By the adjustment of the bearings the greater portion of the error of level can be actually corrected, but it is not possible to make this adjustment as perfectly as the accuracy of astronomical observations requires. Even if the axis were once made level, it would not remain so. Slight changes in the temperature, and possibly changes in the earth itself, alter to a minute extent the piers upon which the bearings rest. The level is thus continually undergoing small fluctuations. The effect of these may be obviated by determining, in the course of the observations each night, the actual inclination of the axis, and then correcting the observations by calculation, so as to reduce them to what they would have been, had the telescope been perfectly adjusted.

§ 38. *Error of Azimuth.*—By correcting the telescope for the error of collimation it is provided that the line of collimation moves in a plane perpendicular to the axis of revolution. The effect of making the line of pivots horizontal is to cause this plane to pass exactly through the zenith. All the planes passing through the zenith are vertical planes, and the last adjustment of the transit instrument consists in so placing it that the particular vertical plane to which the movements of the telescope are restricted shall coincide with the meridian of the place. As the meridian is defined to be the great circle which passes through the zenith and the celestial pole, the adjustment now under consideration will be secured if the plane in which the telescope moves be caused to pass through the celestial pole. For this purpose we have to resort to observations of the stars, as it is by this means alone that this adjustment can be effected.

It is easy to arrange the bearings so that the axes of the pivots shall lie approximately in its correct position due east or west, or, what is the same thing, that the central spider line shall be very close to the meridian. If the Pole Star were actually at the pole, then it would only be necessary to move one of the bearings horizontally northwards or southwards until the Pole Star, when viewed through the telescope, would be seen to coincide with the central wire. As the Pole Star is really a degree and a half from the pole, the operation is not quite so simple. Still the Pole Star (like every other star) crosses the meridian twice in every revolution of the celestial sphere. The times at which the Pole Star is on the meridian can be found from the 'Nautical Almanac,' and the approximate adjustment of the transit instrument may be made by making the observation at that time, and then moving the bearing until the Pole Star is seen to coincide with the central wire.

Though this adjustment is susceptible of a considerable degree of accuracy, yet, just as in the case of the level, it is hopeless to attempt to place the instrument absolutely correctly. Nor, if it were once placed in the true position, would it be safe to calculate that that position would be retained. It is therefore necessary, on each occasion when the telescope is used for accurate observations, to have the means of computing to what extent the axis of revolution really departs from the true east and west position. This is called the *error of azimuth.*

For the determination of the error of azimuth we require the assistance of a good clock, and we choose for the observations some conspicuous star situated near the pole, preferably the Pole Star itself. Let A C B D (Fig. 59) represent the apparent path of the Pole Star in its diurnal motion, the point o, which is the centre of the circle A C B D, being of

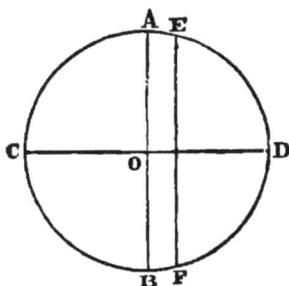

course the celestial pole. The meridian passes through the pole, and we may denote that small portion of the meridian which lies within the circle by the line A B. The telescope having been approximately adjusted, the vertical circle which the axis of collimation describes will pass very close to the pole, and we may denote that portion of it which lies within the path of the Pole Star by the line E F. The diameter of the small circle A B is less than three degrees of arc on the celestial sphere, and consequently we may often regard A B and E F as straight lines, though of course they are really parts of great circles. In a period of one sidereal day the Pole Star moves completely round the circle A C B D; this motion is described uniformly, so that the time taken to move from A through C to B is equal to the time taken to move from B through D to A. The telescope being directed to the Pole Star, the time is to be noted when the star crosses the central wire of the system; the star is then at the point E. After an interval somewhat exceeding half a sidereal day the Pole Star will again be visible at its lower culmination, and will again cross the central wire of the system at the point F. The instant of crossing is to be again noted by the help of the clock. In another period of somewhat less than half a sidereal day, the star will have returned to upper culmination, and the clock time is to be taken again when the star is at E. If the telescope were correctly adjusted, so that the vertical circle described by the line of collimation coincided with A B, then the time interval between the observed upper and lower culmination would be equal to that between the lower and upper; if, however, the telescope be not perfectly adjusted, there will be a discrepancy between these two intervals, arising from the fact that the arc E C F

FIG. 59.

exceeds the arc E D F. It is thus easy to see that by these observations we have the means of ascertaining if the telescope be correctly placed, while if the telescope be not

FIG. 60.

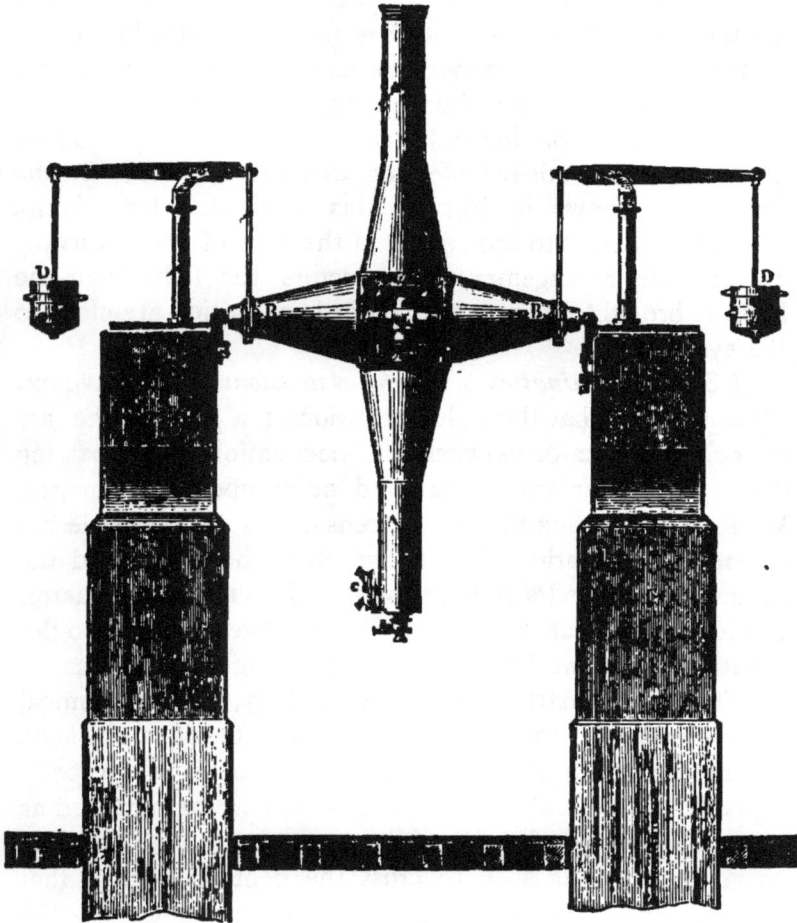

correctly placed the observations give the means of ascertaining how much that place is erroneous.

The general appearance of the transit instrument as it is actually used in astronomical observations is shown in Fig. 60. The bearings of the pivots repose on massive masonry piers C, C. In order to ensure that the form of the pivots shall not be injured by friction, the weight of the telescope and axis is to a large extent relieved by the counterpoises D, D, which, by means of hooks B, B furnished with friction rollers, support the instrument. Just sufficient pressure is left on the pivots to ensure that they shall work steadily in the bearings. When the telescope is used at night it is necessary to make provision for illuminating the spider lines, which would otherwise be invisible against the dark sky. There are various methods for effecting this illumination. In the instrument shown in Fig. 6o this is effected by a lamp, which is attached to the centre of the tube of the telescope. By suitable arrangement of reflectors the light from the lamp is brought down to the spider lines, which are close to the eye piece.

§ 39. *Determination of Right Ascensions.*—We have explained (§ 34) that the right ascension of a star is the arc on celestial equator between the declination circle passing through the star and a standard point upon the equator. We may also define the right ascension to be the angle between the declination circle drawn through the star and the declination circle drawn to the standard point of the equator. If, therefore, we can measure the angle between these two declination circles we have ascertained the right ascension.

The transit instrument and its auxiliary, the astronomical clock, enable us to measure the angle between the two declination circles. When the instrument is properly adjusted, the central wire of the system may be regarded as lying in the plane of the meridian ; so that the moment the image of a star is seen to cross the central wire, at that moment the star is actually crossing the meridian. For the present we may suppose that the right ascensions are to be measured from the declination circle which passes through some bright star—for example, Sirius—and to illustrate the method of observing we shall show how to find the right ascension of some other star—for example, Vega.

The first thing to be done is to make the clock keep

accurate *sidereal* time. For this purpose the transit instrument is to be directed to any convenient star, and the time shown by the clock is to be noted when the star passes across the central wire of the instrument. At its next transit the same star is to be observed, and if the clock be going correctly it will show an interval of twenty-four hours (sidereal time) between the two observations. This interval is equivalent to 23^h 56^m 4^s of ordinary civil time. The choice of a star for this purpose is almost immaterial; the interval will be the same for all stars, so that if the clock be correct for one star it will be correct for all. It is, however, convenient to choose a star which is at a considerable distance from the pole, for then its apparent motion is rapid and the moment of its transit across the central wire can be ascertained with accuracy. On the other hand, it is not convenient to choose a star which is low down at the time of transit. Observations made near the horizon are always liable to be influenced more or less by atmospheric disturbances, which do not affect to so great an extent the observations which are made when the star has a suitable altitude.

If the astronomical clock do not show the correct interval of twenty-four hours of sidereal time between the two transits of the star, then the length of the pendulum must be altered by screwing the bob up if the clock be going too slow, or screwing the bob down if the clock be going too fast. It will not be possible to make the clock go with perfect accuracy, but it is made to go very nearly right, and the amount which it gains or loses in twenty-four hours is determined. This error is called the *rate* of the clock, and the test of a good clock is sought not in the amount of the rate, but in the uniformity with which the rate is maintained. When the rate has been determined, then the time indicated by the clock at which a transit or other phenomenon occurs can be corrected, so as to render the observation as serviceable as if the clock had actually no rate.

When Sirius approaches the meridian the transit instrument is to be pointed so that the star enters the field of view, and at the moment when it crosses the central wire the time shown by the clock is to be noted. After a time Vega approaches the meridian; the instrument is to be pointed at Vega, and again the clock time is to be noted as Vega crosses the central wire. The interval between the two observations (corrected, if necessary, for the rate of the clock) will give the difference between the right ascension of Sirius and of Vega; for when Sirius was on the meridian the declination circle passing through Sirius must have coincided with the meridian, and when Vega was on the meridian the declination circle passing through Vega must have coincided with the meridian. The celestial sphere turns round in twenty-four hours; therefore a declination circle is carried through an angle of 360° in twenty-four hours, so that in one hour the declination circle moves through an angle of 15°. If we turn the time interval between the two observations into an equivalent angle at the rate of 15° for each hour, we obtain the angle between the declination circle passing through Sirius and that through Vega, which is of course the difference between the right ascensions of the two stars.

It is generally more convenient to speak of right ascensions as time intervals in hours, minutes, and seconds than to turn them into arcs expressed by degrees, minutes, and seconds. The reduction from one form of expression to the other is exceedingly simple. Thus we have—

$$15° \text{ correspond to } 1 \text{ hour.}$$
$$15' \quad \text{,,} \quad 1 \text{ minute.}$$
$$15'' \quad \text{,,} \quad 1 \text{ second.}$$

We have still to explain the true origin from which astronomers measure right ascensions. We may anticipate a future chapter so far as to state that there is on the celestial sphere a certain important point called the *vernal equinox.* Like every other point on the heavens, the vernal equinox crosses the meridian in virtue of the diurnal motion. The astronomical clock should be so adjusted that its hands should indicate o^h o^m o^s at the time when the vernal equinox is crossing the meridian. If the clock do this correctly, then *the time shown by the clock when any star crosses the meridian (allowing, if necessary, for the error of the clock) is the right ascension of that star.*

We shall suppose that the astronomical clock is going correctly, so that it has neither error nor rate, and also that the transit instrument has been correctly adjusted. The following is then the actual course of observations by which the right ascension of a star or any other celestial body is to be found :—

The observer first points the telescope at the right elevation. To enable this to be done a small graduated circle (shown in Fig. 60) is attached to the side of the telescope near the eye piece. The arm carrying the index is furnished with a spirit level, and when the index is directed to the proper point on the graduated circle the telescope must be moved until the level is horizontal. In this way the observer is able to ensure that the object which he is anxious to observe shall enter the field of the telescope. When the telescope has been set the observer takes his seat at the eye end, and after a glance at the clock commences to count the seconds. If the star be towards the south, then the image of the star enters the field of view, in virtue of the diurnal motion, at the right hand of the observer, and approaches the vertical wires. Let x (Fig. 55) be the position which the star has at one tick of the clock; then by the next tick it will have passed across the wire and be found at y. The experienced transit observer will rapidly estimate to a fraction of a second the instant when the star coincided with the wire, and he will note this down. Without taking his eye from the telescope he repeats this operation for each of the five wires, and he takes the mean of the five observations for the time of transit over the middle wire. By this method the time of transit is asceitained with much greater accuracy than would be practicable with the single observation at the central wire.

In some of the more important observatories this method of taking transits has been replaced by what is called the *chronographic method.* The chronograph consists essentially of a drum covered with paper, which revolves uniformly by clockwork. A galvanic arrangement is provided, by which, when the observer depresses a key, a small mark is made upon the drum. At each vibration of the pendulum of the clock a current is made, which also impresses a mark upon the drum. When the mark made by the observer is compared with the marks made by the clock, the time at which the observer depressed the key can be determined. In taking a transit by this method the observer depresses the key each time the star crosses one of the vertical wires. There is thus a record made upon the drum, from which the true time of transit is ascertained.

§ **40.** *The Meridian Circle.*—To know the position of a celestial body we require to determine both its right ascension and *declination.* We have already explained how the former element is found, and it remains to show how the

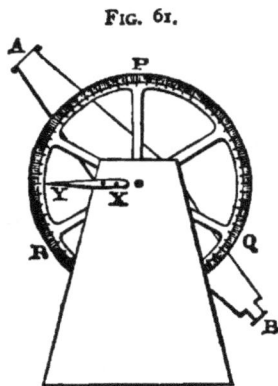

FIG. 61.

latter is ascertained. For this purpose we may employ the instrument known as the *meridian circle,* which, indeed, comprises the properties of a transit instrument with those additional arrangements which fit it for the measurement of declinations.

The principle of the meridian circle is shown in Fig. 61. It consists primarily of a transit instrument A B of the same construction as that already described. Attached to the axis, near each of the pivots, is a graduated circle, the plane of which is perpendicular to the axis of the pivots. One of

these circles is shown in the figure. These circles are fixed
to the axis, so that they revolve when the telescope revolves.
The circles are divided primarily into degrees, the numbers
of which run continuously from 0° to 359°. Each of these
degrees is again subdivided, the extent of the graduation
being limited by the size of the circles. In some of the
best instruments each degree is subdivided into thirty equal
parts, of which each part is therefore equivalent to two
minutes. The reading of these circles is effected by microscopes, which, as explained in § 13, subdivide the circle, so
that the actual readings are recorded in tenths of a second
of arc. It is usual to have four microscopes, at distances of
90°, for the purpose of eliminating the errors arising from the
eccentricity of the circle, as well as certain other errors in
the graduations themselves, which are of a periodic nature.
For the purposes of the general description now about to be
given it will be convenient to speak of the reading of the
circles as if it could be simply performed by a pointer x y
(Fig. 61).

As the diurnal motion of the heavens carries the star
which is being observed across the field of view, the observer moves the telescope so that the line in which the star
moves shall coincide with the horizontal wire c d, Fig. 55.
To enable this adjustment to be made with the delicacy
which is necessary, a slow motion is arranged, by which the
telescope can be moved by a handle in the hand of the
observer. After the star has passed out of the field the
position of the graduated circle is read off, and the result is
recorded to the tenth of a second.

A single observation of this kind is, however, insufficient
to give us any information as to the declination of the star.
We must, further, have the means of pointing the telescope
to some accurately defined spot on the celestial sphere.
Take, for example, some standard star—suppose Sirius—then,
if we observe Sirius in the way we have described, and if
we then observe some other star of which the place is to be

determined, the difference between the readings of the circle in the two cases will be the difference between the declinations of the two objects.

In making these calculations due allowance has to be made for the effect of refraction upon the position of each star. This is done as follows :—When Sirius is observed, the star is elevated by refraction to an amount which is nearly proportional to the tangent of the zenith distance. After the position of the circle has been read off we correct the result to what it would have been had the telescope been *depressed* into the position in which it would have been directed had we been able to view Sirius without the intervention of the atmosphere. In the same way the reading of the circle when the telescope has been directed to the unknown star must also be corrected for refraction, and it is the difference between these two corrected readings which is to be taken as the difference between the declinations of the two stars.

It may, perhaps, be urged that the zenith distance of the object, on which the refraction depends, cannot be known until the observations have been completely corrected, and that thus it would seem that we required to know the zenith distance before we could find the zenith distance. In reply it may be observed that, as the refraction is comparatively small, a small error in the zenith distance will only affect the refraction computed therefrom to an inappreciable extent. We may, therefore, first compute an approximate value of the zenith distance by omitting the refractions. By this approximate value of the zenith distance of the unknown object its refraction may be computed with sufficient accuracy, and then the difference of declinations can be found with precision.

In the list of standard stars published in the ' Nautical Almanac' every year their declinations are recorded for each day. By comparison with the declination of these standard stars we have thus the means of deducing the

declinations of unknown objects from the observations made
by the meridian circle. It is, however, necessary to point
out how the declinations of these standard stars are themselves obtained.

If the pole were a visible point on the celestial sphere,
then we could turn the telescope to that point, read off the
circle, then turn the telescope on the star and read off the
circle again. The difference between these two readings—corrected, of course, for refraction—would be the distance on the
celestial sphere from the pole to the star, or what is called
the *polar distance* of the star. If this polar distance be
subtracted from 90° the difference is the distance from the
star to the celestial equator, or the declination of the star.

There is, however, no star or other mark accurately
placed at the pole, and therefore we have to resort to an
indirect operation for ascertaining declinations. If we were
enabled to point the telescope accurately to the zenith, and
then read off the circles, and if we then turned the telescope
to the star and read off again, the difference between the
two readings corrected for refraction would be the *zenith
distance* of the star. There is, however, no point at the
zenith, just as there was no point situated at the pole, and
therefore we have no means of knowing when the telescope
is pointed precisely towards the zenith. We *have, however,
the means of pointing the telescope accurately towards the
nadir*, and thus we have the means of ascertaining the
distance on the celestial sphere from the unknown object to
the nadir ; then, by subtracting this from 180°, we find the
zenith distance of the object.

The surface of a liquid at rest is a horizontal plane. A
perpendicular to such a plane points upwards towards the
zenith and downwards towards the nadir.

A basin of mercury is placed underneath the meridian
circle, so that when the telescope is pointed with its object
glass vertically downwards the mercury is immediately
underneath it. A lamp being placed near the eye end of

the telescope, a simple contrivance enables a little light to be admitted, which illuminates the spider lines. On looking through the eye piece we are then able to see the spider lines, and also their reflections from the surface of the mercury.

To show how by this means we are enabled to point the telescope so that its line of collimation is accurately directed towards the nadir, we refer to Fig. 62. Let A B be

FIG. 62.

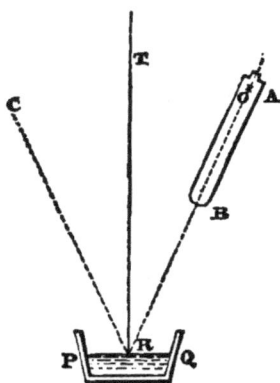

the line of collimation of the telescope, P Q the surface of the mercury, and the cross at o the intersection of the horizontal and vertical spider-lines. The rays of the light, diverging from the illuminated wires at o, pass down the tube of the telescope to the object glass at B. As a beam of parallel rays falling from a star on the exterior of the object glass are turned into a beam of rays which converge to the point o, so a beam of rays diverging from o and falling on the object glass emerge therefrom as a parallel beam, and in this condition they encounter the surface of the mercury. By the laws of reflection of light the beam, after reflection from the mercury, will travel in a direction parallel to the line R C, so that R C and B R make equal angles with the vertical line R T perpendicular to the surface of the mercury. In the circumstances which are represented in the figure it is obvious that the beam, after reflection, will pass away from the telescope altogether, so that the reflected image of the wires will not be seen.

But suppose that the telescope were placed with its axis of collimation exceedingly close to the vertical line R T; then the beam of rays, after reflection from the mercury, would in a great part fall again upon the object glass of the telescope:

These rays were parallel before reflection from the mercury, and they will remain parallel after reflection; they are therefore in a fit condition for coming to a focus again at the object glass of the telescope. On looking into the telescope we then see the system of spider lines, and together with them the images produced by the reflection. The images are somewhat fainter than the original lines, but are still quite distinct. By means of the slow motion of the telescope the relative position of the lines and reflected images can be altered, so that the horizontal line can be made absolutely *coincident with its reflected image.* When this is so, the axis of collimation of the telescope must be directed perpendicularly to the surface of the mercury, and must therefore point exactly to the nadir. While the telescope has this position, the circle is to be read off, and the reading thus obtained gives a standard by which the zenith distances of other stars may be ascertained. The difference between the position of the circle when the telescope is directed to the nadir, and the position when the telescope is directed to the star, is the apparent nadir distance of the star. The supplement of this distance, duly corrected for the effect of refraction, is the true zenith distance of the star at the place of observation.

§ **41.** *Latitude.*—The zenith distance of a star is not so convenient as the declination when the position of the star is to be recorded in a catalogue. This arises from the circumstance that although the zenith distance of the same star is almost absolutely constant at the same spot on the earth's surface, yet, when the locality of the observer is changed, it generally happens that the zenith distance of the stars is also changed. This does not arise from any change in the positions of the stars themselves on the celestial sphere, but it is due to the fact that the position of the zenith on the surface of the celestial sphere depends upon the position of the observer on the earth. It is, consequently, more convenient to refer the position of the stars to points

H

fixed on the celestial sphere than to those which depend upon the place from which they are to be observed. It is, therefore, necessary to explain how, when the zenith distance of the star has been measured, the declination of the star can be computed.

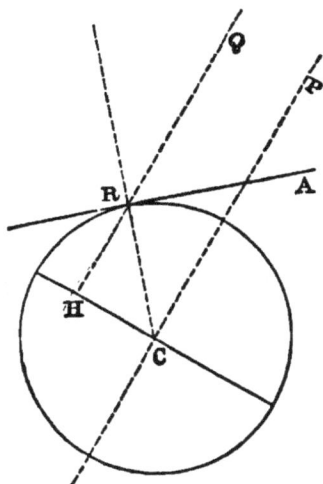

Let c represent the centre of the earth, and let R be a point on the earth's surface. Through c draw a line c P pointing to the pole of the heavens. This line cuts the surface of the earth in the two points which are known as the north and south poles respectively. Through the centre of the earth draw a plane c H perpendicular to the axis c P. This plane cuts the surface of the earth in a great circle which is known as the *earth's equator*. If we regard the earth as spherical, then a line drawn from any point R on the surface of the earth to the earth's centre makes with the equator an angle R C H, which is the *latitude* of the point R.

Fig. 63.

An observer stationed at the point R will see the pole of the celestial sphere in the direction R Q, which is parallel to c P.

A tangent plane drawn through the point R perpendicular to the radius c R forms the plane of the horizon. This plane cuts the plane containing c R and c P in the line R A. This line R A is in fact the direction of the north point of the horizon drawn from the situation of the observer. The angle between R Q and R A is the *altitude of the pole above the horizon.*

Since the revolution of the celestial sphere does not alter the position of the pole, the altitude of the pole is a constant at any point on the earth. The actual altitude of the Pole Star varies through a range of nearly three degrees.

As the angle A R C is a right angle, it is obvious that the two angles Q R A and H R C taken together must also make up a right angle. In the triangle H R C, since H R is parallel to C P, the angle at H must be a right angle, and therefore the sum of the angles H R C and H C R is also a right angle. It follows that the angle Q R A is equal to the angle R C H, but Q R A is the altitude of the pole and R C H is the latitude, whence we have the following very important proposition :—

The altitude of the pole above the horizon is equal to the latitude of the place.

§ 42. *Phenomena dependent on Change of Place.*—The proposition just proved will explain the series of changes in the appearance of the heavens which are presented to a traveller who makes a considerable change in his latitude. We may, for the sake of illustration, suppose that the traveller starts from the North Pole and travels along a meridian to the South Pole. When at the North Pole, it is plain that his latitude is 90°, and, therefore, the altitude of the celestial pole will be 90°. In this case the celestial pole will be actually coincident with his zenith. The horizon of the observer being at all points 90° from the celestial pole in his zenith, must in this case be the celestial equator. The diurnal motion of the celestial sphere will cause the stars to revolve in small circles which are all parallel to the horizon. The phenomena of the rising or the setting of the stars would be unknown. Each star preserves constantly the same altitude above the horizon. A star on the horizon would never ascend higher, and a star below the horizon would never cross the horizon and come into view. The observer would thus never have an opportunity of beholding more than one-half of the celestial sphere, so that (assuming the stars to be uniformly spread over the celestial sphere) about half the celestial objects in the universe would never be seen by him.

If the traveller shift his station to a latitude which, for

the sake of illustration, we shall suppose to be 45°, the apparent movements of the celestial sphere will have become much modified. The pole of the heavens will no longer be at the zenith, but will have an altitude equal to the latitude, i.e. 45°. If a star be situated within a distance of 45° from the pole, then this star will obviously never be able to pass below the horizon, for the polar distance of the star is constant, and the nearest point of the horizon to the pole is 45°. All stars within 45° from the pole will, therefore, be continually above the horizon, and they are called *Circumpolar Stars.*

In the case of a star visible at the latitude of 45° which had a polar distance exceeding 45° the phenomena of rising and setting will be observed. If the polar distance, for example, be 60°, i.e. if the declination of the star be 30°, then the star will set by crossing below the horizon at one of the two points of the horizon which are at a distance of 60° from the pole. A star situated on the equator and, therefore, at a distance of 90° from the pole, will cross the horizon at rising and setting in those two diametrically opposite points which are 90° from the pole.

Stars having south declination will also be seen, provided that their declination does not exceed a certain amount. This amount may be determined as follows. As the nearest point of the horizon to the pole is at a distance of 45°, so the most distant point of the horizon from the pole is 180° −45°=135°. It therefore follows that if a star is to be visible at that latitude, the polar distance of the star must not exceed 135°, or, what is the same thing, the south declination must not exceed 45°.

Continuing his journey, we may suppose that the traveller reaches the equator, where, his latitude being zero, the pole has sunk to the horizon. All the stars rise and set perpendicularly to the horizon, and the phenomena of circumpolar stars are unknown. In this position all the stars in both hemispheres will be visible, and each one will continue above the horizon for half a sidereal day. There is thus a remarkable contrast between the appearance of the celestial sphere as seen from the equator and as seen from either of the poles. In the former case all the celestial sphere is above the horizon in the course of a revolution. In the latter only one-half the celestial sphere is ever visible.

As the traveller enters the southern hemisphere of the earth, the north pole of the heavens sinks below his horizon and the south pole begins to rise. There does not happen to be any bright star indicating the position of the south pole in the convenient way in which the Pole Star shows the position of the pole in the northern hemisphere. At the south latitude 45° the phenomena already described at the latitude of 45° north are presented in an inverse manner, the circumpolar stars being now the stars which surround the south pole.

It will be useful to state more fully the conditions under which a star or other celestial object is visible at a given latitude at one or both of its culminations. Since the elevation of the pole is equal to the latitude of the place, it follows that the angular distance between the zenith and the pole is equal to the complement of the latitude of the place, or to what is called the *colatitude.*

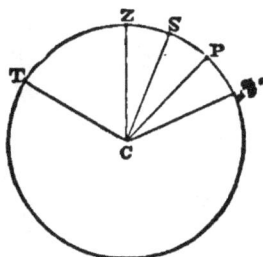

FIG. 64.

In Fig. 64 let P denote the pole, z the zenith, and s, s′ the positions of a star at upper and lower culminations respectively. Let ϕ be the latitude, and δ the declination of the star.

Then the zenith distance of the star at its lower culmination is z s′, but

$$z\,s' = z\,p + p\,s'$$
$$= 90 - \phi + 90 - \delta = 180 - \phi - \delta.$$

In order that a star may be visible, it is necessary that its zenith distance at the time of observation be less than

90°. Hence for a star to be visible at the moment of lower culmination we must have

$$180 - \phi - \delta < 90°$$
or
$$\delta > 90° - \phi.$$

We thus have the following result :—

When the north declination of a star exceeds the colatitude, both culminations of the star will take place above the horizon of a station in the northern hemisphere.

Next suppose a southern star at T. The zenith distance of the star is Z T, but

$$Z T = P T - P Z$$
$$= 90 + \delta - (90 - \phi)$$
$$= \delta + \phi.$$

Hence when the upper culmination of the star takes place above the horizon, we must have

$$\delta + \phi < 90°$$
or
$$\delta < 90 - \phi ;$$

and we have the following result :—

When the south declination of a star is less than the colatitude, then the upper culmination of the star will take place above the horizon of a station in the northern hemisphere.

For example, the latitude of Greenwich is 51° 28' 38"·4, and therefore the colatitude is 38° 31' 21"·6. The upper culmination of every star in the northern hemisphere will be seen at Greenwich, as well as of all the southern stars whose south declination does not exceed −38° 31' 21"·6.

All stars are visible at both culminations at Greenwich of which the declinations exceed +38° 31' 21"·6.

These statements require some modification on account of the effects of refraction, but this need not be considered at present. It should also be observed that, owing to the loss of light incurred when the rays have to pass through a great thickness of the atmosphere, stars cannot be seen when close to the horizon. The actual rising or setting of a star is perhaps never witnessed.

§ **43.** *Determination of Latitude.*—Since the altitude of the pole above the horizon is equal to the latitude, the angular distance from the zenith to the pole is equal to the complement of the latitude, or the *colatitude*. By suitable observations made with the meridian circle the colatitude can be ascertained, and thence the latitude is known. We select for this purpose a circumpolar star, preferably the Pole Star itself, and we determine the zenith distance of the star at its upper and then at its lower culmination.

Let z s, z s′ (Fig. 64) be the zenith distances at upper and lower culmination respectively, then if refraction be omitted, we have

$$z \, \textsc{p} = \tfrac{1}{2} \, (z \, s + z \, s'),$$

or the *colatitude is the arithmetic mean of the zenith distances of a circumpolar star at upper and lower culmination.*

Owing, however, to the effect of refraction, the actual distances z p found by different circumpolar stars will be seen to vary. A star, for example, which, at its lower culmination, passed close to the horizon would be largely affected by refraction, while at its upper culmination it would be but very slightly affected ; the Pole Star would be affected very nearly equally on both occasions. We have seen that the refractions of stars are nearly proportional to the tangent of the zenith distance. They are therefore equal to this tangent multiplied by a certain coefficient, termed the coefficient of refraction, which has to be determined by observation. It is by observations of the kind we are describing that the coefficient of refraction is to be determined. This coefficient must have such a value that, when the corresponding corrections are applied to the zenith distances, the resulting values of the colatitude shall be all equal from whatever star they have been derived. In this way the colatitude has been ascertained, and therefore also the latitude.

The declination of a star can now be readily found. For the declination δ is the complement of the polar distance, which is equal to the zenith distance at upper culmination added to the colatitude. If z be the zenith distance, and λ the latitude, we have therefore

$$90° - \delta = z + 90° - \lambda$$

whence
$$\delta = \lambda - z$$
or
$$\delta + z = \lambda$$

We therefore see that—

The zenith distance of a star at upper culmination added to its declination is equal to the latitude,
and also it can be easily shown that—

The zenith distance of a star at lower culmination added to its declination is equal to the supplement of the latitude.

§ 44. *Numerical Illustration.*—In applying the correction for refractions, it is necessary to take account of the temperature of the air (t degrees Fahr.), as well as its barometric pressure (h inches of mercury). The following formula, which has been derived from the observations we have just described, fairly represents the refractions except in the neighbourhood of the horizon.

$$\text{Refraction (in seconds)} = 992'' \frac{h}{460 + t} \tan z.$$

For the sake of illustration we shall show how the latitude of Dunsink Observatory and the declination of the star γ Draconis have been ascertained by observations of this star at its upper and lower culmination respectively.

At the lower culmination the nadir distance of the star was observed, and appeared to be

$$104° \quad 56' \quad 46''\cdot8.$$

The zenith distance is therefore

$$75° \quad 3' \quad 13''\cdot2.$$

Twelve hours later γ Draconis crossed the meridian again, and its south zenith distance on that occasion was found to be

$$1° \quad 53' \quad 18''\cdot6.$$

We have first to correct these apparent zenith distances for the effect of refraction, so as to bring them to what they would have been had the atmosphere been absent. It appears that at the time of lower culmination the temperature was 40°, and the barometric pressure is 29''·86. Under these circumstances the refraction is

$$3' \quad 41''\cdot9.$$

The corrected value of the zenith distance at the lower culmination is

$$75° \quad 6' \quad 55''\cdot1.$$

The observations at the upper culmination are also affected to a small extent by refraction, the amount being

$$1''\cdot9.$$

Applying this correction, we have for the zenith distance at upper culmination the result

$$1° \quad 53' \quad 20''\cdot5.$$

We therefore have the following two equations from which the latitude λ and the declination δ are to be ascertained :—

$$180-\lambda=75° \quad 6' \quad 55''\cdot1+\delta$$
$$\lambda=1° \quad 53' \quad 20''\cdot5+\delta,$$

whence we deduce

$$\lambda=53° \quad 23' \quad 12''\cdot7$$
$$\delta=51° \quad 29' \quad 52''\cdot2.$$

By repeating these observations, using different stars for the purpose, the latitude can be determined with great accuracy, and in the case of those observatories which are furnished with large meridian instruments the latitude is known accurately to a small fraction of a second.

§ 45. *Star Catalogues.*—The results of the observations on the positions of the stars are given in what are called *Star Catalogues.* These catalogues contain the appropriate designation of the star, either by a special name, or more usually by the constellation in which it lies, together with the number or letter by which the star is distinguished. Then follow columns containing the right ascensions and declinations of the stars, and it is usual to place the stars in the catalogue in the order of their right ascensions. Such catalogues of stars are of the greatest service in astronomy. When once the place of a star has been accurately found and recorded, that place becomes a faithful point of reference from which other measurements can be made. For example, suppose a comet appears which cannot be conveniently observed by a meridian instrument; it is only necessary to select some star in the neighbourhood of the comet, whose place is known by the catalogue. It is then easy to determine the difference of the right ascensions and declinations of the star and the comet by means of an equatorial telescope, which can be directed to whatever point of the heavens the comet may happen to be situated. These differences being determined, the absolute position of the comet is deduced, because the right ascension and declination of the star is known from the catalogue.

Among the numerous catalogues of stars which have been made, we shall enumerate a few of the most celebrated.

Lalande's Catalogue contains the places of 47,390 stars observed in 1790 and following years. These were published in the year 1847 at the expense of Her Majesty's Government.

The British Association Catalogue contains the right ascensions and north polar distances of 8,377 fixed stars. This catalogue has the advantage of not having been derived solely from the labours at a single observatory, but it represents the mean result of all the best authorities as to the positions of the stars. It includes nearly all stars not less than the sixth magnitude, as well as a number of other stars which are considered from one cause or another to possess a special interest.

In these catalogues the right ascension of each star is given to the hundredth part of a second of time, and the polar distance is recorded to the tenth of a second of arc.

The Great Bonn Catalogue, made by Argelander, contains approximate places of 324,188 stars situated between the pole and two degrees of south declination. This includes all stars in the region thus specified, which are not less than the ninth magnitude, and also a good many which are between the ninth and the tenth.

CHAPTER IV.

THE SUN.

§ 46. *Apparent Motion of the Sun.*—Passing from the consideration of the apparent diurnal rotation of the heavens, which is, as we have seen, due to the rotation of the earth upon its axis, we must proceed to consider the movements of those heavenly bodies which do not occupy a fixed position with regard to the stars and constellations. Among this class of bodies we can have no hesitation in regarding the sun as the most important as far as the inhabitants of the earth are concerned. We shall therefore begin by studying the nature of those movements on the surface of the celestial sphere which are performed by the sun.

It is, in the first place, obvious that the sun partakes in the apparent diurnal rotation of the heavens which we have already considered. Like the stars, the sun rises in the east, and having run its course across the heavens, it descends again to set towards the west. When, however, we study the apparent movements of the sun with more attention, we see that the motions of the sun are not simply due to the apparent motion of the entire celestial sphere, but that the sun actually possesses a certain motion on the sphere, which is blended with the apparent motion which it also receives by the diurnal rotation of the sphere. The actual movements of the sun are, however, so slow when compared with the movement arising from the apparent diurnal motion, that it requires some little attention to perceive them. The difficulty would be greatly lessened if it were possible to see the stars on the celestial sphere in the neighbourhood of the sun. No doubt it is possible by the aid of telescopes to see bright stars in the daylight, but these stars to be visible must be at least 15° from the sun, otherwise the overwhelming brilliancy of the sun would obliterate the feeble rays from the star. There are, however, several indirect methods by which we can easily see that the circumstances attending the apparent motions of the sun are different from those connected with the stars. In the first place, if we note carefully the point of the heavens at which a *star* first becomes visible after it has risen, and if we compare this position with the surrounding terrestrial objects, we shall find that the point of the horizon at which the star actually rises is constant, so that it remains from one end of the year to the other at the same distance from the north or south point. This constancy of the point of rising is an obvious consequence of the diurnal motion of the heavens, when it is remembered that the star remains in a constant position on the surface of the celestial sphere. The rising of the sun is, however, very different from that of a star ; for the point of the horizon at which the sun rises changes periodically. This is seen by noting the position of the sun at sunrise with regard to the adjacent terrestrial objects. If the distance of the point at which the sun rises from the southern point of the horizon be measured, it is found that this distance fluctuates between certain limits, and the period in which these changes are accomplished is one year. Precisely similar phenomena are witnessed at sunset; the point at which the sun descends below the horizon oscillates between certain limits. This is seen in a very striking manner at certain places on the western coasts of Europe, where the sun appears to set in the sea, and where the point at which it sets is easily defined by the aid of islands which are seen near the horizon.

There is also another point in which the movement of the sun is seen to differ from the apparent movements which the stars make in virtue of the diurnal rotation of the heavens. When a star culminates, the altitude of that star is a maximum. It is easily perceived that this maximum altitude of a fixed star is always the same, whatever may be the season of the year at which it is observed. This is, however, not the case with the sun. At or near noon each day the sun culminates, but everybody knows that the altitude which the sun has at culmination is far from constant. In winter the sun is low when it culminates, while in summer it is high. In the arctic regions this feature of the sun's movements is so strongly marked that in midwinter the sun actually does not come above the horizon at all, while in midsummer the sun never sets. These points of contrast between the movements of the sun and those of the stars can only be explained by the supposition that the position of the sun on the celestial sphere is constantly changing. We can confirm this by observing the heavens in the west as soon after sunset as the decreasing twilight has rendered the stars visible. If we note the positions of the stars with respect to the terrestrial objects, and if the observation be repeated after an interval of two or three weeks, it is seen that the stars which at the first observation were considerably above the western horizon at sunset have now closely approached it. Again repeating the observation after the same interval, it will be found that the stars already mentioned are now actually below the horizon at sunset. After the lapse of some time it will be found that these stars, having ceased to be visible in the west after sunset, will become visible in the east shortly before sunrise. These phenomena can only be explained by the supposition that the sun is in motion on the surface of the celestial sphere.

It will be instructive for the learner to observe these phenomena for himself, and it may facilitate his doing so if we describe the appearance presented by a certain region of the heavens at the different seasons. We shall again take for the purpose of illustration the well-known group in the constellation Taurus known as the Pleiades (§ 26), and we shall describe the position occupied by the Pleiades at eleven o'clock P.M. on the nights of January 1, March 1, May 1, July 1, September 1, November 1. If the weather be unfavourable on any of the nights we have named, then the next fine night *at the same hour* will answer instead. We shall also assume that the observer is situated in the northern hemisphere, at about the latitude of the British Islands.

At eleven P.M. on January 1, the Pleiades may be found high up in the sky, a little to the west of south. At the same hour on the evening of the 1st of March they will be visible rather low in the west. On the 1st of May they are not visible ; on the 1st of July they are not visible. On the 1st of September they are visible low in the east. On the 1st of November they are high in the heavens, a little to the east of south. On the next 1st of January they will be in the same position as they were on the last, and in the course of the following year the same cycle of changes which we have described will be repeated.

It would therefore seem as if the Pleiades were at first gradually moving from the east to the west, that then they dipped below the horizon, and after a short time reappeared again in the east, so as to regain at the end of a year the position they had at the beginning.

The reader will, it is hoped, not confuse the *annual* motion which we are here considering with the apparent diurnal motion previously discussed (§ 26). In the apparent diurnal motion the phenomenon is observed by looking out at different hours on the same night. To observe the apparent annual motion we must look at the same hour on different nights separated by a considerable interval.

We shall now endeavour to explain the apparent annual motion of the Pleiades. We have chosen the hour at eleven P.M. because by this time in the summer months the heavens are sufficiently dark to enable the stars to be visible. We might, however, have traced the same series of changes had we adopted any other constant hour, and used a telescope whenever the daylight would not permit the stars to be seen without its aid. It will also be understood that the Pleiades have merely been taken as an illustration: many other objects would have answered equally well.

We must first ask, What does eleven P.M. really mean? It will be necessary to anticipate to some trivial extent certain points to be more fully discussed hereafter; we may, however, assume that the sun crosses the meridian each day very near noon, and that consequently, at eleven P.M., the sun has passed the meridian about eleven hours previously. We find that at eleven P.M. on March 1, the Pleiades are farther from the meridian than they were at eleven P.M. on January 1. But as the sun is nearly at the same distance from the meridian in the two cases, it follows that the Pleiades must be nearer to the sun on March 1 than on January 1. It is, therefore, plain that the relative positions of the sun and the Pleiades on the surface of the heavens must be changing. By comparing the sun in the same way with any other stars, it is found that the stars to the east of the sun appear to be approaching the sun. But we have already noticed that the positions of the stars *inter* .

se do not change, and therefore we are obliged to come to one of two conclusions : either, firstly, that all the stars in the universe have an annual motion from east to west relatively to the sun, which remains fixed, or that the sun has an apparent annual motion from west to east, while the stars remain fixed. We cannot hesitate to choose the latter alternative as the correct one. The apparent diurnal motion we have already explained to be due to the real rotation of the earth upon its axis. Were this rotation to cease, the stars would appear immovable, but the apparent annual motion of the sun among the stars would remain as before. In the discussions of the motion of the sun, which we shall now commence, we shall only consider the apparent annual motion of the sun among the stars, for, of course, that part of the sun's apparent motion which is due to the diurnal motion which it shares in common with the stars needs no other explanation than the rotation of the earth upon its axis.

§ 47. *Observations of the Sun.*—We shall first consider how the position of the sun may be ascertained by observations made with instruments capable of giving numerical results. The instruments first to be described are of exceedingly simple construction, and though the results they give are only to be regarded as coarse approximations, yet in the early epochs of astronomy it was by their means that acquaintance was made with the apparent movements of the sun, and that the laws of those movements were discovered.

The most simple instrument of the class now under consideration depends upon the shadow which the sun casts on a fixed object. As the sun moves, the shadow moves in a corresponding manner, and by the measurements of the position of the shadow on the plane the movements of the sun can be deduced. This instrument in its simplest form is represented in Fig. 67. A rod A B is fixed vertically, so as to project from the horizontal plane B A C. On this plane a line A C is traced north and south for the purpose of facilitating the measurements.

The shadow of the vertical rod A B occupies the position A D on the horizontal plane, and by the position and length of this shadow the position of the sun can be ascertained. The plane B A D which contains the rod and its shadow is vertical, and the intersection of this plane with the horizontal plane makes with the line A C an angle equal to the azimuth of the sun. Thus the azimuth of the sun can be ascertained by measurement of the angle B A C. It is also

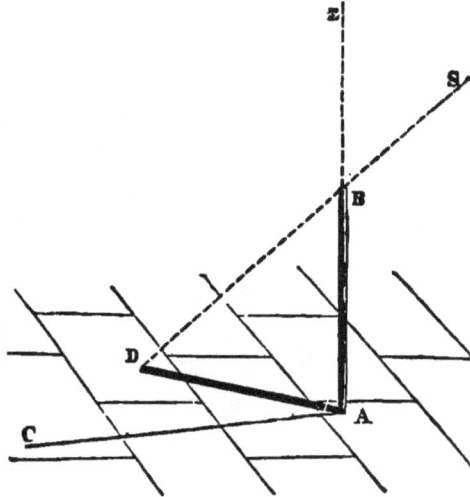

easy to obtain the zenith distance of the sun. For the line joining the end of the shadow D to the top B of the rod must point directly towards the sun, and as the line A B points directly towards the zenith, it follows that the angle Z B S or A B D must be equal to the zenith distance of the sun. The angle B A D being a right angle, it follows that the tangent of the zenith distance is equal to A D ÷ A B, and, as A B is constant, it appears that the tangent of the zenith distance of the sun is proportional to the length of the shadow. At noon each day the zenith distance of the sun is at its smallest value, and the shadow of the sun, which is then on the line A C, has its smallest value also. It is, however, found that the length of the shadow at noon is different on different days. In midsummer the length of the shadow has its minimum value, and as the season advances so does

I

the length of the shadow, until in midwinter the length attains its greatest value. This shows that the zenith distance of the sun at noon has its least value in midsummer, and its greatest value in midwinter, and by measurement of the length of the shadow and dividing the result by the length of the vertical rod, the tangent of the zenith distance can be computed.

The accuracy of this instrument is greatly impaired by the circumstance that the apparent size of the sun is so considerable. Each luminous point on the surface of the sun emits rays of light, and will throw a shadow of the vertical rod. We thus see not a single shadow, but a great number of shadows partially superposed on each other. This composite nature of the shadow entails a certain degree of ambiguity and want of sharpness in the shadow itself, and is a very serious obstacle in the way of exact measurement. So crude an instrument as this is therefore quite unsuited for the exact measurements required in modern astronomy.

It has been sought in some degree to rectify the uncertainty attendant upon measurements of the shadow by using

FIG. 66.   FIG. 67.

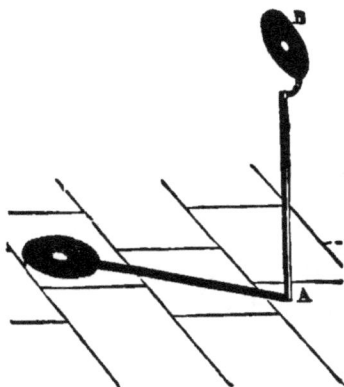
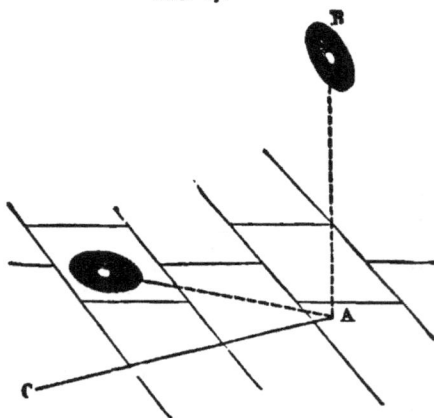

a modified form of this instrument, shown in (Fig. 66). The vertical rod A B (Fig. 66) carries at its summit a plate B pierced with a small orifice. The shadow of this plate is thrown on the horizontal plane, and the measurements are made from the point A to the centre of the illuminated disk (Fig. 67). In this way it is found that measures can be made with a considerably greater degree of precision than is attainable when the extremity of the shadow of the rod is the point whose distance is to be measured. It might be thought that if the orifice were made exceedingly small the point in the centre of the shadow of the disk would also be an exceedingly small point ; but in this case again the apparent size of the sun intervenes, and prevents the point in the centre of the shadow from being so sharply defined, as it would be were the apparent diameter of the sun smaller than it actually is. All instruments of this class are, however, replaced for accurate work by suitably mounted telescopes, which enable measurements to be made that are incomparably more accurate than anything which can be attempted by the contrivances we have just been describing.

Before the indications of these accurate instruments can make known to us the precise place of the sun, it is necessary to have some acquaintance with the apparent shape and form of the sun's disk as we see it. We shall therefore first turn our attention to this subject. The disk of the sun on the celestial sphere appears, at first sight, to be a circle : it is, however, necessary to make the shape of this disk the subject of exact measurement, in order to show how far the disk is exactly circular, and also to determine the angular value of its radius.

§ 48. *The Micrometer.*—The instrument which is used in these investigations is generally attached to the eye end of a telescope, and is employed for measuring the dimensions of the images of the objects in the field of view. There are several different contrivances used : we shall describe the form which is known as the *parallel wire micrometer*.

In the focus of the object glass of the telescope are

stretched three spider lines. Two of these are parallel, and the third is at right angles to them. Each of the parallel wires is mounted in a frame, by which it is enabled to be moved parallel to itself in the plane of the focus, and the movements of the two wires are completely independent. The movement of each wire is effected by a screw called the *micrometer screw*, and it is on the accuracy with which this screw is made that the utility of the micrometer mainly depends. If the screw be strictly uniform in the shape of its thread, then the distance through which the wire is moved is always proportional to the number of whole revolutions and parts of a revolution which the screw has made. Means are provided by which the entire number of revolutions by which the screw has been turned from a standard position can be ascertained. The head of the screw is also usually subdivided into hundredths, and by estimation each hundredth part may be subdivided into tenths. It is therefore possible to record the distance through which the screw has been moved from a standard position accurately to the thousandth part of a revolution. The single wire which intersects the movable wires passes through the centre of the field of view. It has no motion of the kind we have just described, but, in common with the two movable wires and the eye piece, is capable of being rotated around the axis of the telescope. There is also a graduated circle attached to the micrometer, so as to enable the angle, through which the micrometer has been rotated, to be ascertained.

To illustrate the method of using this instrument we shall suppose the case of three stars in the field of view, and we shall describe how the relative positions and distances of these stars are to be ascertained. To enable the measurements to be made with facility and accuracy the micrometer ought to be attached to a telescope, which is mounted equatorially (§ 27), and which is furnished with a clock-work apparatus. When the clock is going the stars will remain constantly in the field of view, and the attention of the observer will not be distracted by the diurnal motion, which, if uncompensated by the action of the clock, would entirely preclude any accurate results. The circles which are attached to the telescope are not employed in connection with the measurements now to be described. In fact, in an equatorial instrument, these circles are generally only intended for the purpose of pointing the telescope in the right direction, so as to bring the right object into the field of view, but they are not generally intended to serve any purpose in the making of exact measurements.

For convenience we speak of the three stars as A, B, C, and we propose to measure the sides and angles of the triangle A B C. The triangle which we see is of course not the actual triangle, which has the three stars at its vertices : it is only the projection of this triangle on the surface of the celestial sphere which is amenable to our measurements. Thus the triangle with which we are dealing is really a spherical triangle, but as the field of view is always only a very minute portion of the entire celestial sphere, it follows that the portion of the celestial sphere visible in the telescope differs but little from an exact plane, and may generally for our present purposes be simply regarded as a plane. We have then to measure the angles subtended by the sides of the plane triangle, A B C, by the aid of the micrometer. We proceed as follows. Turn the micrometer, and of course its wire system also, so as to place the single wire approximately parallel to one of the sides of the triangle (suppose A B) ; then, by means of the slow movements which are provided for the purpose of slightly changing the telescope in right ascension and declination, bring the telescope into such a position that the stars A and B are close to the single wire. Then adjust the single wire so that it passes symmetrically across the two stars. The two movable wires are next to be placed so that one of the wires is exactly on A and the other on B. These adjustments having been carefully made, the micrometer is to be read off. First the position of the graduated circle is to be noted, and then the positions of the screws, as well as the indications of their graduated heads, are to be recorded. Without altering the position of the single wire, the two movable wires are to be brought into coincidence, and the indications of the screws are again to be read off. It is thus known how far each of the wires has been moved from the point where they coincide to the position which it occupied when it was coincident with the star, and thus the entire distance between the two stars, *expressed in terms of the revolutions of the screws*, is ascertained.

The whole micrometer is now turned so as to bring the single wire across A C, and the operations already described are to be repeated. The difference between the readings of the graduated circle in the two positions of the micrometer will give the angle B A C, while the readings of the screws gives the distances A B and A C, so that the whole triangle is ascertained. The results may be verified by actually measuring the distance between B and C, and comparing it with the distance ascertained by calculation.

It remains, however, to show how the apparent distances thus found, which are expressed by the arbitrary value of the revolutions of the screw, can be evaluated in angular magnitude. In fact, what we want actually to find is the angle which two stars—A and B, for example—subtend at the eye. It therefore becomes necessary to ascertain the number of seconds of arc which are equivalent to one revolution of the screw. We may put the question now under consideration in a somewhat different manner. Suppose that two stars on the surface of the celestial sphere were separated by a distance equal to one revolution of the micrometer screw : it is required to find how many seconds of arc these stars subtend at the eye. There are different methods by which this important problem may be solved. One of these methods depends upon direct measurement, while others are founded upon astronomical observations. We shall proceed to describe the two processes, taking first the method of direct measurement.

By means of an accurately graduated scale we can measure the actual distance in inches or fractions of an inch at which the wires are separated. We are then to measure the distance from the wires to the optical centre of the object glass, which is in fact the focal length of the object glass. The angle subtended at the object glass by the two points in which the fixed wire intersects the movable wires, is obviously equal to the angle which the stars subtend at the centre of the object glass when the stars are situated at the two points of intersection. But the angle subtended by the two points at the centre of the object glass is easily found when the measurements have been made. The distance between the points divided by the focal length gives the circular measure of the angle, and multiplying the circular measure by 206265, we obtain the required angle in seconds. The difficulty attending this method of determining the value of the micrometer screw principally arises from the uncertainty with which the measurements of the distance of the wires and the focal length of the telescope are affected. Any error in the determination of the distance between the wires will produce a corresponding error in the determination of the value of the screw. This source of uncertainty may to a large extent be obviated by separating the wires as far as possible, so that the distance between them shall be fifty or more revolutions of the screw. Any error that is made in the measurement of the distance will thus be subdivided by fifty, when we come to the evaluation of a single revolution, and thus when two stars, tolerably near together, are measured, the error arising from using the incorrect value of the micrometer screw will be intrinsically insignificant. The measurement of the focal length of the telescope is also a matter of some difficulty. It is no doubt easy to define the focal length to be the distance from the optical centre of the object glass to the point where all the constituents of a beam of parallel light falling on the object glass converge. The optical centre, however, is not a point which is accessible to measuring instruments, nor is it easy to find at what point of the thickness of the object glass it is really situated. There is thus no insignificant degree of uncertainty about the actual focal length of a telescope. These causes render the method of determining the value of the micrometer screw by direct measurement to be of comparatively little reliability and it is seldom resorted to in practice.

With a telescope which is really well and steadily mounted, the value of the micrometer screw is most conveniently and accurately found by observations of the Pole Star. For this purpose a clock or chronometer accurately rated to sidereal time is required. The micrometer is first to be adjusted so that the Pole Star, when carried by the diurnal motion, will run accurately along the fixed wire. The two movable wires are then to be placed at a convenient distance apart, and the telescope is to be firmly clamped in right ascension. The time shown by the clock is to be noted when the Pole Star crosses the first of the wires, and then when it crosses the second wire. The difference between the two recorded times, corrected, if necessary, for the rate of the clock, gives the time occupied by the Pole Star in passing from one of these wires to the other. This determination should be repeated several times, and the mean of the several results chosen as the final value. It is desirable, for the reasons already explained, to have the interval between the two wires as great as possible. On the other hand, the success of the method depends entirely upon the telescope remaining absolutely fixed while the Pole Star is crossing the interval between the two wires. If this distance be too great, some uncertainty is apt to arise, particularly when the telescope is not mounted in a very substantial manner. When the observations are finished, it is usually necessary to make some trifling allowance on account of refraction. What we actually observe is the time in which the image of the Pole Star as deranged by refraction moves over a certain distance. What we want to find is the time in which the image of the Pole Star would have moved over the same distance had there been no such thing as atmospheric refraction. We do not enter into the details of the method by which the small correction which the time is to receive is computed ; indeed, except for the more refined branches of observational astronomy, such accuracy is not required. We shall therefore proceed at once to show how from the observed time which the Pole Star requires to move over a certain number of revolutions the value of a single revolution is to be ascertained.

We have already explained how the declination of a celestial object is determined, and we may therefore assume that the declination of the Pole Star is known. The Nautical Almanac contains the apparent declination of the Pole Star for every day in the year, and we cannot do better than adopt the declination there given as the basis of the calculations which are to be made. From the declination the polar distance can be ascertained, and in this way the dimensions of the small circle on the celestial sphere, in which the Pole Star moves in its diurnal motion, is known. The Pole Star performs its journey round this circle in twenty-four hours of sidereal time. As we know the length of the circumference of the small circle expressed in degrees; minutes, and seconds, we are able to ascertain the number of degrees, minutes, and seconds that the Pole Star travels in a given time. We are, therefore, able to find the distance in minutes and seconds which the Pole Star will accomplish in the time which it takes to move from one of the wires to the next, and thus we are enabled to ascertain the minutes and seconds of arc on a great circle by which those two wires are separated. As we also know the number of revolutions of the micrometer screw which corresponds to the interval between the two wires, we are enabled to deduce the value of one revolution.

A word should be added as to the grounds on which the Pole Star is chosen for the purpose of these observations. There is really no particular virtue about the Pole Star for this purpose, save that it is near the pole. Any other star tolerably near the pole would do equally well. For some reasons, indeed, a star a little more removed from the pole than the Pole Star is perhaps preferable. But a star very far from the pole moves across the field of view so rapidly that a small error in the determination of the time of transit from one wire to the other will be a comparatively large fraction of the total time observed ; by choosing a star close to the pole, the interval of time will be greatly increased, and the effect of an error in the time will have a much less appreciable effect.

The Pole Star is attended by a small companion star, which is visible in a small telescope and is a very conspicuous object in a large telescope. In the latter case it is often preferable to use the companion of the Pole Star, rather than the Pole Star itself. The moment when the small star crosses the wire can generally be more accurately noted than in the case of the larger star. In this case, of course, it is the declination of the companion which must be employed in the calculations instead of that of the Pole Star.

Another method of finding the values of the micrometer screw has also much to recommend it. This consists of measuring with the micrometer the distance of a pair of stars which have already been measured by a previous observer, and of which the distance is therefore known. For the application of this method, certain stars in the Pleiades (§ 26) are especially suitable. The arcual distances of several of the stars in this group have been accurately determined by Bessel, and by these the value of the revolution of a screw in any micrometer can be readily ascertained. It is only necessary to choose two of Bessel's stars which are at a suitable distance for the particular micrometer under consideration, and then to measure them with the micrometer. By comparing this with Bessel's measures, the number of seconds of an arc of the great circle on the celestial sphere which corresponds to a certain number of revolutions and parts of a revolution of the micrometer screw is determined. From this of course it is easy to ascertain the number of seconds of arc in one revolution.

The value of a revolution of the micrometer screw depends, as we have seen, upon the focal length of the telescope as well as upon the actual linear distance between the wires. The latter generally, and perhaps the former also, depends to a certain extent upon the temperature of the instrument. The effect of temperature upon the value of the screw can only be obtained empirically. This is done by using any of the methods we have described, and noting the temperature by a thermometer. These observations being repeated at different seasons, when the temperature is different, enable a table to be formed which will give the value of the screw for any given temperature.

§ 49. *Apparent Diameter of the Sun.*—Having determined the value of the micrometer screw, we shall now show how it may be used for measuring the apparent dimensions of the sun on the surface of the celestial sphere. For this purpose the micrometer should be fitted to a telescope mounted equatorially, and the telescope being directed to the sun, should be clamped in declination while the clock movement carries it on in right ascension. The single wire should be placed on that diameter of the sun which it is proposed to measure, and the movable wires are then to be brought so as to form tangents to the disk of the sun at the extremities of the diameter. It is important in placing the wires to observe that the screws are always turned in the same direction ; for example, it is a convenient rule whenever a wire is being set to draw it up to its final position, so that its motion is towards the head of the screw. If this precaution be not taken, irregularities will be apt to arise from the presence of a certain degree of shake in the bearings of the screw. After the two movable wires have been set, the positions of the screws are to be read off and recorded. We may then proceed in two different methods : the simplest method is that already described, of bringing the two wires into coincidence, reading off the screws, and then calculating the distance in arc through which each screw has been moved. This method is, however, very inferior to the somewhat more tedious plan of *reversing* the wires, i.e. moving the wire from one extremity of the diameter to the other, and *vice versâ*, and then reading the screws again. It can be proved that the required length of the diameter is equal to the arithmetical mean of the distances over which the two screws have been moved. After a measure has been made in this way, it is desirable to turn the entire micrometer through 180°, and then repeat the measurement and adopt the mean of the results as the final value of the sun's diameter. If increased accuracy be desired, it is well to make two settings of the wires before reversal, then two settings after reversal, and the same series again after the micrometer has been turned through 180°.

§ 50. *The Heliometer.*—Another instrument which is used for the purpose of measuring the diameter of the sun is termed the heliometer, and is founded upon quite a different principle to the parallel wire micrometer. The heliometer is essentially an achromatic telescope, the object glass of which is divided in half by a plane which passes through the optic axis of the telescope. If the disk in Fig. 68 represent the object glass, then the diameter of the disk shown in the figure is the line in which the object glass is cut. One half of the object glass A may be fixed in the tube of the telescope, but the other half B is mounted in a frame so that it moves in a plane perpendicular to the optic axis of the telescope, and is thus enabled to assume the position shown in Fig. 69. The movement of B is effected by a screw which has a graduated head, and also an arrangement by which the entire number of revolutions can be counted. It is thus possible to measure with the greatest accuracy the precise distance through which the half of the object glass has been moved, and it is by the indications of this screw that the measurements are effected by the heliometer.

To understand the action of this instrument let us first suppose the heliometer to be directed to some small celestial object—a fixed star, for example—and let us see what effect the movement of the half of the object glass will have on the appearance of the field of view. When the two halves

FIG. 68.

FIG. 69.

of the object glass are together in the position shown in Fig. 68, then the instrument performs exactly in the same way as an ordinary astronomical telescope ; but when the two halves of the object glass have the relative positions shown in Fig. 69, the condition of affairs is altered. It is to be remembered that the function of an achromatic object glass is to convey to a single point all the rays of a parallel beam falling upon it near the direction of its optic axis. Every part of the object glass bends the rays which it receives in the required direction. It therefore appears that if parts of the object glass were covered over, the rays which fell on the uncovered portions would be brought to a focus and form an image of the object, which would only differ from the image formed by the whole object glass in being proportionally less brilliant. When, therefore, the object glass has been divided in the heliometer, each portion of the glass forms its own image of the object. If the two halves are in the position of Fig. 68, the two images made by the two halves are coincident; but when the halves are in the position of Fig. 69, the two images are separated by a distance which depends upon the distance through which one half of the glass has been displaced relatively to the other.

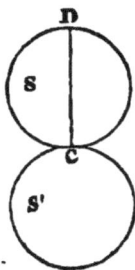

We can now understand how this instrument can be applied to the measurement of the apparent diameter of the sun. When the two halves are coincident, the image of the sun appears single in the telescope, because in this case the separate images formed by the two halves of the object glass are superposed; but when the halves are separated, then the image of the sun produced by the moving half of the object glass withdraws from the other, and a double image of the sun is the result. The observer at the eye piece having the movement of the half object glass under his control, gradually .increases the separation, until the two images of the sun s and s' are in contact, and in fact just on the point of being separated. It is then clear that the image of the sun has moved through a distance equal to D C, or the apparent diameter of the sun. The actual distance through which the half object glass has been moved is ascertained from the screw, and hence we learn the apparent diameter of the sun in terms of the revolutions of the screw.

FIG. 70.

By this means we are enabled to determine the apparent diameter of the sun in terms of the revolutions of the screw, which moves the half of the object glass. It remains to point out how the actual distance in arc of the celestial sphere, which corresponds to a revolution of this screw, can be ascertained. This may be found by drawing a circle of white on a black ground and observing this circle with the heliometer, when placed at a considerable distance. If the diameter of the circle is known, and also its distance from the telescope, then the angle which a diameter of the circle subtends at the object glass of the telescope can be calculated. By the screw of the heliometer, the white circle can be measured in the same manner we have just described with reference to the sun. We have thus the means of finding the actual value of a revolution of the micrometer screw, and therefore of determining the apparent diameter of the sun.

Either the wire micrometer or the heliometer may be employed to determine the shape of the sun's disk as it appears to us. For this purpose it is only necessary to measure different diameters of the sun drawn through the centre in various directions. It appears from these measures that the lengths of the various diameters are all equal, and therefore the measurements show that the sun's disk is not appreciably distinct from a perfect circle. The diameter of this circle is not always the same. In midwinter its apparent diameter is 32' 36", and in midsummer 31' 32" : these are the extreme limits within which the fluctuations of the diameter are confined. We shall generally not be far wrong in assuming the sun's diameter to have its mean value of 32'. We shall subsequently explain the cause of these apparent fluctuations in the diameter, and it is only necessary here to remark that they are not due to any intrinsic changes in the body of the sun itself.

It is very instructive to contrast the circular figure of the sun with the elliptical figure of the earth. If the sun really presented to us the figure of an ellipse similar to that which is produced by a section of the earth drawn through its polar axis, there would then be a difference of no less than 7" between the greatest diameter of the sun and the least diameter. A difference so great as this would be readily appreciated in measurements so exact as those which can be made with the heliometer or the parallel wire micrometer. We are therefore led to the conclusion that even if the figure of the sun (or rather that projection of the figure which we see in the celestial sphere) be elliptical, its ellipticity must be very much smaller than the ellipticity of the section of the earth which is obtained by a plane drawn through the polar axis.

§ 51. *Effect of Refraction.*—To observe that the apparent disk of the sun is circular, the opportunity should be chosen when the sun is situated at a considerable altitude above the horizon. When the sun is very low down, the refraction of the atmosphere surrounding the earth distorts the apparent disk of the sun, so that it is no longer circular, but assumes approximately the form of an ellipse with its smaller axis vertical. It is not difficult to explain how refraction causes this curious transformation in the apparent shape of the sun. It will be remembered that the effect of refraction on a celestial body is to raise the apparent place of that body up towards the zenith. It therefore follows that the entire disk of the sun is raised up by refraction, so as to appear higher in the heavens than it would do if we could see the sun without the intervention of the atmosphere. But besides this effect of refraction in bodily raising the whole image of the sun, it has also the effect of slightly distorting that image. The upper edge of the sun is nearer to the zenith than the lower edge, the difference in the zenith distances being of course equal to the diameter of the sun, when the highest point of the sun is compared with the lowest point. The refraction, however, increases with the zenith distance, and hence it follows that while both the upper and lower edges of the sun are raised by refraction, the lower edge is raised more than the upper edge ; consequently, in the refracted image of the sun, the vertical distance, between the highest point and the lowest point, must be less than it would be in an unrefracted image. Refraction, however, can have no appreciable effect in altering the length of the horizontal diameter of the sun. But if the vertical diameter be shortened while the horizontal diameter is unchanged. the image of the sun has ceased to be circular, and has assumed a form which is nearly that of an ellipse.

When the sun is actually at the horizon, the effects produced by atmospheric refraction are of a very marked character. Let us take the case when the sun appears to be on the point of setting, and select for special consideration that particular phase in which the lower edge of the sun *appears* just to have come in contact with the horizon.

The appearance of the sun in these circumstances is represented in Fig. 71 ; the horizontal line is the horizon, and the elliptical figure s′ denotes the apparent shape of the sun, the lower edge being in contact with the horizon. At the moment when the sun appears to have this distorted shape, and to be just in contact with the horizon, the real position of the sun is altogether below the horizon at s.

FIG. 71.

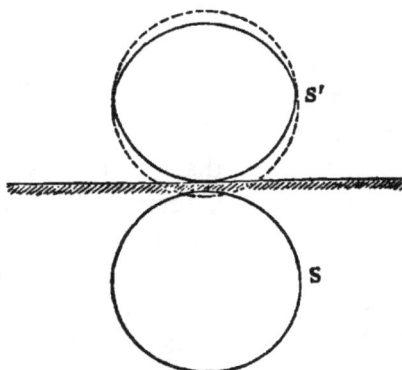

The atmospheric refraction has both raised the sun above the horizon to the position s′, and also altered its shape. The amount of the distortion of the figure when the sun is in this position is very considerable ; in fact, the vertical diameter is shortened to the extent of one-sixth part of its total amount. This distortion, though necessarily always present, decreases very rapidly indeed, as the altitude of the sun above the horizon is augmented. At the altitude of 45° the sun's vertical diameter is only diminished by $1''$, which is about one three-hundredth part of the effect which is produced near the horizon.

When the sun is in the vicinity of the horizon, either

shortly after rising or shortly before setting, it is often remarked that his apparent size seems considerably greater
than when he is high in the heavens at noon. This is
really only an illusion, but none the less it is worthy of a
word of explanation. Measurement lends no countenance
to this illusion. We have already seen that the horizontal diameter is unchanged as the sun approaches the horizon, while
the vertical diameter is actually shortened. The apparent
dimensions of the sun, when tested by measuring instruments,
are therefore less near the horizon than high up, though
to the unaided eye the reverse seems the case. The
explanation of this illusion is found in the circumstance
that near the horizon we tacitly compare the size of the sun
with the various terrestrial objects in its vicinity. We cannot then fail to observe that the sun is more distant from
us than any of the terrestrial objects. Ordinary experience
shows us that if two objects subtend the same angle at the
eye, the more distant of the objects must be larger than the
other. Hence, seeing that the sun is more distant than
any terrestrial object, and yet observing that the apparent
dimensions of the sun are considerable, we naturally consider the sun to be larger than in the case where, as it is
high in the heavens, we have no terrestrial objects with which
to compare it.

§ 52. *Right Ascension and Declination of the Sun.*—As
the apparent disk of the sun is circular, it is convenient to
define the position of the sun on the celestial sphere by the
position of the centre of the circle. We may therefore
speak of the right ascension and declination of the sun,
meaning thereby the right ascension and declination of the
sun's centre. The centre of the sun, being only marked out
by geometrical considerations, is not a visible point capable
of being observed by the telescope, like a star. We are
therefore obliged to resort to indirect means to ascertain the
position of the sun's centre.

For this purpose the meridian circle which we have

described in § 40 is very suitable. It is, however, necessary, when employing a telescope to observe the sun, to take precautions which will counteract the effects of the excessive heat and light of the sun. For this purpose it is usual to place a diaphragm over the end of the telescope, which intercepts a portion of the sun's rays before they fall upon the object glass. The total light and heat received at the eye piece is thus reduced to a fraction of what it would be were the telescope to be used in its ordinary condition. But even with this precaution the light and heat, though greatly reduced, are still too considerable to allow the un-protected eye to be placed at an ordinary eye piece. Further protection is afforded by using an eye piece modi-fied in a suitable manner. The most simple arrangement for this purpose is to cover the eye piece with a piece of glass which, though very deeply coloured, is still sufficiently transparent to allow a portion of the sun's light to pass through it. One disadvantage of this contrivance is that the observer retains no power of adjusting the light so as to make the image more or less brilliant according as circum-stances may require. This difficulty has been obviated by using eye pieces of a more elaborate construction, one of which may be here referred to.

It is well known to those who are acquainted with the properties of light, that when a beam of rays falls upon a plane glass surface at a particular angle, the rays which are reflected from that surface are *polarised*. If a polarised beam fall perpendicularly upon a plate of the mineral called *tourmaline*, the amount of light which will pass through the tourmaline depends entirely upon the position of the plane of polarisation of the rays with respect to a certain axis of the tourmaline. When this axis is parallel to the plane of polarisation, the tourmaline is comparatively transparent. When this axis is normal to the plane of polarisation the tourmaline is almost opaque. By rotating the plate of tour-maline in its plane, the quantity of light which is transmitted can be varied at will. This property of polarised light is applied in the construction of a solar eye piece. The rays from the sun, after passing through the object glass and down the tube of the telescope, fall on the glass plane near the eye piece. A great portion of those rays pass directly through the glass and never reach the eye of the observer. A portion of the rays are, however, reflected from the plate, and the plate being placed at the proper angle, the reflected beam is polarised, and in this state falls upon the eye piece. Between the eye of the observer and the first lens of the eye piece a plate of tourmaline is interposed. This is so mounted as to be capable of being rotated, and by its aid the observer has the most complete control over the degree of illumination of the image.

With his eye protected by one of the methods we have described, or by other analogous contrivances, the observer proceeds to determine the right ascension and declination of the sun by observations made with the meridian circle. As the centre of the sun is not a recognisable point, the observations are always made by noting the contact of the circular margin of the sun's disk, or what is technically called the *limb* of the sun, with the wires stretched in the focus of the telescope. As the sun enters the field of view, the observer counts the seconds in the same way as if he were observing a star, and he notes the times when the preceding limb touches the first and subsequent wires, from which he can ascertain the moment when the preceding limb of the sun crosses the central wire ; then allowing for the errors of the instrument, he deduces the right ascension of the sun's preceding limb. The observer also notes the contacts of the following limb of the sun with the successive wires, and in that way the right ascension of the following limb is determined. As the disk of the sun is known to be circular, we can find at once the right ascension of its centre by taking the mean of the right ascensions of the preceding and following limbs. At the same time as these observations are made, the meridian circle will also enable the declination of the sun's centre to be determined. While the sun is passing across the wires, the telescope is to be moved so as to bring the horizontal wire to touch the upper limb of the sun. The four microscopes are then to be read off rapidly, and ere the sun has passed out of the field the telescope is to be moved so as to bring the horizontal wire into coincidence with the sun's lower limb, and the microscopes are to be read again. From the indications of the microscopes the declinations of the upper and lower limbs of the sun are to be ascertained in the way already explained (Chap. III.), due allowance being made for the effect of refraction. The mean of these observations gives the declination of the sun's centre.

If we determine the right ascension and declination of the centre of the sun on different days, we shall find that neither of these quantities remains constant. The right ascension, for example, is continually increasing. Expressing the right ascension in arc of a great circle, we find that the daily increase is about 1°. If the right ascension be expressed in the more usual method as an interval of time, then it appears that the daily increase of the sun's right ascension is about 4 minutes. Nor are the variations in the declination more difficult to recognise. The northern declination of the sun in the spring gradually increases until it reaches a maximum in midsummer of about 23° 27′; after attaining this value, the declination decreases, reaches zero in the autumn ; then the centre of the sun crosses the equator, till in midwinter the southern declination has attained the same value as the northern declination had in midsummer. Again the sun approaches the equator, and crossing it, the declination repeats the cycle of changes to which we have referred.

The changes in right ascension and declination of the sun stand out in marked contrast to the fixity of the right ascension and declination of the stars : they indicate that the sun is not at rest upon the surface of the heavens, but that it has, or appears to have, a motion which is constantly changing its position with respect to the stars and constellations.

As the right ascension of the sun is increasing, the return of the sun to the meridian, when expressed in sidereal time, is later and later each sidereal day. The difference being on an average about four minutes, the interval between two successive transits of the sun's centre across the meridian is four minutes longer than the sidereal day. Thus, if a star came on the meridian to-day at the same moment as the sun's centre, when the star reached the meridian to-morrow, the sun would have moved away from the place it originally occupied, and its centre would not cross the meridian until about four minutes after the star.

§ 53. *Mean Time and Sidereal Time.*—The interval between two successive transits of the sun is called an *apparent solar day* ; and as the sun is of such transcendent importance to the earth, it is necessary to regulate our ordinary avocations by solar days rather than by sidereal days. There is, however, an apparent difficulty in adopting the solar day as the measure of time. The interval between two consecutive returns of the centre of the sun to the meridian is not exactly constant. The sun, in fact, does not augment its right ascension with uniformity, but the motion is sometimes more rapid than at other times. The length of the apparent solar day is thus not absolutely constant, though the limits between which it fluctuates are narrow. We therefore introduce the conception of the *mean solar day*, of which the length is equal to the average of a very great number of consecutive *apparent solar days*. This mean solar day is the unit used for all ordinary civil reckoning: its duration is absolutely constant, and generally different from, though always very close to, the length of the apparent solar day.

The mean solar day is equal to $24^h\ 3^m\ 56^s\cdot56$ of sidereal time. The mean solar day is (like the sidereal day) sub-

divided into twenty-four hours, each hour into sixty minutes,
and each minute into sixty seconds.

The reader may, perhaps, think that needless complexity
is introduced by using two different methods of measuring
time. He might suppose that simplicity would be gained
either by making all astronomical measurements in ordinary
mean solar time, or possibly he might suppose that the
sidereal day could be used for ordinary civil purposes.
We must therefore show why the two distinct measures of
time must be maintained.

The great convenience which sidereal time possesses for
astronomical purposes arises from the circumstance that
when sidereal time is used each star crosses the meridian
daily at the same moment of time. Thus when the sidereal
clock is going correctly, a star which crosses the meridian
at 4^h 30^m to-day will cross it at the same hour to-morrow,
and every day (subject only to very minute variations,
which need not be further considered at present). The
time of culmination as shown by a sidereal clock is, in fact,
the right ascension of the star. If we were to take the
time from an ordinary mean solar clock, then the time of
culmination of the star would be continually changing, being
about four minutes earlier every day. The inconvenience
arising from this would be intolerable, so that the sidereal
clock must be the regulator of time in the observatory. For
the ordinary purposes of life, on the other hand, sidereal time
would not answer. We are obliged to regulate the time
used in daily life by the movements of the sun. Our notion
of a day is so inseparably connected with the sun that the
length of the day must be the average duration of the apparent diurnal revolution of the sun around the earth. Custom
has decreed that our measurement of time shall commence
from a moment which is always close to the time of culmination of the sun, and at that moment a correct mean
solar clock should show 0^h 0^m 0^s, or, as it is more commonly
called, twelve o'clock. When we speak of any other hour—for

example, 2 P.M.—then everyone knows that the sun has at that hour passed the meridian about two hours previously. But were we to attempt to use sidereal time, endless confusion would be the consequence. By the sidereal clock the sun comes on the meridian about four minutes later every day, so that in a month the time of culmination would have changed by a couple of hours, and in the course of a year the time of culmination would have been at every hour of the twenty-four. We are therefore obliged to use a mean-time clock regulated by the sun for the purposes of civil life, while in the observatory we use a sidereal clock regulated by the stars.

In consequence of the continual increase in the right ascension of the sun, it is found that after the lapse of 365 days, or, more strictly, of 365·2421 days, the right ascension has gained 360°, or twenty-four hours of sidereal time. At the same time it is found that the movements of the sun in declination have also run through their cycle, so that at the close of the epoch just mentioned, which is approximately 365¼ days, the right ascension and declination of the sun have both regained the values they had at its commencement. The sun has, therefore, in the course of this time returned to the point of the heavens from which it originally departed.

§ 54. *The Ecliptic.*—To ascertain what the actual movements of the sun during the interval have been, we make use of a celestial globe. By the observations of the right ascension and declination of the sun's centre made at stated intervals of time, we are enabled to plot down on the celestial globe the actual position which the centre of the sun occupied on the celestial sphere at the corresponding epochs. When this is done we find that all the points, which have been laid down, lie on the circumference of a great circle, and we therefore see that the sun moves in a great circle on the surface of the celestial sphere. This great circle is called the *ecliptic.*

It is desirable to dwell for a moment on the important result which is thus obtained. That the path of the sun on the celestial sphere must be a closed curve of some kind is obvious from the consideration that, at the termination of a period of $365\frac{1}{4}$ days, the sun returns to the position which it originally had at the commencement of that period. The number of closed curves which can be drawn on the surface of a sphere are of course of unlimited variety. If, however, the sun's motion takes place in a plane, then its apparent motion on the surface of the sphere can only be in such a curve as lies on the surface of the sphere, and also in a plane ; but a section of a plane by a sphere is a circle, and as the sun appears to move in a circle, it follows that the sun moves in a plane. There is a very remarkable circumstance with respect to this plane, for as the circle in which the sun moves is a great circle, the plane must pass through the centre of the celestial sphere. The earth is stationed at the centre of the celestial sphere, and hence we see that the apparent motion of the sun takes place in a plane which passes through the earth.

Let the sphere P Q (Fig. 72) represent the celestial sphere, of which the points P and Q are the north and south poles respectively. The great circle C E A E is the celestial equator. The ecliptic is the great circle A B C D, which intersects the equator in the points A C, called the *Equinoxes*. The plane of the ecliptic is inclined to the plane of the equator by an angle which is called the *Obliquity of the Ecliptic*, and is equal

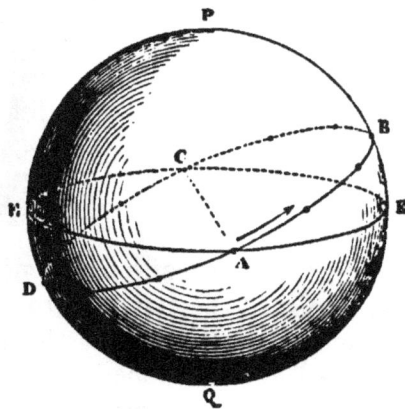

FIG. 72.

to about $23° 27'$. The points on the ecliptic B D, which are

situated at the middle points B and D of the semi-circum-
ferences A B C and A D C, are termed the *Solstices.* The sun
moves round the ecliptic in the direction indicated by the
arrow. He passes through the point A about March 21,
whence that point is termed the *Vernal* equinox. In three
months afterwards, about June 22, he reaches B, the summer
solstice, and C, the autumnal equinox, on September 23, and
D, the winter solstice, on December 22.

  § **55.** *Day and Night.*—To study the variations in the
length of the day, a celestial globe is specially convenient

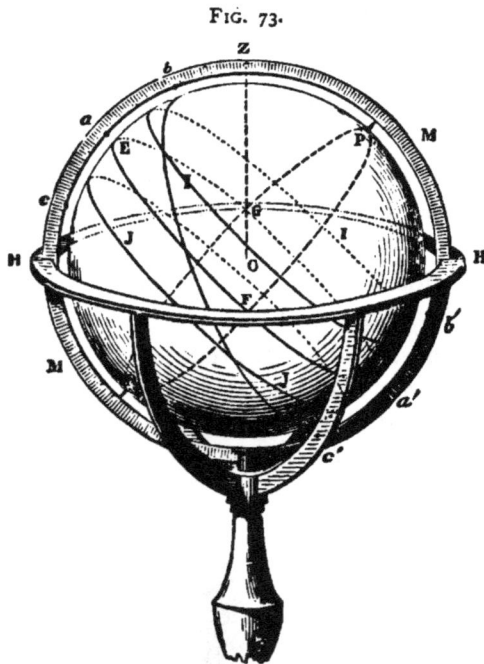

FIG. 73.

(Fig. 73). The earth is supposed to have shrunk into a
small point at the centre O of the celestial globe. The plane
of the horizon, drawn at the place where the observer is
supposed to be stationed, must cut the celestial sphere in a
great circle. This great circle corresponds to the ring H F H,
which is attached to the frame by which the globe is sup-

ported. A celestial object will be visible to the observer if it be on that part of the globe which is above the plane H F H. All objects on the hemisphere below that plane are invisible.

To examine the variations in the length of the day, the globe must first be set correctly for the latitude of the place where the observer is stationed. We have shown in § 41 that the elevation of the pole of the heavens above the horizon is equal to the latitude. To set the globe it is, therefore, necessary to place the axis P Q, about which the globe revolves, so that it makes with the plane H F H an angle equal to the latitude.

Let z be the zenith ; then the great circle passing through z and P is the meridian, which of course cuts the horizon at right angles. The meridian on the globe is represented by the ring H Z M. The great circle É F G, drawn perpendicular to the polar axis P Q, is the equator. Since two great circles must mutually bisect each other, it is obvious that the equator is bisected by the horizon at F and G, and therefore half of the equator is below the horizon and half is above. A celestial object which is situated on the equator will, in the course of one revolution of the heavens, remain above the horizon for precisely the same time that it is below. When the sun is in the equator the length of the day must be equal to the length of the night. If the sun always moved in the equator then the day and night would be constantly equal. The path of the sun lies, however, in the *ecliptic*, which crosses the equator at the vernal and autumnal *equinoxes*. The sun is, therefore, in the equator when, and only when, he is in the act of passing through one or other of the equinoctial points on March 21 or September 23. At either of the dates mentioned the length of the day is equal to the length of the night.

After passing the vernal equinox on March 21 the sun gradually increases his declination. The path of the sun as caused by the *diurnal rotation* ceases to be a great circle after the equinox is passed. It is then a small circle, of which the dimensions gradually decrease until June 22, when the sun having attained its greatest north declination, the small circle which it appears to describe in virtue of its apparent diurnal rotation is a minimum. At a date intermediate between March 21 and June 22 the path of the sun will be the small circle 1 1 (Fig. 73). It is plain that this circle is bisected by the great circle F P G, and therefore the part of the small circle, which is above the plane of the horizon, is greater than that which is below it. When the sun appears to move in this circle it continues above the horizon for a longer period than it is below, and consequently the length of the day exceeds that of the night. The disparity between the length of the day and night will increase as long as the sun increases in north declination, until the difference attains its maximum value on June 22, when the sun, having attained the declination of $23° 27'$, is in the summer solstice. In this case the sun passes the meridian at the point b, the distance $a b$ being equal to $23° 27'$.

The sun having passed the summer solstice, begins to return towards the equator. On September 22 the sun reaches the equator, and the length of the day is again twelve hours, and equal, of course, to the length of the night. After this date the sun is below the equator, its south declination gradually increasing to the value $23° 27'$, which it has on December 22. The sun each day after passing the autumnal equinox will describe a small circle of the celestial sphere parallel to the plane of the equator, and lying to the south of it. The dimensions of this circle gradually diminish, until the winter solstice is reached. Let J J be the circle described by the sun on a date between the autumnal equinox and the winter solstice. Part of this circle is above the plane of the horizon and part is below, and it is easy to see that the latter part is greater than the former. The plane P F Q must obviously bisect this circle, and the two points of bisection, and therefore more than half the circle, are below the plane of the horizon.

When the apparent diurnal motion of the sun is performed in this circle J J, the sun is below the horizon for a longer period in each revolution than it is above, and hence the day is shorter than the night. The disparity between the length of the day and that of the night goes on augmenting until the winter solstice, when the shortest day is attained. In the movement of the sun from the summer solstice to the winter solstice we see that the length of the day changes from its greatest value to its least. After passing the winter solstice the same series of changes in the relative lengths of the day and night is repeated in the inverse order, and in the course of 365¼ days, or one year, a complete cycle of changes has been gone through and a new cycle is commenced.

There are, however, remarkable phenomena connected with the recurrence of day and night in countries near either of the poles, that merit special attention.

The globe is set in Fig. 74 so as to correspond to a considerable northern latitude, which we shall suppose is not less than 66° 33'. The distance of the pole P from the zenith Z is equal to the complement of the altitude of the pole, and as the altitude of the pole is equal to the latitude, it follows that the zenith distance of the pole in the present case is less than 23° 27'. Let us now consider the circumstances under which the sun can rise above the horizon in a case like the present. Whenever the sun is above the horizon, its distance from the zenith must be less than 90°. The zenith distance of the sun is always smaller when the sun is on the meridian than when the sun is either approaching to the meridian or after he is past it. We have, therefore, to see under what circumstances the sun, when on the meridian, is above the horizon, or, what is the same thing, when the meridian zenith distance of the sun is smaller than 90°. To give definiteness to what we have to say on this subject, let us take in the first instance a latitude of 75°, so that the angle between z and P is equal to 15°. The distance from the pole to the equator is equal to 90° : hence the distance in the present case from the zenith to the equator is 90° −15°=75°. If then the sun were on the equator, its zenith

FIG. 74.

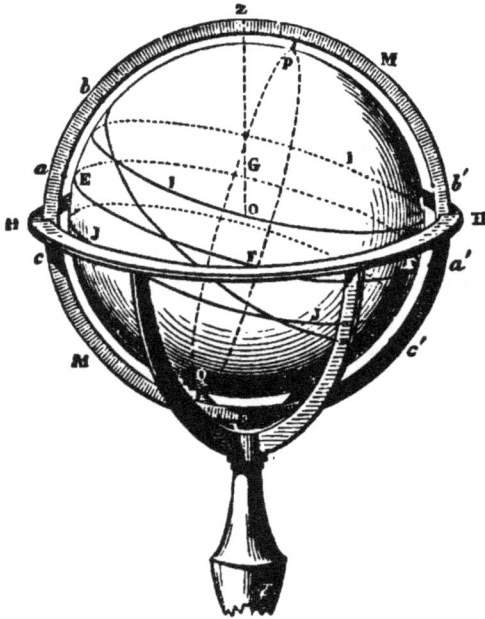

distance at the moment of culmination would be 75°. The sun must then be visible every day, because the zenith distance at culmination is less than 90°. As the sun moves from the equinox towards the summer solstice, the meridional zenith distance gradually diminishes. When the sun is at the solstice, its declination is 23° 27′. It is, therefore, north of the equator by this amount, and consequently the zenith distance at culmination is 75°−23° 27′=51° 33′. The sun crosses the meridian twice daily, viz. at noon and at midnight ; in the former case the zenith distance has its smallest value, and in the latter the zenith distance has its greatest

value. We have seen that the southern point E of the
equator is 75° from z. The northern point of the equator
is below the horizon, and its distance from z must be 90°
+15°=105°. On June 22, when the sun is in the summer
solstice, the north declination is 23° 27', and when the sun
is on the meridian at midnight it will be 23° 27' nearer the
zenith than the north point of the equator. It follows that
the zenith distance of the sun at midnight will be 105°
−23° 27'=81° 33'. As this is less than 90°, we learn the very
remarkable fact that the sun under these circumstances does
not descend below the horizon at all, so that, in fact, as long
as this state of things continues we have perpetual day.
Thus is explained the midnight sun in the Arctic regions.

It is plain that as the sun sets every day, even in midsummer, in the latitude of the United Kingdom, while it
does not set at midsummer in the latitude of 75°, there must
be some intermediate latitude where the sun just reaches the
horizon at midnight in midsummer. The determination of
this latitude is very easily effected by the celestial globe.
The polar distance of the sun is equal to the complement of
the declination, and therefore at midsummer the polar distance is 66° 33'. When the sun is on the horizon at midnight,
the polar distance of the sun must be equal to the altitude
of the pole; but the latter is the latitude, and hence when
the sun is on the horizon at midnight in midsummer the
latitude must be 66° 33'. The parallel which passes through
all the points which have the north latitude 66° 33' is called
the *Arctic Circle.* At all places to the north of this circle
the sun remains above the horizon, even at midnight, for a
certain portion of the summer. The number of days in the
year during which the midnight sun is to be seen increases
as the latitude increases, and may be found as follows.

After the sun has passed the vernal equinox its declination gradually increases, and the amount by which the
diurnal circle dips below the horizon gradually diminishes
until a declination is attained, when the diurnal circle I I

(Fig. 74) just grazes the northern point of the horizon. On the day when this declination is attained, the sun's centre will therefore not set. The next day the sun's centre will at midnight be some distance above the horizon, and the midnight sun will be continually witnessed until, after passing the summer solstice, the sun has returned so far towards the autumnal equinox that he is again performing his diurnal motion in the circle I I. At midnight the centre of the sun will then just graze the horizon, and on the following revolution the centre will pass below the horizon. When the equinox has been reached, the length of the night has become equal to that of the day, and the length of the latter gradually diminishes until the sun has attained a declination so far to the south that the circle J J, which its centre describes, only just grazes the horizon. In this case when the sun is on the meridian at noon, its centre has just reached the horizon, and on the following day the centre of the sun will not reach the horizon at all. Perpetual night will then reign until the sun, having attained the winter solstice, has moved so far towards the equator that it again performs its diurnal motion in the circle J J. When this point has been attained the centre of the sun just grazes the horizon at noon, and the following day the sun may be seen to rise : the length of the day then gradually increases, and the cycle already described is repeated.

It should also be observed that the limits of latitude in which perpetual day is enjoyed at midsummer are identical with the limits in which perpetual night is found in midwinter. At the winter solstice, the south declination of the sun is 23° 27′. If, therefore, the sun at culmination in the winter solstice be on the horizon, the southern point of the equator will be 23° 27′ above the horizon. The meridional distance from the zenith to the equator will therefore be 90° − 23° 27′ = 66° 33′. It is, however, easy to see that this is the altitude of the pole, and therefore the latitude of the place. It follows that in all localities of which the latitude exceeds 66° 33' there will be perpetual day for a certain period at midsummer, and perpetual night for a corresponding period at midwinter.

To take an extreme case, let us consider the condition of day and night which could be presented to an observer who was stationed at the latitude of 90°, or, in other words, at the pole itself. The pole would then coincide with the zenith, and the celestial globe adapted to these circumstances is

FIG. 75.

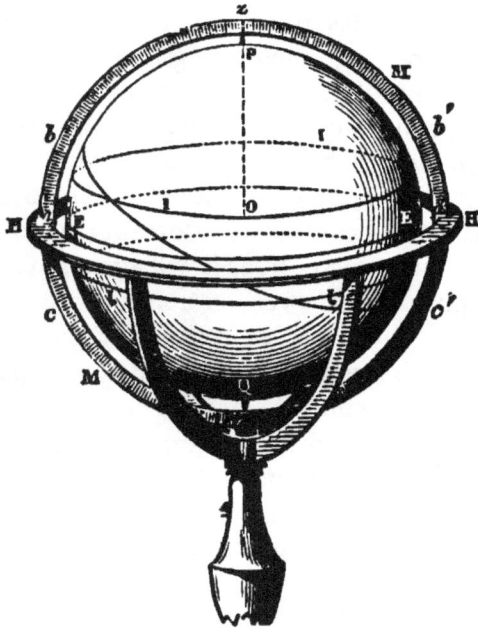

shown in Fig. 75. The equator, being at every point 90° distant from the pole, must now be coincident with the horizon, and the small circles in which the stars revolve will be all parallel to the plane of the horizon. The altitude of a star would then be a constant, and the phenomena of rising and setting stars would be unknown. If the sun were actually fixed in the ecliptic then it would always describe the same small circle, and would be visible or not, according as that point of the ecliptic lay in the half which was above or below the plane of the horizon. As, however, the sun moves in the ecliptic, it will be seen constantly through part of the year, and remain invisible during the remainder. Suppose, for example, that the sun is in the vernal equinox. The centre of the sun is then on the equator, and as the equator is coincident with the horizon, the centre of the sun will lie on the horizon. The diurnal motion will, therefore, carry the sun round the horizon. As the declination increases the sun will gradually become elevated above the horizon, so that it will really describe a sort of spiral on the celestial sphere, each revolution being at a greater altitude than the preceding one, until the summer solstice is attained, when the sun will revolve nearly in a small circle parallel to the horizon, at an altitude of 23° 27'. After the summer solstice has passed, the spiral course of the sun will recommence, but its motion will now be gradually downwards, so that after a number of complete revolutions it will again have reached the horizon. Thus an observer stationed at the north pole will find the sun above his horizon during the whole interval from the vernal to the autumnal equinox. After the latter equinox the sun descends below the horizon, and performs a spiral movement in the southern hemisphere, until, at the winter solstice, it has attained the south declination of 23° 27'. It then commences to return towards the horizon, to reappear in the northern hemisphere after the vernal equinox. At the north pole, therefore (neglecting small variations), one half of the year is day and the other half is night. Finally, let us take the other extreme case of an observer who was stationed at the equator. The globe set for this case is shown in Fig. 76. As the latitude is now equal to zero, the elevation of the pole above the horizon must be zero, and therefore the pole must actually lie in the plane of the horizon. All the small circles which are described by the celestial bodies must, therefore, be perpendicular to the plane of the horizon. It is easy to see that half of each circle lies below the plane of the horizon, and half is above. Thus every celestial object will (so far as the diurnal motion is concerned) be above the horizon for half the day, and below it for the other half. The sun is no exception to this statement : the small circle in which the sun moves each day is bisected by the horizon, and therefore the day and night are of equal length.

At the equinox the sun at rising is stationed precisely at

FIG. 76.

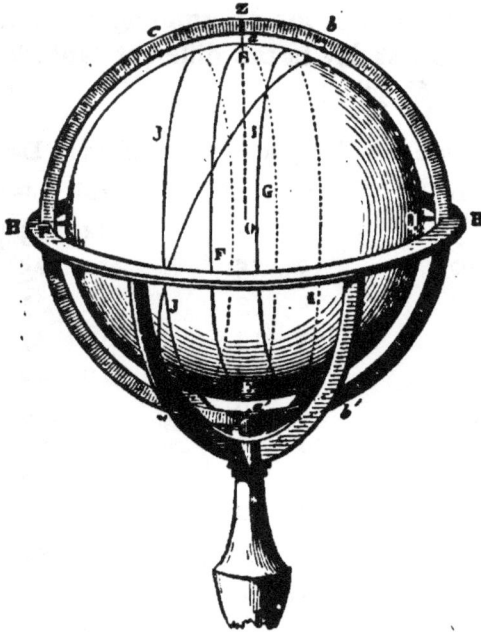

the east point of the horizon ; it then ascends to the zenith, and descends again to set in the western point of the horizon. After the vernal equinox is past, the sun has moved into the northern hemisphere, and the small circle described in the diurnal motion, though still perpendicular to the plane of the horizon, cuts the horizon at points less than 90° from the north point. Thus each day the sun rises somewhat to the north of due east, ascends perpendicularly till it culminates at a point on the meridian to the north of the zenith, then descends to set perpendicularly at a point of the horizon somewhat to the north of west. The magnitude of the diurnal circle gradually decreases until the summer solstice is reached, when the sun at noon is 23° 27′ to the north of the zenith ; the circle afterwards increases till it becomes a great circle passing through the zenith at the autumnal solstice, and then repeats in the southern hemisphere the series of changes that we have described in the northern.

We give here a table for determining the length of the longest and shortest day in the different latitudes, from the equator up to 66° 33′.

Latitude.	Longest Day.		Shortest Day.	
0°	12^h	0^m	12^h	0^m
5	12	17	11	43
10	12	35	11	25
15	12	53	11	7
20	13	13	10	47
25	13	34	10	26
30	13	56	10	4
35	14	22	9	38
40	14	51	9	9
45	15	26	8	34
50	16	9	7	51
55	17	7	6	53
60	18	30	5	30
65	21	9 ·	2	51
66° 33′	24	0	0	0

At localities above the latitude of 66° 33′ the length of the longest day is never less than twenty-four hours, while there is no day at all during a certain period at midwinter. It is, however, of interest to calculate the connection between the latitude and the number of days during which the sun does not set in summer or rise in winter. The results are given in the following table.

North Latitude.	Sun does not set for about	Sun does not rise for about
66° 33′	1 day	1 day
70	65 days	60 days
75	103 ,,	97 ,,
80	134 ,,	127 ,,
85	161 ,,	153 ,,
90	186 ,,	179 ,

§ 56. *Zones into which the Earth is divided.*—The parallels which pass through all the points of latitude 66° 33′ in either hemisphere, con-veniently mark off two important regions in the vicinity of the north and south poles respectively. These parallels are repre-sented by A A′ and B B′ (Fig. 77), and the regions which they define are A P A′ and B Q B′. The northern circle A A′ is called the *Arctic Circle*, while B B′ is the *Antarctic Circle*. The regions with-

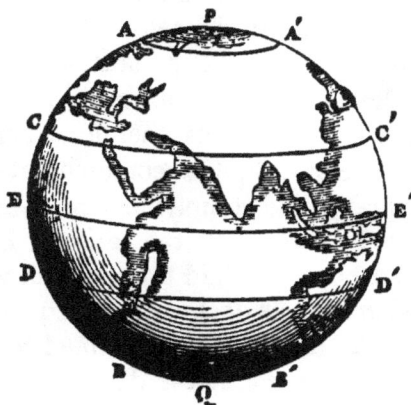
FIG. 77.

in these circles are characterised by the presence of con-tinual day at midsummer, and continual night at midwinter. They are the coldest portions of the earth, and are known as the *North Frigid Zone* and *South Frigid Zone* respectively.

Another very natural division of the earth is presented by the equator E E′, which divides the portion of the earth between the Arctic and Antarctic circle into equal parts. There is also a natural division by means of the circles C C′ and D D′, which are known as the Tropic of Cancer and the Tropic of Capricorn. It will be necessary to explain the significance of these circles.

The sun at midday attains its greatest elevation above the horizon, so that the zenith distance of the sun is then a minimum. At midsummer the sun at noon is nearer to the zenith than it is at any other season. In European latitudes it is always seen that even when the sun has its greatest altitude it is still at a considerable distance from the zenith. The question, therefore, arises as to whether there are any localities on the earth's surface at which the sun, when its altitude is greatest, is actually situated at the zenith.

The altitude of the pole being equal to the latitude, the angular distance from the pole to the zenith is equal to the complement of the latitude, or to what is called the co-latitude. If, therefore, the sun at its moment of culmination be situated in the zenith, the polar distance of the sun must be equal to the co-latitude. The polar distance of any celestial object is, however, equal to the complement of its declination. It is, therefore, evident that when the sun is in the zenith, the complement of its declination must be equal to the complement of the latitude, but this can only be the case when the latitude is equal to the declination; and hence we find that the sun will culminate in the zenith whenever its declination is equal to the latitude.

As the ecliptic is inclined to the equator at the angle of 23° 27′, the greatest declination which the sun can have is 23° 27′, and this is only attained when the sun is at the summer or winter solstice. As the sun can only attain the zenith when its declination is equal to the latitude, and as the greatest declination is +23° 27′, it follows that the sun cannot attain the zenith of localities which are situated at a greater distance north and south of the equator than 23° 27′. The north and south parallels, which correspond to this latitude, bound the regions in which it is possible for the sun to attain the zenith.

As the sun in the course of the year passes through every

phase of declination between $+23°\ 27'$ and $-23°\ 27'$, so it
follows that the sun will appear during the year at, or close
to, the zenith of every place contained between these two
parallels.   It would not be correct to say that the sun would
actually culminate in the zenith, for it will generally happen
that at the time when the sun passes close to the zenith, its
declination on two consecutive days will be one a little
greater and the other a little less than the latitude.   The
sun will, therefore, on one day culminate a little on one
side of the zenith, and the following day it will be a little on
the other side.

The circles c c' and D D', bounding this very naturally
marked region on the surface of the earth, are known as the
tropics.   The name is given to them from the circumstance
that an observer on one of these circles will, as midsummer
approaches, find the sun daily approach nearer and nearer to
the zenith at the moment of culmination, until on midsummer
day the culmination will be exceedingly close to the zenith.
The centre of the sun will then be actually in the zenith if
it should happen that the moment of culmination is identical
with the moment when the sun has actually attained its
greatest north declination.   After this point has been
attained, the sun will then begin to *turn* back, for at the
next culmination it will be a little south of the zenith in the
northern hemisphere, or north of the zenith in the southern
hemisphere.

A resident, either on the equator or at any station intermediate between the two tropics, will have the sun in his
zenith twice each year.   Suppose, for example, that the lati-
titude of the place be $+10°$.   At the vernal equinox the
sun's declination is zero, but it gradually increases, and,
when it attains the value of $10°$, being then equal to the
latitude, the culmination must take place at the zenith.
After the declination of $10°$ is passed, the sun gradually
moves to the summer solstice, when its declination is $23°\ 27'$,
and then, in commencing to return towards the equator,

the declination diminishes down to 10°, and, therefore, the culmination is again at the zenith. We thus see that a station in the northern hemisphere which has a latitude less than 23° 27′ will have the sun in its zenith on two occasions between the vernal and the autumnal equinox.

At a place situated on the equator, the latitude is of course zero, and hence the declination of the sun must be zero when it culminates at the zenith. This is only the case at the vernal and autumnal equinox. It is easy to show that in the southern hemisphere the sun is twice near the zenith, between the autumnal and the vernal equinox, provided the latitude is less than 23° 27′ south of the equator.

The regions between the equator and the tropic of Cancer on the one hand, and the equator and the tropic of Capricorn on the other, are known as the *north* and *south torrid zones* respectively. The region between the tropic of Cancer and the arctic circle is the north temperate zone, while that between the tropic of Capricorn and the antarctic circle is the south temperate zone. Each of the two hemispheres into which the equator divides the earth is thus conveniently divided into three zones, and the boundaries are indicated in a very natural manner, by the apparent annual movements of the sun.

§ 57. *Effect of Refraction on the Length of the Day.*—In what we have hitherto stated with respect to the variation in the length of the day, in different seasons of the year and at different latitudes on the earth, we have overlooked the effect of atmospheric refraction in changing the apparent place of the sun. It is, therefore, necessary to point out briefly the modifications which must be introduced into our results when refraction has to be taken into account. In the tables given on pages 148–9, it has been assumed that the day commences at the moment when the centre of the sun is actually on the horizon, as it would be seen were there no atmosphere. The refraction of the atmosphere tends to raise all objects towards the zenith, and at the horizon this is no less than 33'. In fact, as we pointed out in § 51, the sun's centre appears to be on the horizon, while *in reality* the sun's centre has to ascend 33' in order to be on the horizon. It therefore follows that, as the sunrise is apparently accelerated and sunset retarded, the length of the day really appears longer than calculations show, which neglect the effect of the atmosphere.

At the north pole the effect of refraction upon the length of the day is indicated in a somewhat remarkable manner. We shall only consider the position of the *centre of the sun*, so that some modifications are required when the dimensions of the sun are taken into account. When the sun, in passing from the winter solstice to the vernal equinox, has approached so near to the latter that its south declination is 33', then the refraction of the atmosphere raises the sun's centre to the horizon, and the perpetual daylight of summer has commenced. Again, when the sun, after passing the summer solstice, is again descending to the horizon, the refraction appears to retard its setting, so that it is not until the sun's centre has really passed below the horizon to the distance of 33' that the phenomenon of setting is witnessed. Remembering that for an observer stationed at the north pole the celestial equator is coincident with the horizon, we see that the arctic day will be enjoyed during the whole time that the sun moves from 33' south declination to $+23° 27'$ north declination at the solstice, and then back again to 33' south.

§ **58.** *Twilight.*—The atmosphere has also an effect upon the duration of daylight, which arises from a different source. After the sun has set, his rays still traverse the higher regions of the atmosphere, and illuminate either the air or the particles which the air retains in a state of suspension. This light from above diffuses a certain amount of illumination on the surface of the earth, which gradually decreases in intensity as the sun sinks below the horizon. The twilight in the evening is generally known as *dusk*, while that which precedes the rising of the sun is known as the *dawn*. Were it not for the effect of the air, the brightness of the day would totally cease when the last portion of the sun's disk disappeared below the horizon at sunset, and the daylight would only commence when the disk commenced to ascend above the horizon at sunrise.

The twilight either at dusk or dawn is only seen when the sun is within a certain distance of the horizon. The exact amount of the distance appears to be dependent upon the state of the atmosphere. It may, however, be generally stated that the distance is about 18°. The dusk is thus usually visible until the centre of the sun has been carried by the diurnal motion to a perpendicular distance of 18° below the horizon. Similarly the earliest glimpses of dawn may be caught when the sun, in his approach to the horizon, has attained a distance of only 18° therefrom. It will of course be understood that the distance from the horizon which is here alluded to is the distance of the sun from the nearest point of the horizon, or the arc which is intercepted by the plane of the horizon on the great circle drawn from the zenith to the sun. This will not, of course, be generally coincident with the actual course in which the sun is apparently moving by the diurnal motion. The sun will generally have to move over a distance considerably greater than 18° before it reaches the horizon after the first glimpses of dawn have been perceived.

The duration of the twilight either at dusk or dawn will be found to vary very considerably at different seasons, as well as at different latitudes. To take a simple case, let us suppose the observer to be stationed at the equator, and let us calculate the duration of twilight when the sun is at either of the equinoxes. Under these circumstances, as the sun is actually situated in the celestial equator, and as the equator passes through the zenith, the diurnal motion of the sun will be effected in the great circle which passes through the zenith and the eastern and western points of the horizon—the circle, in fact, which is often known as the *prime vertical.* Before sunrise the sun will ascend perpendicularly to the horizon, and after sunset it will descend perpendicularly below it. The twilight will therefore continue in this special case during the time when the sun moves through an arc of 18° on the celestial sphere, in virtue of the apparent diurnal motion. As the diurnal motion completes its revolution in twenty-four hours, a point on the equator which moves through 360° in one revolution must move through 15° in one hour. To move through an arc of 18° a time of 1ʰ 12ᵐ will therefore be required. Hence it appears that under the circumstances we have described the twilight at dusk and at dawn will last for a period of 1ʰ 12ᵐ.

At other places or seasons the duration of twilight can be found by the celestial globe with sufficient accuracy for most purposes. First set the globe to the latitude of the place by placing the axis of the celestial sphere at the correct inclination to the horizon. Then find on the globe the point on the ecliptic which the sun occupies on the day in question, and place this point on the horizon. The angle through which the globe must be rotated in order to bring the point from the horizon to a distance of 18° below it, converted into time at the rate of 15° to an hour, is the duration of twilight.

It is well known that in our latitude, at midsummer, the night is never perfectly dark even at midnight. Looking towards the northern horizon, the midnight twilight is seen to come from that region, and its origin can easily be explained. When the sun crosses the meridian at midnight, its distance below the horizon is greater than when the sun is at any other point of its diurnal path; if, therefore, the depression of the sun below the horizon at midnight be not greater than 18°, the sun will, during the entire night, be within 18° of the horizon, and consequently the twilight will never cease.

At midsummer, the sun's declination being 23° 27', the polar distance of the sun is 90°—23° 27'=66° 33'. If, therefore, the sun be 18° below the horizon, the elevation of the pole above the horizon will be 66° 33'—18°=48° 33'. It therefore appears that when the latitude of the place is 48° 33', the sun at midsummer will only just attain the distance below the horizon at which twilight ceases. At any latitude exceeding 48° 33' we shall, then, have twilight at midnight in midsummer. Take, for example, the latitude of 53°. The elevation of the pole above the horizon is therefore 53°, and consequently the distance of the midsummer sun below the horizon at midnight is 66° 33'—53° =13° 33'. As this is less than 18°, the twilight will still be seen at midnight.

Similar considerations will also enable us to find the number of days at a given latitude during which the midnight twilight will be seen. At the latitude of 53°, which we may adopt as before, for the sake of illustration, the elevation of the pole is, of course, 53°. If the sun comes to the meridian at midnight at a point which is 18° below the horizon, the distance from the pole to the sun is 53+18°=71°. The declination of the sun is equal to the complement of the polar distance, and hence the declination is 90—71=19°. The midnight twilight will therefore be perceived whenever the declination of the sun exceeds 19°. To find the number of days during which the midnight twilight continues, it is therefore only necessary to find the number of days in which the sun moves from the declination 19° to the summer solstice, and then back again to the declination 19°. In the particular case now before us, the number of days is about 74.

The effect of twilight in mitigating the darkness of the long arctic winter should also be noted. In midwinter the sun does not ascend above the horizon in the arctic regions, but if the sun approaches the horizon to a distance less than 18°, the twilight of course becomes visible. At the winter solstice, the south declination of the sun is 23° 27'. If the sun at culmination be 18° below the horizon, it follows that the equator must cut the meridian at an elevation of 23° 27' — 18° = 5° 27' above the southern horizon; but it is easy to show that this must be the complement of the elevation of the pole above the horizon. The whole semicircumference of the meridian is made up of the elevation of the pole, the altitude of the point on the equator, and the distance from the pole to the equator. As this last is equal to 90°, the two former must be complementary. The elevation of the pole at that latitude in the arctic region, where the midday sun in winter just comes near enough to the horizon to afford a glimpse of twilight, is 90° — 5° 27' = 84° 33'; at lower latitudes the number of days in which twilight is witnessed in midwinter gradually increases. Within the distance of 5° 27' from the pole, there are a certain number of days every winter, during which not even a glimpse of twilight announces the fact that the sun has reached the meridian at noon. This number gradually increases from the latitude of 84° 33' to the pole.

As the sun approaches the winter solstice, its south declination gradually increases, until about the 13th of November the Sun has descended 18° below the equator. Assuming that the atmospheric conditions of twilight at the pole resemble those which are found in latitudes to which access is obtained, it would appear that the twilight at midday will cease on the 13th of November, and will continue absent until the sun, having passed the winter solstice again, reaches the distance of 18° from the horizon on January 29.

§ 59. *Changes of Temperature on the Earth.*—The earth is constantly radiating out heat from its surface to the celestial spaces, and tends to grow cooler and cooler by the continual loss of heat which is thus sustained. On the other hand,

the rays from the sun, besides their effect in illuminating the
earth which we have already considered, convey to us
heat which counteracts the continuous loss of heat to which
the earth is exposed by the effect of its radiation. During
the night this radiation is continually cooling the earth, and,
as there is no counterbalancing heat then derived from the
sun, the temperature goes on decreasing after sunset. When
the sun rises, its rays begin to warm the earth, and though the
loss by radiation still continues, yet the increase of heat due
to the sun soon counterbalances the loss arising from radiation, so that the temperature gradually rises. The efficiency
of the sun rays in warming the earth increases as the sun
approaches to the meridian, and attains a maximum when
the sun is on the meridian at noon. As the sun commences
to decline towards the western horizon in the afternoon, the
amount of heat received from his rays by the earth decreases.
Still, so long as the amount which is received exceeds that
which is lost by radiation towards the celestial spaces, the
temperature of the earth goes on increasing. Thus the
greatest temperature in the day is attained not actually at
noon, but some time afterwards. When the amount of heat
received just equals the amount lost, then the maximum
temperature of the twenty-four hours is reached. The loss
then exceeds the gain, and consequently the temperature
falls.

The series of changes in the temperature which occur
daily would be reproduced each day in precisely the same
manner if the apparent diurnal motion of the sun were
always performed in the same course. But as the sun has
an apparent annual motion on the surface of the celestial
sphere, so there are corresponding modifications in the path
of the diurnal motion from one day to another. For the
sake of illustration we may take the case of the northern
temperate region. As the length of the day commences to
increase in spring, the quantity of heat received from the sun
between sunrise and sunset increases also. This is due not

only to the longer continuance of the sun above the horizon, but also to the increased elevation of the sun, which causes its rays to descend more nearly perpendicularly on the earth. The temperature is of course still diminished by radiation from the earth during the night, but the decreasing length of the night combined with the increase of heat during the day soon makes the gain to exceed the loss, and consequently the temperature steadily rises. As the summer is neared, the daily gain in temperature approaches a maximum, so that on midsummer day the region receives more heat during the day, and loses less during the night, than on any other day in the year. It does not, however, follow that the hottest period of the year is precisely at midsummer day; just as in the case of the changes of temperature during the day it appears that the maximum of temperature does not coincide with the moment when the most heat is being gained, so in the case of the annual changes of temperature the day when most heat is received is not necessarily the hottest day. At midsummer the gain during the day largely exceeds the loss during the night, and though the daily gain is less on subsequent days, yet the total heat must be increasing so long as there is any gain at all. The greatest temperature will, therefore, be attained at the moment when the gain during the day is equal to the loss; this will be some little time after midsummer. On subsequent days the loss will exceed the gain, and consequently the temperature will fall; but just as the greatest heat is not attained precisely at midsummer, so the greatest cold is not attained precisely at midwinter. At midwinter no doubt the loss of heat by radiation exceeds the gain during the day by a greater amount than at any other season, but so long as the loss is in excess of the gain, so long will the temperature continue to decrease; nor will the decrease be arrested until, by the gradual lengthening of the days during the approach of spring, the amount of gain during the day will precisely counterbalance that lost by radiation. When this point is attained, the temperature will have sunk to its lowest amount for the year ; this point is reached about a fortnight after the winter solstice. What has been said as to the diurnal and annual variations of temperature may also be applied to the southern temperate region, remembering that midsummer here is midwinter there, and *vice versa.* A somewhat different account must be rendered of the variations of temperature between the tropics on the one hand, or inside the arctic circles on the other.

On the equator the length of the day is equal to the length of the night, whatever be the point of its apparent path in which the sun is situated : the changes of temperature corresponding to different periods of the year are therefore much less marked than is the case in the temperate latitudes, where the relative lengths of the day and night undergo considerable fluctuations. Any changes in the temperature at the equator are therefore limited to the effects which can be produced by the fluctuations of the sun's zenith distance. At the equinox, when the sun is in the equator it passes through or close to the zenith at the moment of culmination. At the solstices the sun is at culmination 23° 27' to the north or south of the zenith. As the sun is therefore shining more perpendicularly upon the earth at the period of the equinoxes, these periods are, so far as this is concerned, the hottest parts of the year. At regions in the tropics not situated on the equator there is of course some difference between the lengths of the days and nights, and a corresponding fluctuation in the temperature, but this is much less marked than in places situated in the temperate regions. In the arctic regions the circumstances are modified by the fact that during part of the year the sun remains entirely above the horizon, while during part of the year it is as continuously below.

§ 60. *Mean Temperature.*—The mean temperature of any spot on the earth depends chiefly upon the latitude. At the equator the mean temperature is a maximum, and it gradually decreases as the latitude increases until the pole is attained. The variation in the mean temperature is not to be explained by the relative proportions of daylight and night : the number of hours of sunlight in the year is pretty much the same whatever the latitude may be. Thus at the equator we have twelve hours of sunlight out of every twenty-four, and therefore the sun is above the horizon at the equator for half the whole number of hours in the year, and below it for the other half. This is the same as we find at the pole, for the sun is there continuously above the horizon for six months, and then continually below for six months. At the intermediate latitudes, the great length of the days in summer is counterbalanced by the great length

FIG. 78.

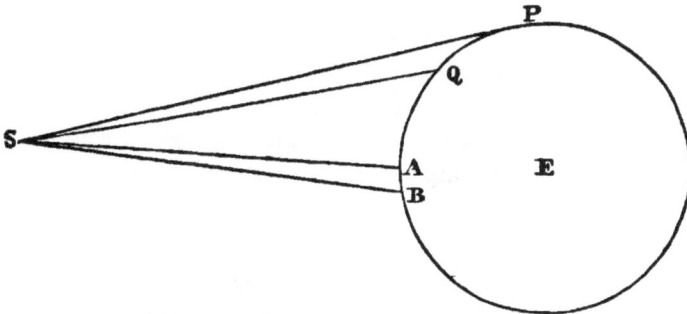

of the nights in winter, so that here again we find that for one half of the year the sun is above the horizon, and for one half below. It is, therefore, necessary to search for some other reason for explaining the great variations in the mean annual temperature at the different latitudes. The explanation will be found in the different elevations which the sun attains in different latitudes, combined with the fact that the efficiency of the sun's rays for warming the earth depends upon the *angle of incidence.* This latter remark may be explained by Fig. 78. Let s represent the sun, and E the earth ; then the rays from the sun diverge in various directions S P, S Q, S A, S B, and it may be assumed that the heat dis-

M

tributed in the various directions is constant. If we conceive a very large sphere to be drawn of which s is the centre, then all the heat from the sun will fall upon the interior of this sphere. This heat is uniformly spread over the surface, so that a given area receives the same quantity of heat whatever be the situation which it occupies on the surface of the sphere. For simplicity we may suppose the area under consideration to be a small circle, and the quantity of heat which falls on that circle will be proportional to the area of the segment of the sphere which it cuts off. If the circumference of the small circle be regarded as the base of a cone of which the centre is at the point s, then the cone contains a portion of the sun's rays which bear to the total radiation the same proportion which the segment of the sphere bears to the entire area of the sphere.

Let two right circular cones be drawn with equal vertical angles, and let the section of these cones, by the plane of the paper, be P S Q and A S B (Fig. 78), the former falling on the earth near the pole, and the latter in the vicinity of the equator. Since the angles at the vertices of these two cones are equal, it is evident that an equal portion of the sun's radiation passes through each, and that, consequently, a certain area of the earth, of which P Q is a section, receives the same quantity of heat as the area near the equator of which the section is A B. It will at once be seen that the area which the cone P S Q cuts on the earth's surface is much larger than the area cut out by the cone A S B. The same quantity of heat which falls on a certain region near the equator has, therefore, to be distributed over a much larger area near the pole; hence the temperature, per unit of area, must be less at the neighbourhood of the pole than in the neighbourhood of the equator. More generally it appears that the temperature decreases as we pass from the equator towards the poles.

§ 61. *Effect of the Atmosphere.*—All the considerations which have here been brought forward with reference to the efficiency of the sun in heating the earth, must receive certain modifications when the existence of the atmosphere surrounding the earth is taken into account. The heat of the sun acting on one region of the earth raises the temperature of that region. The air which is in contact with the earth becomes warm, and as it expands by heat must necessarily become lighter. Thus the air over a heated region of the earth ascends, and its place is supplied by cooler air rushing in along the surface of the earth. These currents of air, which rush towards the heated regions, form what we know as *winds.* Their regularity is, however, greatly modified by the configuration of the land and other circumstances. It would appear that the influence of the winds, so far as their effect on the temperature is concerned, is entirely of a moderating character. The hot region is cooled by the cool air rushing in upon it, while the heated air, which ascends, carries the heat with it to some less favoured region of the earth.

So considerable are the effects produced by the action of the winds, that the law of the diurnal changes of temperature which we have laid down, and still more the law of changes of temperature in the course of the year. are often seriously incorrect. These laws are. therefore, to be regarded as merely expressing the mean or average state of things from which the irregularities produced by the influence of the atmosphere have been eliminated.

§ 62. *The Origin of Right Ascensions.*—We are now in a position to explain how the point is to be found on the equator from which right ascensions are to be measured. The two most important fixed circles of the celestial sphere are unquestionably the *equator* and the *ecliptic*, for though of course, at a given place, the meridian or the horizon are of very great significance, yet these are not fixed circles on the sphere, as their relation to the celestial objects changes every moment. As the right ascensions are measured on the equator, it is natural to make the measurements start from one of the two most remarkable points on the equator, these being *the two equinoxes* or the intersections of the equator with the ecliptic. There still remains a choice as to which of the two equinoxes shall be selected as the zero of right ascension. Custom has settled on the *vernal equinox*, through which the sun's centre passes about March 20 every year, when changing its declination from south to north.

It has already been explained (§ 39) that when the sidereal clock is going correctly, the right ascension of any object is the sidereal time at which that object crosses the meridian. The determination of the origin from which right ascensions are measured is, therefore, the same problem as the setting of the sidereal clock so that it shall show correct time. We shall, therefore, first direct our attention to the problem of ascertaining both the *rate* of the sidereal clock and its *error*.

The sidereal day is the interval between two successive upper culminations of the same star. Let us, for the sake of illustration, fix our attention upon the bright star Sirius. The interval between the upper culmination of Sirius on January 1 and January 2, 1877, is 24^h 0^m 0^s·007 of sidereal time. Is this interval constant or not? If we repeat the observations on March 1 and 2, we find for the interval 23^h 59^m 59^s·985. It is no doubt true that each of these quantities only differs by an extremely small fraction of a second from twenty-four hours, but still there is a difference. Let us now compare the interval between two successive culminations of another bright star, Vega, with what we have already found for Sirius. It appears, when the observations are made, that the interval between the upper culmination of Vega, on January 1 and January 2, 1877, is 24^h 0^m 0^s·013.

We have already explained how the successive returns of the stars to the meridian, in the apparent diurnal motion, are really due to the rotation of the earth upon its axis. It may be reasonably objected to this explanation that, if it be true, we ought to find the interval between two successive returns to the meridian the same for all the stars and constant for each one ; yet the observations appear to show a difference in the case of Sirius and Vega, and that even the period is not constant in either star. The reply to this objection is that, if we could see the *real* culmination of the stars, the interval between two successive culminations would be exactly the same for each star, and constant for each one, but that the *apparent* culminations, which alone we can see, are affected by certain sources of error now wholly, or in great part, understood. When due allowance is made for the effect of these errors, in modifying the time of culmination. it is found that the interval between two successive culminations is the same for each star, and that it is constant for centuries.

By observing a pair of consecutive culminations of the same star, the rate of the clock can be accurately determined. It remains now to show how the hands are to be set so that the time indicated by the clock shall really correspond with the right ascension of the celestial bodies on the meridian. As the right ascension of the vernal equinox is to be zero, it follows that a correct sidereal clock should show 0^h 0^m 0^s when the vernal equinox is on the meridian. What the sidereal clock does actually show, when the vernal equinox is on the meridian, is the error of the clock. These observations would be comparatively simple were the vernal equinox a point so marked that its presence could be detected in the telescope, or if a fixed star happened to be situated at the equinox. As these conditions are not fulfilled, it is necessary to resort to indirect means to ascertain when the vernal equinox is actually culminating.

Let s s' (Fig. 79) be the ecliptic, and M M' the equator in the neighbourhood of the vernal equinox A. The sun's centre at culmination on the day previous to its passage through the vernal equinox is at s, and on the following day it is at s′ ; at some epoch between these two culminations the centre of the sun must actually have been in the point

FIG. 79.

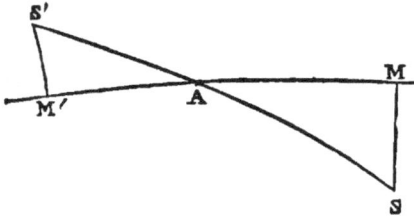

A. It will render the process of calculation which is employed more easily understood, to give in detail a series of observations for the purpose of determining the error of the clock on a special occasion. The instrument used is a meridian circle, and the observations are presumed to have been made on March 19 and 20, 1877. At the culmination of the sun on March 19, the declination of its centre was ascertained in the way already described, and was found to be −0° 23′ 21″·9. The next day the sun's centre has passed to the north of the equator, and its declination as determined by the meridian circle at culmination is +0° 0′ 20″·4. If from s an arc s m of a great circle perpendicular to the equator be drawn, then the arc s m is equal to 23′ 21″·9. Similarly the arc s′ m′ is equal to 0′ 20″·4. We may for our present purposes assume that the triangles A M S and A M′ S′ are plane triangles, and as the angles at M and M′ are both right angles, the two triangles are obviously similar. We therefore have the relation

$$\text{A M} : \text{A M}′ :: \text{S M} : \text{S}′\,\text{M}′.$$

The arc M M′ intercepted on the equator by the two declination circles through the sun, is the difference between the right ascensions of the sun on the two consecutive days of observation. But this difference is known from the observations with the meridian circle. It is only necessary to know the rate of the sidereal clock, and observe the time when the centre of the sun crosses the meridian on each of the two days. The difference between these times is independent of the error of the clock, which is what we want to find. On the days in question, it will be found that the difference between the two times, after making allowance for the rate of the clock, is $3^m 38^s\cdot52$. This is, therefore, the length of the arc M M' on the equator, which is of course equal to the sum of the arcs A M and A M'. We can therefore find the arcs A M and A M' measured in time, as we know both their sum and their ratio. Thus we have :—

$$\text{A M} = 3^m\ 38^s\cdot52 \times \frac{23'\ 21''\cdot9}{23'\ 42''\cdot3}$$

$$= 3^m\ 35^s\cdot39.$$

Similarly we have

$$\text{B M} = 0^m\ 3^s\cdot13.$$

Thus we see that the difference between the right ascension of the sun and the equinox is $3^m 35^s\cdot39$ on March 19, and $0^m 3^s\cdot13$ on March 20. In the former case the equinox comes to the meridian after the sun, and in the latter the equinox precedes the sun. Let us, then, suppose that on March 19 the clock showed $23^h 56^m 49^s\cdot81$ at the moment of culmination of the sun's centre. Then the clock would show at the moment of culmination of the equinox

$$23^h\ 56^m\ 49^s\cdot81 + 3^m\ 35^s\cdot39 = 0^h\ 0^m\ 25^s\cdot20.$$

This proves that the clock is $0^m 25^s\cdot20$ fast, for if it had been correct it should have shown $0^h 0^m 0^s$ at the moment of culmination of the equinox.

The method of determining the clock error which we have just described can only be conveniently applied when the sun is in the immediate neighbourhood of either of the equinoxes. Of course the time shown by a correct sidereal clock, when the *autumnal* equinox culminates, is equal to twelve hours, so that either equinox will answer for the purpose.

It is no doubt true that any two complete observations of the sun at culmination theoretically afford the means of determining the error of the clock. For let s and s' (Fig. 80) be the positions of the sun at the moments when the observations are made, then the declinations s' M' and s M are determined by the meridian circle, while with the help of the clock of which the rate is known the difference in right ascension of the sun on the two occasions is known. This difference converted into arc at the rate of 15° to an hour is equal to M M'. We can therefore construct the equinoctial point A, by taking an arc of a great circle of the known length M M', and then drawing the arcs s M and s' M' perpendicular thereto. We thus determine the points s, s', and the great circle through s s', and therefore the point A in

FIG. 80.

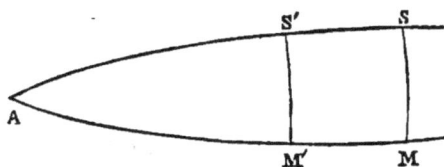

which it cuts the line M M' is known. We have therefore reduced the question to the problem in spherical geometry of calculating the arc A M. This arc is really the right ascension of the sun at the time of the first observation, and by comparison with the observed time the error of the clock is determined. In the neighbourhood of the equinoxes the method just described coincides with that of which we have already given an illustration. The method cannot be satisfactorily applied except near the equinoxes.

The determination of the sun's declination is. like all other measurements, open to some degree of error. It is therefore proper to enquire whether the calculated distances A M will be much affected by errors in the declination. It appears that when s and s' are distant from the equinoxes, a very trifling error in the observed declinations will produce a very large error in the concluded value of A M. This can easily be seen by taking s and s' near the solstices, when it is plain that a very slight alteration in either of the lengths s M or s' M' will produce a very considerable displacement of the ecliptic upon the equator. The alteration in the sun's declination on two consecutive days at the solstice is only about 4″·5; consequently an error in one of the declinations of half a second will be a ninth part of the difference of declinations, and hence the position of the equinox which is concluded from the difference of declinations will be exposed to a very large degree of uncertainty. At the equinox, however, the sun's declination in a single day changes through no less than 23′ 42″·3 = 1422″·3. But there is no reason to suppose that the observed declinations will be more erroneous at the equinoxes than at the solstices. We may therefore assume at the equinoxes as at the solstices that the risk of an error of 0″·5 is incurred. But this error, which in the case of the solstices is one-ninth of the total amount with which we have to deal, is in the case of the equinoxes little more than the three-thousandth part (1 ÷ 2844·6). We hence conclude that the equinoctial determination of the zero of right ascension is far less liable to error than a determination made at any other season of the year.

It is, however, manifest that, as no clock can be trusted to run accurately during so long a period as from one equinox to the next, we must have some practical method for determining the error of the clock at intermediate times. In fact, in an observatory, when great accuracy is required, the error of the clock is of such importance that it requires to be determined carefully every night on which observations of right ascension are made, or indeed several times in the course of a single night. Further, it may be added that the determination of the error of the clock, by the method we have described, can only be done in a well-equipped observatory; and even then it is a laborious operation, and liable to be seriously interfered with by the weather at the critical times during which alone it is possible. On all these grounds, it is, therefore, obvious that some more practical and simple method of determining the clock error is required, though it must be distinctly understood that the method to be described is subsidiary to, and indeed ultimately based upon, the fundamental method of equinoctial observations of the sun.

Suppose that on one occasion of the sun passing through the equinox we have succeeded in determining accurately the error of the clock. We then observe the transit of a fixed star with the same instrument at a time when the error of the clock is certainly known. We can, therefore, find with great precision the right ascension of the star at the epoch in question. The process can be repeated for other standard stars, and then, by observation of any one of these stars at subsequent epochs, the clock error can be ascertained. If the equinox remained absolutely fixed on the surface of the celestial sphere, then the right ascensions of the stars would remain absolutely constant, except in so far as the actual proper motion of the star is capable of producing an appreciable change. The position of the equinox is not, however, constant, as will be explained subsequently, but the nature of its motion is so well understood, that when once the right ascension of a star has been ascertained at any epoch, its right ascension at any subsequent epoch can be calculated. We are thus able to predict, with a high degree of accuracy, the actual right ascensions of a large number of fundamental stars, and it is by the observations of these fundamental stars that the errors of the sidereal clock can be determined. The difference between the right ascension of the star and the time of transit as shown by the clock is the error of the clock.

By referring to the 'Nautical Almanac' the astronomer will always find a standard star which will shortly come on his meridian; he then makes an observation with the transit instrument or the meridian circle, and determines the moment of culmination of this star by his sidereal clock. This is compared with the right ascension given for the star in the 'Nautical Almanac' for the day in question, and the difference between the two is the error of the clock. Thus, for example, we have from the 'Nautical Almanac' the following right ascensions of Vega at the corresponding dates :—

1877. January 1	.	.	.	.	18^h 32^m 44^s·85
„ April 2	.	.	.	.	18 32 47·28
„ July 3	.	.	.	.	18 32 49·46
„ October 2	.	.	.	.	18 32 48·13

Take, for the sake of illustration, an observation of the transit of Vega on July 3, 1877. After the mean of the results obtained at the different wires had been computed, and after allowance had been made for the errors of the instrument, due to level, collimation, and azimuth, it was found that the clock time of crossing the meridian was 18^h 32^m 51^s·42. But had the clock been correct, it must have indicated the true right ascension of the star, which at this date, according to the 'Nautical Almanac,' is 18^h 32^m 49^s·46. It follows that the clock is too fast, and that the time which the clock indicates should receive the correction of -1^s·96 in order to show the true time. The *rate* of the clock may be determined by observing the error which it has at any subsequent period ; for of course any alteration in the *error* of the clock can only be due to the *rate*.

When the right ascensions of objects are about to be determined by the transit instrument, the observer commences by ascertaining the errors of his instrument ; and taking one or two standard stars, he then observes the unknown objects, and after suitably correcting the observations for the instrumental errors, he infers the clock time at which each of the unknown objects crosses the meridian. During the course of these observations, and also at their close, the observer is careful to observe a few additional standard stars ; so that, as he knows accurately the error of his clock at the commencement of the obser-

vations, and also at their close, and perhaps in some intermediate points, he is then able to determine with great
accuracy the clock error which existed at the moment when
each of the unknown objects was crossing the meridian.
By applying this correction to the observed time, the true
sidereal time of transit is obtained, or, in other words, the
right ascension.

It is generally convenient to express right ascensions
in hours, minutes, and seconds of sidereal time. It can,
however, easily be transformed, when necessary, into degrees,
minutes, and seconds of arc, by remembering that fifteen
degrees of arc correspond to one hour of time.

§ **63.** *Celestial Latitude and Longitude.*—When the right
ascension and declination of a celestial body are known,
then the position of the body on the celestial sphere is
determined. It will, however, be readily understood that
for the purpose of indicating the position of a celestial body,
other methods are possible besides that of giving its right
ascension and declination. The celestial pole and the celestial equator are the fixed objects of the celestial sphere by
which right ascensions and declinations are measured. In

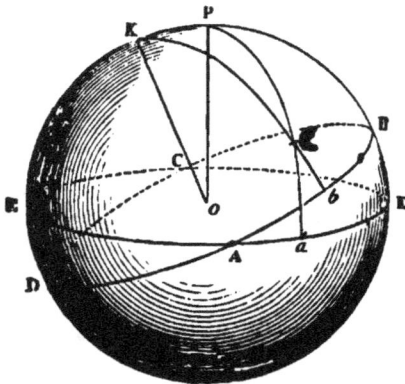

FIG. 81.

the same way the ecliptic
and the pole of the ecliptic
can be used for the pur-
pose of measuring what
is called the *latitude and
longitude.*

Let the circle A B C D
(Fig. 81) be the ecliptic,
and A E C E be the equator.
Let P be the pole of the
equator, and K that of
the ecliptic, while *e* is the
position of a body on the
celestial sphere. Then a great circle P *e*, drawn from P
through *e*, cuts the equator in a point *a*, and the arc A *a* is

the right ascension of *e*, while *e a* is its declination. If, instead of taking P, the pole of the equator, we took K, the pole of the ecliptic, and drew the arc K *e*, which cuts the circle A B in *b*, then the position of the point *e* could be completely specified by knowing the length of the arc A *b*, measured along the ecliptic from the equinox, and also the arc *b e*, measured perpendicularly to the ecliptic. These quantities are known as the *longitude* and the *latitude* respectively. The latitude and longtitude of a celestial body bear to the ecliptic precisely the same relations which the declination and right ascension bear to the equator.

The longitudes are measured round the ecliptic from 0° to 360° in the same direction in which the apparent motion of the sun is performed. The latitude of a point on the ecliptic is, therefore, zero, while its longitude is equal to the arc intercepted on the ecliptic between the point and the equinox. When the right ascension and declination of a celestial object have been observed, then its latitude and longitude can be obtained by calculation.

§ 64. *The Sun's Path in Space.*—What we have ascertained with respect to the apparent motion of the sun shows us that its centre moves along a great circle of the celestial sphere. It is, however, necessary to observe that this does not necessitate that the actual path of the sun's centre, through space, shall be performed in a circle, of which the earth is centre. We have, hitherto, avoided any reference to the actual distance at which the sun may be situated when at the various points of its path. If these distances were always the same, then no doubt the path of the centre of the sun would really be a circle, but the apparent motions of the sun, so far as they have been explained up to the present, are quite consistent with the supposition that its path is widely different from a circle. We have shown that the centre of the sun is always to be found on a great circle of the celestial sphere which is called the ecliptic. This has been proved by observing the right ascension and declination of the sun at different periods, and then plotting the observed places on the celestial globe, or making calculations which are equivalent thereto. But merely observing the right ascension and declination of an object does not define the actual position of an object in space ; it merely points out the *direction* in which the object is seen. In fact, any number of objects situated at even enormous distances apart will have the same right ascension and declination, provided that they all lie upon one straight line which is directed to the earth. All, therefore, that our observations of the sun have hitherto shown us, is the *direction* in which the sun was seen on each occasion. It therefore appears that the centre of the sun is always to be seen in a direction which can be found by joining the earth's centre to a point on the ecliptic. As the ecliptic is a great circle of the celestial sphere, of which the earth's centre is the centre, all the lines drawn from the centre to points on the ecliptic will lie in a plane, and, therefore, the *centre of the sun must always move in the same plane which passes through the centre of the earth.* All, therefore, that the observations hitherto described have taught us, is that the apparent motion of the sun is performed in this plane. The actual form of the apparent path might, for anything we have yet proved, be any closed plane curve which contained the earth in its interior.

To ascertain the actual shape of the curve, it is necessary to know the distance at which the sun is situated at any point of its path, as well as the right ascension and declination which point out its direction. We shall subsequently explain the methods by which the actual distance of the sun is to be determined. For the present, it will be sufficient to point out how the relative distances of the sun at different seasons can be ascertained.

It will of course be admitted that the actual dimensions of the sun are constant, so that any changes in his apparent dimensions, or the angle which the diameter of the sun subtends at the earth, is to be attributed to a change in the distance between the earth and the sun. At a superficial glance the dimensions of the sun appear to be the same at all seasons of the year ; it would, therefore, appear at first sight as if the distance from the sun to the earth were constant, so that consequently the apparent path of the sun must be a circle. When, however, the apparent dimensions of the sun are accurately measured by means of the micrometer we have already described, it is found that though the dimensions are nearly constant, yet that there is an appreciable degree of variation, according to a regular law which is given in the following table :—

Apparent Diameter of the Sun.

Jan. 1	. .	32′ 36″·4	July 1	. .	31′ 32″·0
Feb. 1	. .	32 31·8	Aug. 1	. .	31 35·8
March 1	.	32 20·6	Sept. 1	. .	31 47·0
April 1	. .	32 4·0	Oct. 1	. .	32 2·6
May 1	. .	31 48·0	Nov. 1	. .	32 19·2
June 1	. .	31 36·4	Dec. 1	. .	32 31·6

It appears from this table that the apparent diameter of the sun attains its greatest value at the end of December and the beginning of January, its value being then 32′ 36″·4. After that date the apparent diameter begins to diminish, and reaches its lowest value about the 1st of July, being then 31′ 32″·0 ; after this it commences to increase again, so as to regain its maximum value in the course of the next winter. It appears from § 11 that the apparent diameter is inversely proportional to the distance, and hence from this table we can conclude, with some degree of accuracy, the distances which the sun has on the different dates. Thus it appears that the sun is closest to the earth in midwinter, and most distant from the earth at midsummer, the ratio of the greatest and least distance being

$$\frac{32' \ 36''\cdot4}{31' \ 32''\cdot0} = \frac{1956\cdot4}{1892\cdot0}$$

The distance of the sun from the earth, therefore, changes in a year through about one-thirtieth part of its mean value. It is thus plain that the orbit of the sun cannot be accurately a circle with the earth situated at its centre. The variation in the distance at the different seasons can, however, be explained to a certain extent by the supposition that the orbit of the sun is a circle, if we suppose that the earth is not situated at the centre of the circle. Let A S B

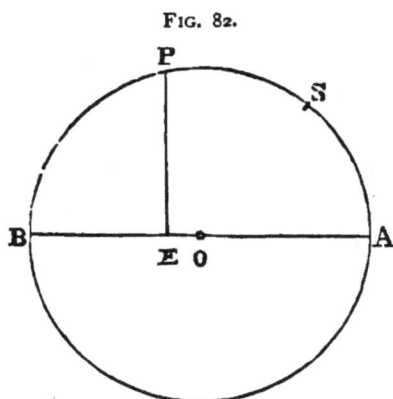

FIG. 82.

(Fig. 82) represent the orbit of the sun on this hypothesis, let o be the centre of the circle, and let E be the position of the earth. If, then, the distances E A and E B be connected by the relation $\dfrac{\text{E A}}{\text{E B}} = \dfrac{1956\cdot4}{1892\cdot0}$, the sun when at A or at B would exhibit to an observer at E an apparent semi-diameter conformable to what observation shows to be the actual value; the ratio of the distance E O to O A can be easily found, for we have obviously

$$\frac{\text{AO} + \text{EO}}{\text{AO} - \text{EO}} = \frac{1956\cdot4}{1892\cdot0},$$

whence $\dfrac{\text{EO}}{\text{AO}} = \dfrac{1}{60}$ very nearly.

To test this supposition with regard to the motion of the sun, we shall calculate the apparent diameter of the sun on this hypothesis at an intermediate position P. We shall take the point P so that the angle A E P is a right angle. We have then, from a well-known proposition in geometry—

$$\text{EP}^2 = \text{EA} \times \text{EB}.$$

If we denote the distance E A by 1956·4, then we have

$$E P^2 = 1956·4 \times 1892·0,$$
whence E P = 1923·8.

Assuming, therefore, that the apparent motion of the sun can be faithfully represented by supposing the orbit to be a circle, in which the earth is situated excentrically at a distance from the centre equal to the sixtieth part of the radius, we must have for the apparent diameter at P the relation .

$$\frac{\text{diameter at P}}{\text{diameter at A}} = \frac{1956·4}{1923·8,}$$

whence we deduce the diameter at P to be 1923″·8 = 32′ 3″·8. The sun arrives at P about three months after it . leaves A, and its diameter as given by the table on April 1 is 32′ 4″·0. It would therefore seem that the apparent changes in the sun's magnitude could be fairly represented by the suppositions we have made. It will, however, be shown subsequently that the orbit is not quite circular, but that it is really an ellipse which approaches so closely to a circle that the difference cannot be satisfactorily established by the observations we have been describing.

§ 65. *Velocity of the Sun.*—The rate at which the sun performs its motion in its orbit must next engage our attention. In the course of an entire year, the sun returns to the point of the celestial sphere from which it started, and therefore in a little more than 365 days must have described an arc of 360°. The average daily motion of the sun must therefore be somewhat less than one degree. But it is found that the rate of motion is not constant. The sun at certain parts of its path is moving more rapidly than at other parts. If we observe the angular distance through which the centre of the sun actually moves on certain days of twenty-four hours distributed through the year, we shall find that the

N

alterations in its rates of motion are very appreciable. This
will be made clear by the following table.

Date.	Arc described in one Day.	Date.	Arc described in one Day.
Jan. 30 . .	1° 0′ 53″·8	July 29 . .	0° 57′ 23″·3
March 1 . .	1 0. 9·1	Aug. 28 . .	0 58 0·7
March 31 . .	0 59 9·6	Sept. 27 . .	0 58 56·9
April 30 . .	0 58 11·5	Oct. 27 . .	0 59 57·4
May 30 . .	0 57 29·0	Nov. 26 . .	1 0 46·6
June 29 . .	0 57 11·8	Dec. 31 . .	1 1 10·1

It appears from this table that the sun attains its greatest
arcual velocity about the end of the year, when it moves
over no less than 1° 1′ 10″·1 in a day of twenty-four hours.
This, it will be observed, corresponds with the season at
which the apparent diameter of the sun attains its greatest
value. The smallest value of the sun's apparent arcual
velocity is on July 1, when the distance moved over in one
day is 57′ 11″·5. This date corresponds with that at which
the apparent diameter of the sun has its smallest value.
This coincidence suggests that the sun may be moving
uniformly, and that the variations in its apparent velocity are
only due to the circumstances of the eccentric position
which the earth occupies in the orbit. It is easy to test
this supposition by the table which we have just given. If
the sun were really moving uniformly, then the apparent
velocity would be inversely proportional to its distance.
If, therefore, we assume the orbit to be that represented
in Fig. 82, we must have

$$\frac{\text{E A}}{\text{E B}} = \frac{1° 1′ 10″·1}{57′ 11″·5} ;$$

from this it is easy to compute that

$$\frac{\text{E O}}{\text{A O}} = \frac{1}{30} \text{ very nearly.}$$

This result is fatal to the hypothesis on which the calculations were commenced. If we suppose that the sun were to move uniformly in a circle in which the earth was placed eccentrically, then, according to the observed variations of the sun's apparent diameter, the distance of the earth, from the centre of the circle, should be the *sixtieth* part of the radius ; according, however, to the observed values of the arcs through which the sun moves at the seasons of greatest and least apparent velocity, the distance of the earth from the centre of the circle should be about the *thirtieth* part of the radius. These two results are so inconsistent, that we are obliged to discard this hypothesis with respect to the apparent motion of the sun. If the sun's motion be uniform, it cannot be circular ; or if it be circular, it cannot be uniform : as a matter of fact the apparent motion of the sun is neither circular nor uniform.

The real path in which the sun appears to move, as well as the law according to which its velocity changes in different parts of its orbit, can be determined by what are known as Kepler's laws. These laws were originally discovered by observations of the planet Mars, and we shall return to their consideration on a subsequent occasion. For our present purpose it will be sufficient to assume—

1st. That the apparent path of the sun around the earth is a plane ellipse with the earth situated in one of the foci.

2nd. That the area described by the radius vector, drawn from the earth to the sun, in a given time, is proportional to that time.

These two laws have been so abundantly verified that we may adopt them without the slightest hesitation. By their aid it will be very easy to render a satisfactory account both of the variations in the apparent diameter of the sun and the alteration in its velocity.

Let M *b* N (Fig. 83) be the ellipse in which the sun moves, let M N be the major axis of this ellipse, and let T be the focus in which the earth is situated. When the sun is at

N, its distance from the earth attains its greatest value and the angular dimensions are a minimum. On the other hand, when the sun is at M, the distance is a minimum, and therefore the angular diameter is a maximum. By the method we have already explained, we can deduce the rela-

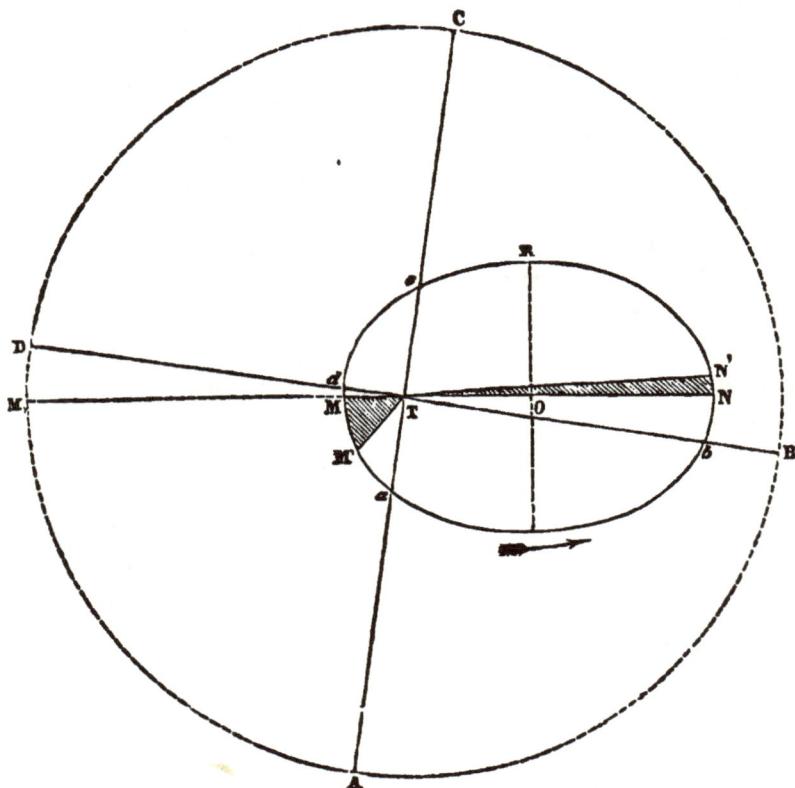

FIG. 83.

tive magnitudes of T N and T M, and thence the ratio of O T to O N when O is the centre of the ellipse. This is called the *eccentricity of the ellipse*, and, as already explained, it is equal to 1÷60.

The elliptic motion will therefore explain the alterations in the apparent diameter of the sun. It remains to be shown how the second law, discovered by Kepler, will explain the changes in the velocity of the sun's motion.

Let M' be the position of the sun when near one extremity of the ellipse, and N' a position near the other extremity. Let these points be so chosen that the time occupied by the sun in moving from M to M' is equal to that required to move from N to N'. According to Kepler's second law, the areas swept over on each occasion by the radius vector will be proportional to the time, and consequently the area M T M' will be equal to N T N'. This being the case, it is obvious that the arc M M' must be considerably longer than the arc N N' in order that the shaded areas represented in the figure shall be equal. It thus appears that the sun, when in the neighbourhood of M, must move more rapidly than when in the neighbourhood of M'. We can also bring this easily to the test of numerical calculation. If we take the points M' and N' so near to M and N respectively that the arcs M M' and N N' may be practically considered as straight lines, then from the equality of areas we have—

$$\text{M M}' \times \text{T M} = \text{N N}' \times \text{T N},$$

whence

$$\frac{\text{M M}'}{\text{T M}} : \frac{\text{N N}'}{\text{T N}} :: \text{T N}^2 : \text{T M}^2.$$

But the two first quantities denote the angles M T M' and N T N' when these angles are both indefinitely small, and hence—

$$\frac{\text{angular velocity at M}}{\text{angular velocity at N}} = \frac{\text{T N}^2}{\text{T M}^2};$$

but we have already seen that $\dfrac{\text{T N}}{\text{T M}} = \dfrac{61}{59}$,

and hence $\dfrac{\text{angular velocity at M}}{\text{angular velocity at N}} = \dfrac{16}{15}$ very nearly.

This closely coincides with the ratio of the greatest and least angular velocities actually found, and it can also be shown that the sun's angular velocity at any other points of its path is accurately expressed by this law.

§ 66. *Changes in the Length of the Seasons.*—The elliptic path of the sun lies, of course, in the plane of the ecliptic. This plane cuts the celestial sphere in a great circle A B C D (Fig. 83). In this figure A C is the line of equinoxes and B D the line of solstices. These lines are accurately drawn with regard to the ellipse, and it will be noticed that the major axis of the ellipse is nearly coincident with the line of solstices. When the sun is seen in the vernal equinox, it is referred to the point *a* on the celestial sphere, though it really occupies the position A. In moving from the vernal equinox to the summer solstice the sun moves from *a* to *b* along its real path. The duration of spring is therefore equal to the time the sun takes to go from *a* to *b*. In the summer the sun travels from *b* to *c*, in autumn from *c* to *d*, and in winter from *d* to *a*.

A curious inequality in the length of the different seasons follows from the eccentricity of the sun's orbit. It will be noticed that during the spring and the summer the sun is in that portion of its path where it moves least rapidly, while during the autumn and the winter it moves most rapidly. The changes in the lengths of the seasons which are thus produced are not considerable, though still sufficient to be appreciable. The true lengths of the seasons when this circumstance is attended to are :—

				Days	Hours	Min.
Spring .	.	.	.	92	20	59
Summer	.	.	.	93	14	13
Autumn	.	.	.	89	18	35
Winter.	.	.	.	89	0	2

The duration of spring and summer taken together is found to be 186 days 11 hours 12 minutes, while the duration of autumn and winter together is 178 days 18 hours

37 minutes. During spring and summer the sun is in the northern hemisphere, while in autumn and winter he is in the southern hemisphere. It therefore appears that in each year the sun remains about eight days longer in the northern hemisphere than in the southern.

§ 67. *Dimensions of the Sun.*—The sun, though really a globe, appears to the unaided eye in the form of a circular disk, of which the diameter subtends at our eyes an average angle of about 32′. The mean distance of the sun from the earth, as concluded from various data by Newcomb, is between 92,200,000 and 92,700,000 miles. Adopting a mean value, then it is easy to calculate that the diameter of the sun is about 860,000 miles—that is, about 110 times the diameter of the earth. The volume of the sun is therefore about 1,330,000 times as great as the volume of the earth. A distance of 447 miles on the surface of the sun subtends an angle of one second at the earth. A spider line used in a large telescope will cover a portion of the sun which may be a quarter of a second in breadth : thus a strip of the sun's surface over 100 miles wide is hidden behind the spider line. The diameter of the earth subtends at the sun an angle of about $17''\cdot52$. To subtend an angle so small as this, a globe one foot in diameter would have to be moved to a distance of upwards of two miles. The radius of the earth subtends of an angle $8''\cdot76$.

§ 68. *Sun Spots.*—When the sun is observed through a telescope, proper precautions having first been taken to diminish by means of coloured glass, or some similar contrivance, the intense brilliancy of its image, a remarkable phenomenon is often noticed. The brilliant surface of the sun is often found marked over with black spots such as are represented in Fig. 84. The central portion of the spot is intensely dark, and it is surrounded by a margin which, though much darker than the general surface of the sun, is not nearly so dark as the centre of the spot. Sometimes the spots attain

very considerable dimensions : occasionally a spot will attain
a diameter of thirty or forty seconds.   It will thus be obvious

FIG 84.

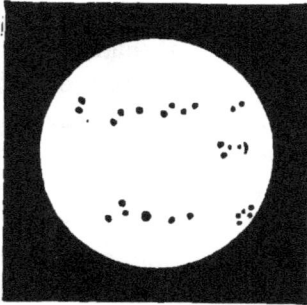

that the real dimensions of the
spots are enormous, because, as
a second corresponds to a distance of 447 miles, it is evident
that a spot thirty seconds in
diameter must have a real diameter of 13,400 miles.   At their
first appearance the spots are
generally found on the eastern
edge of the sun ; they then
pass over the surface of the
sun, and in about fourteen days
they disappear on the western side.

From observations of these spots we are led to the

FIG. 85.

very remarkable conclusion that the sun
is actually rotating.   It might at first be
supposed that the spot was a body situated
above the surface of the sun, and actually
revolving around the sun ; but a little reflection will show that this supposition
is inconsistent with the observed facts.
Let s represent the sun, let T be the position of an observer on the earth, and
suppose that a spot is caused by a body
revolving around the sun in the orbit
A B C.   When the object reaches A, the
observer will for the first time see the spot
on the surface of the sun, and it will continue visible on the surface until it arrives
at B.   During the rest of the journey of
the object through the arc B C A it will
not be visible on the surface of the sun.
It therefore appears that the object will
be projected against the sun's disc while traversing the arc

A B, but will not be so projected while traversing the arc B C A.   Assuming, therefore, that the object moves with approximate uniformity, and observing that the arc A B is only a very small part of the total orbit, it follows that the time during which the object is invisible must greatly exceed the time during which it is visible. This, however, is not the case. Each spot is visible, projected on the sun's surface, for a period of fourteen days, and it remains invisible for the same period.   This can only be explained by the supposition that the spot is actually on the sun himself.   Observing also that the time of revolution of the spots is nearly the same for each spot, and that the same motions are participated in by the bright objects on the surface of the sun known as *faculæ*, it appears that the only reasonable supposition we can form is that the sun rotates upon its axis, carrying with it the spots and the faculæ.

As the sun always appears circular, and as it rotates upon its axis, the actual shape of the sun must, undoubtedly, be spherical.   The most careful observations have not afforded reliable indications of any ellipticity in the figure of the sun.

When a spot is carefully observed for some days, it is often seen to undergo changes in its form and size which cannot be accounted for by the foreshortening which is due to its approach to the limb of the sun.   Occasionally a spot is seen to become smaller and smaller, or, indeed, to disappear altogether, while new spots are often formed in places where previously no trace of a spot existed.   The changes in the form of a spot generally take place gradually, but sometimes this is not so, and marked changes have occasionally been noted in the lapse of a few hours or less.

The apparent time of rotation of the sun on its axis, as concluded by the reappearance of the spot, is therefore liable to be affected by the actual motion of the spot.   An accurate determination of the periodic time of the sun's rotation is, therefore, at best, only to be made by taking the mean result of a number of observations made with spots on different parts of the surface.

The spots are not distributed uniformly over the surface of the sun. They are principally found on two zones, one on the northern hemisphere, and the other on the southern. These zones lie between 10° and 35° of heliocentric latitude.

Fig. 86.

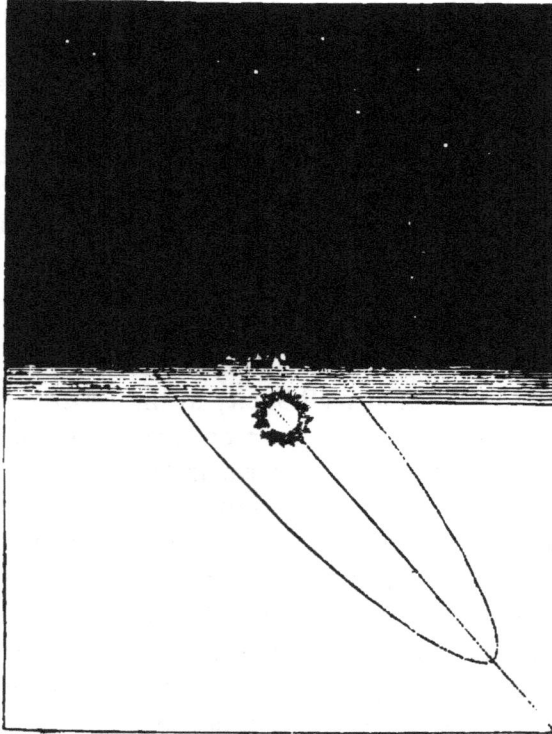

The numbers of the spots on the surface of the sun undergo regular periodic changes in amount. Schwabe, who observed the sun regularly, from the year 1826 to 1868, has discovered that the sun spots have very well marked periods of maxima and minima, which succeed each other at intervals of about eleven years. The number of spots, for example, increases for about five or six years to a maximum, and then decreases for almost five or six years to a minimum.

§ 69. *Zodiacal Light.*—At certain seasons of the year the phenomenon known as the zodiacal light may be seen in the west shortly after sunset. The zodiacal light is a faint luminosity of a triangular shape, shown by the shaded portion of the adjoining figure (Fig. 86). As the sun gradually descends lower below the horizon, this phenomenon disappears; it is noticed that the direction in which the light is most extended coincides with the ecliptic, and that consequently the sun is situated with regard to the zodiacal light in the manner shown in the figure. It is generally supposed that the matter whose luminosity gives rise to the zodiacal light is of lenticular form, and that it is a sort of appendage placed around the sun. Whatever be the materials of which this lenticular object is composed, they must be of extreme tenuity, inasmuch as faint stars are easily visible through its entire thickness.

CHAPTER V.

MOTION OF THE EARTH AROUND THE SUN.

§ 70. *Revolution of the Earth.*—It has been already shown, in discussing the apparent diurnal revolution of the heavens, that the true explanation of that motion is to be sought in the fact that the earth revolves on its axis, and thus produces the same effect as would be observed if the celestial sphere actually rotated in the opposite direction. We have now to consider whether the apparent annual motion of the sun around the heavens is to be regarded as real, or whether it may not also be found to arise from an actual movement of the earth. In the first place it is to be noted that the apparent movement of the sun could be explained by the supposition that the earth is revolving around the sun, and not the sun around the earth; this may be rendered clear by an illustration. Imagine a large open circular space to be entirely surrounded by a forest, and in the centre of the space a single tree to remain. If you were to walk round this tree, you would at each point of your path see the tree projected against the trees of the forest, but the position of the projection will be continually changing, and by the time you have regained your original position, the tree will in succession have appeared in line with each of the trees in the ring bounding the open space; if you could be unconscious of your own motion, the observed facts would suggest that the central tree was actually revolving round you. To apply this illustration to the present case, we suppose the central tree to be analogous to the sun, and the distant trees of the forest to be analogous to the stars. What we actually observe is the apparent motion of the sun among the stars, but this is really due to the motion of the earth, of which we are unconscious.

It will be easy to show that the apparent elliptic orbit of the sun about the earth in one of the foci could really be produced by an actual motion of the earth in an elliptic orbit, with the sun in one of the foci. In considering this subject we may overlook the effect of the earth's rotation upon its axis, because the effect of this rotation has already been completely examined. Let s, s', s'', s''' (Fig. 87) represent the apparent orbit of the sun around the earth in the focus T. Suppose the ellipse be rotated through two right angles around the point x, which bisects s T, until the ellipse assumes the position T, T', T'', T'''. Then the lines T s', T s'', T s''', in the original ellipse, will be transferred to the positions s T', s T'', s T'''. Since the ellipse has been turned round through two right angles, every straight line in the figure will also have been turned through two right angles, and consequently each line in the second ellipse will be parallel to the corresponding line in the first ellipse. It thus appears that the lines T s', T s'', T s''' will be respectively parallel to the lines s T', s T'', s T''' ; observing that the stars are enormously distant compared with the dimensions of the ellipses now under consideration, then the apparent motion of the sun among the stars can be explained by supposing either that the sun moves in the orbit s, s', s'', s''', around the earth T at rest, or that the earth moves in the orbit T, T', T'', T''', around the sun s at rest. For example, when the sun is at s', and the earth at T, the sun appears to have a certain position with respect to the fixed stars ; but this position will be the same as if the sun had been at s, and the earth at T, for as the lines T s'

Fig. 87.

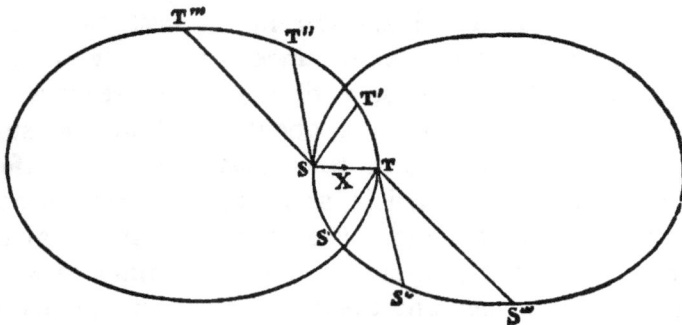

and T's are parallel, and as the celestial sphere may be regarded at an infinitely great distance, the two lines T s' and T' s will really intersect the celestial sphere in the same point. In the same way the apparent position of the sun at s'', seen from the earth at T, coincides with the position in which the sun at s is seen from the earth at T'', and similarly for the other points of the orbit. It is thus equally easy to explain the apparent movements of the sun by the hypothesis that the sun is at rest in the focus, and that the earth moves around the sun in an ellipse, or by the hypothesis that the earth is at rest in the focus, and that the sun moves round the earth in an ellipse. It is, indeed, easy to see that on either supposition the law of the description of equal areas in equal times will be fulfilled. To decide which of these two hypotheses we shall accept, we must introduce other considerations.

It is in the first place to be observed that the diameter of the sun is more than one hundred times as large as the diameter of the earth, and that the bulk of the sun is stupendously greater than the bulk of the earth. It therefore seems more natural that the earth, being a small body, should revolve around the sun, being a large body, rather than that the converse should take place.

There is, however, other evidence on this subject, derived from a study of the planets. Let us take, for example, the planet Venus. Shortly after sunset at the proper season, this object appears like a brilliant star in the west. On subsequent evenings the angular distance between Venus and the sun gradually increases, until the planet reaches its greatest elongation, when it is about 47° from the sun. Venus then begins to return towards the sun, and after some time ceases to be visible, as its light is overpowered by the brilliancy of the sun. Ere long the planet may be seen in the east shortly before sunrise. The time between the rising of the planet, and the rising of the sun, gradually increases, until Venus again reaches the greatest distance from the sun, after which it commences to return again, passes the sun, and may be seen at evening in the west as before. It appears from these observations that Venus is continually moving from one side of the sun to the other, and is never found at a distance of more than 47° from the sun. The question remains as to how these movements are to be explained. It is noticed that when Venus is at its greatest distance from the sun, its apparent movement is much slower than when it is nearer the sun. It is further shown by the telescope that when Venus is at its greatest angular distance from the sun, the disk is seen to be half illuminated, like the moon at the quarter. The planet is also, on very rare occasions, seen actually to pass between the earth and the sun, the phenomenon being known as the transit of Venus. From all these facts it is inferred that the planet Venus is really a dark globular body, which moves around the sun in 224 days in a certain orbit. As the sun executes its apparent motion among the stars, the planet seems to accompany it, alternately appearing to the east and the west of the sun, and never more than 47° distant therefrom. Precisely similar, though on a smaller scale, are the apparent motions of Mercury. This planet does not go so far from the sun as Venus, its greatest elongation being only 28°. The time in which Mercury revolves round the sun is 87 days.

We thus see that there are two planets, Mercury and Venus, which certainly appear to move round the sun. If we compare Mercury or Venus with the earth, we find some striking points of resemblance. All three bodies are approximately spherical, and they are all dependent upon the sun for light. It is therefore not at all unreasonable to enquire whether the analogy between the three bodies may not extend farther. We have already seen that the phenomenon of the apparent annual motion of the sun could be explained by supposing that the sun is really at rest, and that the earth moves round the sun. When we combine this fact with the presumption afforded by the analogy between the Earth and Mercury and Venus, we are led to the belief that the Earth, Mercury, and Venus are all bodies of the same general character, and all agree in moving around the sun, which is the common source of light and heat to the three bodies.

Nor are other confirmations wanting of the same important truth. By careful observations of the fixed stars, a very beautiful phenomenon, known as the aberration of light, has been discovered. This receives a most satisfactory explanation when the revolution of the earth around the sun is admitted, while it would be wholly inconsistent with the hypothesis that the sun revolves around the earth.

The theory of universal gravitation affords so satisfactory an explanation of many most remarkable phenomena connected with the motions of the heavenly bodies, that not a doubt can remain of its truth in the mind of any person capable of understanding the subject. Yet the theory of universal gravitation is indissolubly connected with and identified with the theory that the earth revolves around the sun.

We shall, therefore, assume that the earth really describes an elliptic orbit around the sun, though this statement is indeed not absolutely accurate. Even if all the other planets were absent, all we could say would be that the earth and the sun each describe elliptic orbits about their common centre of gravity. Owing to the vast preponderance of the mass of the sun on the mass of the earth, the centre of gravity of the two bodies is comparatively close to the centre of the sun, and may for most purposes be regarded as absolutely identical with the centre of the sun. The problem of the earth's motion becomes still more complicated when the actions of the different planets upon the earth and sun are taken into account. Owing to these disturbances, the ellipse which the earth describes is gradually modified both in form and position within narrow limits. We may, however, generally suppose that the earth revolves in an ellipse in one of the foci of which the sun is situated.

When the earth occupies the point т (Fig. 87), it is then nearer to the sun in the focus s than it is in any other part of the orbit, and this point is termed perihelion. The opposite extremity of the major axis of the ellipse is termed the aphelion, and the earth when in that position is at its greatest distance from the sun. The mean distance of the earth from the sun is an arithmetical mean between the aphelion distance and perihelion distance, and is equal to the semiaxis major of the ellipse.

The axis of rotation of the earth is, of course, carried

round the sun once in the year by the motion of the earth itself. It is exceedingly important to observe that, during the translation of the axis, the direction of the axis remains constantly parallel to itself. This is shown in Fig. 88, which represents the earth in four different parts of its orbit, though it is impossible in a figure of this kind to maintain the actual proportions of the respective sizes and distances. The lines N S in the four different positions of the earth denote the axes of rotation, and these four lines are parallel.

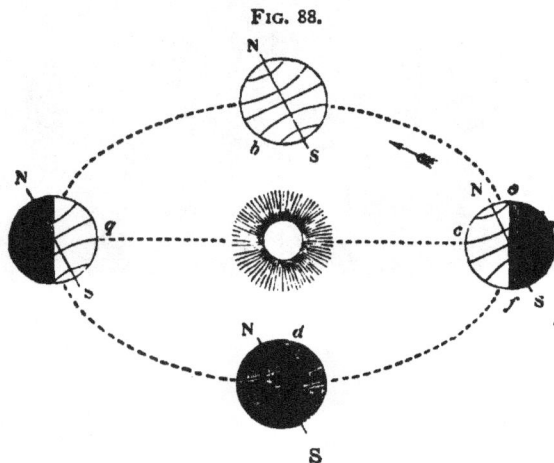

Fig. 88.

As the whole dimensions of the orbit of the earth are quite insignificant compared with the distances of the stars, the direction of the axis of rotation of the earth points always to the same point on the celestial sphere, i.e. to the point we have already determined as the celestial pole.

For a more detailed exposition of the circumstances attending the revolution of the earth around the sun, we may refer to Fig. 89, in which the position of the earth is given for each month of the year. The positions of the earth are also shown at the summer and winter solstices as well as at the vernal and autumnal equinoxes. When the earth is at the summer solstice, then all the region within the arctic circle remains constantly on the illuminated side of

O

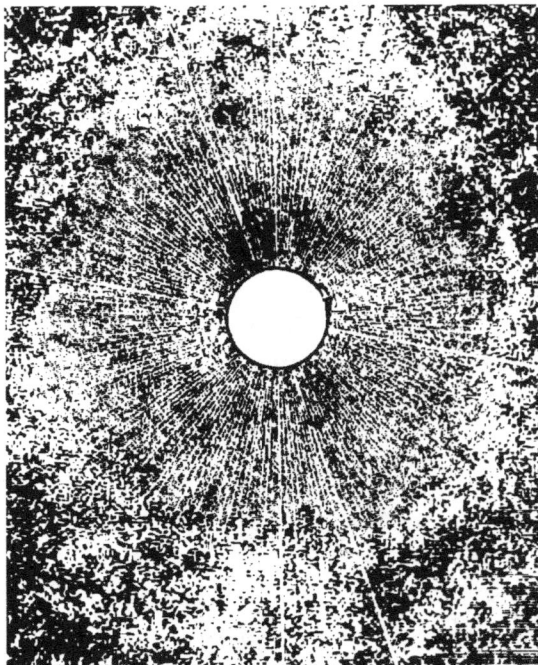

the earth, while at the winter solstice the region within the arctic circle is in continual darkness. At the equinoxes it will be observed that the boundary of light and shade passes through the pole.

§ **71.** *Precession of the Equinoxes.*—Although we are practically correct in asserting that during the revolution of the earth around the sun, the direction of the axis of rotation remains constant, yet when the direction of this axis is compared with the position which it occupied at a remote interval of time, it is seen that certain well-marked alterations in the position have taken place. These changes are principally manifested to us by an apparent alteration in the places of the stars. We have already explained how the position of a star is determined by its right ascension and declination. Both these measurements are made with respect to the celestial equator. If the equator of the earth were constantly and absolutely parallel to itself, then the celestial equator would remain constant also ; but if the earth's equator were in motion, then the celestial equator would undergo corresponding alterations in its position with reference to the fixed stars. If we were unconscious of any motion of the earth, then the alterations in the celestial equator would produce changes in the places of the stars. The extent of these changes can be determined by observation, and we can thence deduce the actual changes of the equator, and therefore of the direction of the axis about which the earth is rotating. We shall first point out the nature of the changes in the apparent positions of the stars which observation has revealed to us, and which are to be attributed to the changes in the position of the earth's axis of rotation.

Take for example the most brilliant star in the heavens, Sirius, and determine the right ascension of that star a intervals of time differing by ten years. We find from actual observations the following values :—

Mean Right Ascension of Sirius.

		Hours	Min.	Sec
January 1,	1847,	6	38	25
,,	1857,	6	38	51
,,	1867,	6	39	17
,,	1877,	6	39	44

We thus see that on the four dates here referred to, the mean right ascension of Sirius has perceptibly altered. It will also be perceived that these changes have taken place with considerable uniformity, the average increase of the right ascension being about 2·65 seconds per annum. Though this may seem to be a small quantity, yet from the circumstance that the change is always taking place in the same direction, its amount soon becomes appreciable, and in the course of centuries attains a magnitude which could hardly be overlooked even by the rudest methods of observing.

To understand the importance of this phenomenon, it is necessary to bear in mind that right ascension is measured from the vernal equinox, which is one of the points of intersection of the ecliptic with the celestial equator. It appears that on January 1, 1847, the point on the celestial sphere where the vernal equinox was at that time situated, crossed the meridian 6^h 38^m 25^s before Sirius crossed the same meridian. Thirty years later, on January 1, 1877, the vernal equinox crossed the meridian 6^h 39^m 44^s before Sirius. This statement admits of but a single interpretation. The angular distance between Sirius and the vernal equinox, as measured on the celestial sphere, must be greater in 1877 than it was thirty years previously. This phenomenon has been called the *Precession of the Equinoxes*, because, comparing the equinox with Sirius, the former comes a little earlier on the meridian in 1877 than it did in 1847. The years we have chosen are merely for the value of illustration : whatever two years were taken, it would be found that in the latter of these two years, the distance of Sirius from the vernal equinox (measured in the same direction on the equator) would be greater than the former.

Nor is there any special feature connected with the star Sirius, which renders it peculiarly fitted for the exhibition of this phenomenon. Had any other star been chosen, it would have been found that the distance between the vernal equinox and the star was continually changing, though it may be with a different velocity, and in some cases even with a different direction, to the changes we have noted in the case of Sirius.

From these observations we are obliged to admit that the positions of the equinoxes, and the positions of the stars on the surface of the celestial sphere, are constantly in a state of change. As the equinoxes are merely the intersection of the ecliptic and the equator, it becomes necessary to admit that either the equator or the ecliptic, or both, must have a certain motion on the heavens relatively to the stars, or that the stars must have a motion relatively to these circles.

The solution of the problem is much narrowed by observing that, at any rate, there is no motion of the ecliptic with respect to the stars at all adequate to account for so considerable a phenomenon as the precession of the equinoxes. Each year the sun passes through the same constellations in its annual progress, and the path of its track as marked out by the stars along the line is sensibly constant, or at all events is not endowed with any change large enough to be seriously considered for the present purpose. If we were enabled to see the stars in the vicinity of the sun (which under ordinary circumstances his bright light overpowers,) we should find, for example, that every 23rd of May the sun passed between the Pleiades and the Hyades; that every 21st of August the sun was to be found very close to Regulus (α Leonis), while on every 15th of October he passed a little above Spica (α Virginis). The track of the sun among the stars is practically invariable, and hence to account for the apparent motion of the equinox we must suppose that the ecliptic is at rest, and that the equator is in motion.

We have hitherto only referred to the changes in the right ascensions of the stars ; before proceeding further we must ascertain whether there are not corresponding changes in the declinations or the polar distances of stars. It will be found by comparing observations of the same star, made at considerable intervals of time, that the polar distances are changing. Thus, for the case of Sirius, we have at the dates already given :—

Mean Polar Distance of Sirius.

January 1,	1847,	106°	30′	37″
,,	1857,	106	31	24
,,	1867,	106	32	11
,,	1877,	106	32	56

It thus appears that the angular distance of Sirius from the pole of the celestial sphere is constantly increasing. In the lapse of thirty years, this increase has attained to the substantial amount of upwards of two minutes, at the uniform rate of about 4″·6 annually. Similar observations made with respect to other stars show, in general, an alteration of the polar distance which will be found to be uniform in the same stars for hundreds of years. It follows that either the celestial pole must be moving relatively to the stars, or that the stars must be moving relatively to the celestial pole. Which alternative will it be the most reasonable to adopt ? Let us remember that the configurations of the stars *inter se* are constant, and that the constellations of the present day are the same as the constellations thousands of years ago. As, therefore, the stars do not move perceptibly when compared together, it is much more natural to suppose that the changes of he pole with respect to the stars are really due to the motion of the pole and not to any motion of the stars.

To this conclusion we are also conducted by the observed changes in right ascension of the stars, for if the equator be in motion—and we have seen that it is—then the pole, which is merely the point on the celestial sphere 90° from the pole, must be in motion also.

Having thus ascertained that the pole has a certain slow movement, the next point to be considered is the exact determination of the rate of that movement, and of the path along which the pole moves. We shall show that the observed facts can be explained by the supposition that the celestial pole is moving on the surface of the celestial sphere in a small circle with a uniform velocity. It remains to determine the pole of this small circle, and the length of the arc which is equal to its radius.

We have already explained how the obliquity of the ecliptic is to be determined : it is, in fact, merely the greatest declination of the sun at midsummer (§ 54). We here give the obliquity of the ecliptic determined by this method in the four years already referred to.

June 21, 1847	.	.	.	23° 27′ 23″·56
„ 1857	.	.	.	23 27 37·12
„ 1867	.	.	.	23 27 13·85
„ 1877	.	.	.	23 27 26·51

It would not be correct to say that these different values of the obliquity are absolutely identical. Yet the variations in its value are extremely small. In fact, the mean of the four values just given is—

$$23° \quad 27′ \quad 25″·26,$$

and the difference between this quantity and the greatest or least of the four observed values is only about one seven-thousandth part of the total amount. For our present purpose we may overlook these small discrepancies, and suppose that the obliquity of the ecliptic is constant. We are thus led to the conclusion that, notwithstanding the equator is continually in motion among the stars, it still preserves constantly the same inclination to the plane of the ecliptic.

Let A B C D (Fig. 90) denote the ecliptic, and let E E be the equator. Then regarding A B C D as fixed, the equator moves so as to occupy the successive positions E′ A′ E′, E″ A″ E″, &c., while the inclinations at A, A′, A″ remain constant. Let K be the pole of the ecliptic, and let P be the pole of the equator in the position E A E, while P′ and P″ are the positions of the pole on the equator when in the position E′ E′ and E″ E″ respectively. The angle between two great circles on the sphere is equal to the angle between their poles, hence the inclination of the ecliptic to the equator is equal to the arc between the poles of the ecliptic and the equator.

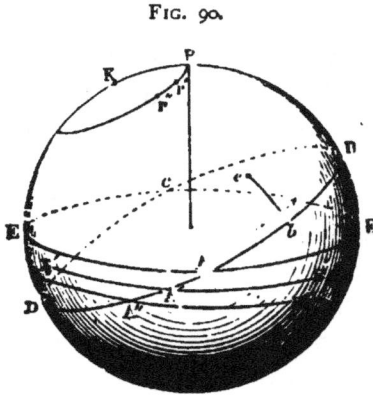

FIG. 90.

Thus in the three positions of the equator, represented in Fig. 90, the angles at A, A,′ A″ are respectively equal to the arcs K P, K P′, K P″, and as the angles at A, A′, A″ are all equal, it follows that the arcs K P, K P′, K P″ must be all equal; the points P, P′, P″ must, therefore, lie upon the circumference of a small circle of which K is the pole; and hence the pole of the equator describes a small circle about the pole of the ecliptic, the radius of which is equal to the obliquity of the ecliptic.

The effect of the precession of the equinoxes on the position of a star is more simply seen when we refer the place of the star to the ecliptic by its latitude and longitude.

Thus if *e* be the position of a star (Fig. 90), and *e b* be the arc let fall from the star, perpendicular to the ecliptic, *e b* is the latitude, while A *b*, being the distance from the equinox to the foot of the perpendicular, is the longitude of the star. The effect of the precession of the equinoxes is here manifested only by the change in the point from which the longitudes are measured : thus without any real change in the place of the star, its longitude, which was originally A *b*, increases to A' *b*, and subsequently to A'' *b*, and so on indefinitely. The latitude, on the other hand, remains constant. It is further to be observed that this increase in the longitude must necessarily be the same for all stars.

§ 72. *Numerical Determination of the Precession.*—To determine the numerical value of the precession of the equinoxes, it will be sufficient to have ascertained for any one star the annual change which precession has produced in the longitude of that star. In making this calculation it will be desirable to have observations of the star separated by as long an interval of time as possible. It appears from the observations of Hipparchus that in the year 128 B.C. the longitude of the star Spica Virginis was about 174°. By the observation of Maskelyne, the longitude of this star in 1802 is 201° 4' 41''. The difference between these two results is due to the alteration in the point from which the longitude is measured, which has taken place during the interval of 1,930 years.

We may also use for this purpose two observations separated by a comparatively small interval when those observations have been made with the accuracy attainable with modern instruments. Leverrier has found that the annual alteration of the intersection of the equator with the ecliptic, measured along the ecliptic, amounts to 50''·24. From this it is easy to calculate that the equinox will move completely round the ecliptic and regain its original position in a period of about 26,000 years.

It appears from these considerations that the actual path

of the celestial pole among the stars is a small circle, of which the radius is 23° 27', and that the duration of the revolution is 26,000 years.

In the course of ages the precession of the equinoxes is calculated to produce the most marked changes upon the relation of the pole to the constellations. At present it so happens that the bright star, called the Pole Star, is about $1\frac{1}{2}$° from the pole. The movement of the pole in the small circle which it describes about the pole of the ecliptic, is at present diminishing the distance between the pole and the Pole Star. The approach will continue until the year 2120, when the distance will not be much more than half a degree. The distance will then increase until, in the lapse of 13,000 years, it will amount to about 47°. Long ere this happens, however, the pole will have become so far from the Pole Star as to deprive the latter of the great utility which at present it possesses.

It appears that the movement of the north pole is calculated to bring it near to the star Vega (α Lyræ) in about 12,000 years, from which star the pole is at present distant by about 51°. The distance will diminish to a minimum of about 5°, and Vega will then by its brilliancy, as well as by its proximity to the pole, be able to fulfil many of the purposes which render the present Pole Star so convenient.

§ 73. *Gradual Displacement of the Perihelion of the Earth's Orbit.*—Just as we have seen that the plane which contains the orbit of the earth has a gradual slow motion in space, so it is found that the elliptic orbit of the earth has itself a slow motion in the plane of the orbit, so that the major axis of the orbit occupies successively different positions.

The movements of the earth's orbit are determined by observations of the movement of the sun, for, as we have seen, the apparent orbit of the sun round the earth is similar to that of the earth round the sun. If, therefore, observation has detected a motion in the solar perigee, that really indicates a movement in the major axis of the earth's orbit.

It is found by comparing old observations with modern observations that the longitude of the sun's perigee changes by 61‴·72 per annum (Leverrier). If this augmentation of the longitude of the solar perigee had been equal to 50′·24, it would have been completely explained by the alteration in the equinoxial point which we have just considered. Under these circumstances the position of the solar perigee among the stars would have been constant. As, however, the motion of the longitude perigee exceeds the precession of the equinoxes by 11‴·48, it follows that the position of the perigee is displaced along the ecliptic to the extent of 11‴·48 annually, and that the motion is direct.

§ 74. *The Aberration of Light.*—The phenomenon which is known as *aberration of light* is one of the most interesting discoveries which have ever been made in astronomy. It had long been surmised that if the earth really moved round the sun, as was admitted in the Copernican theory, that then certain apparent movements ought to be observed in the places of the stars, and that from such movements the distances of the stars would be determined. With the view of detecting the changes whose existence was thus surmised, the celebrated astronomer Bradley commenced a series of observations with an instrument specially arranged for the purpose. The star which he chose was γ Draconis, which culminates very near the zenith, at the latitude of Greenwich, and the place of which is, therefore, but little deranged by refraction. The observations were commenced in 1725, and continued with great assiduity at all available opportunities. Bradley soon found that certain changes were taking place in the positions of the stars, but that these changes were of quite a different character from those of which he was in search. It appeared, for example, that the star γ Draconis was in March 1726 about 20″ more to the south than it was in December 1725, while from March to the following September the same star moved towards the north through a distance of not less than 39″. In

December 1726 the star had returned to the same position
which it had one year previously. It thus appeared that the
zenith distance of this star oscillated through a distance of
about 20″ on each side of its mean value. It can, however,
be shown that these changes could not be accounted for
merely by the displacement of the earth in its orbit, and
could not, therefore, be the parallax of which Bradley was
in search. It was not until similar movements had been
detected in the case of many other stars that Bradley was
enabled by a happy conjecture to give it a satisfactory explanation. The movements of the stars were found to be a
consequence of the fact that the velocity of light, though
exceedingly great, is still not incomparably greater than the
velocity with which the earth is travelling in its orbit round
the sun. The phenomenon which has been thus so happily
explained, is called the *aberration* of light. This discovery,
though it relates to magnitudes so small as only to be perceptible in very accurate measurement, is yet of so delicate
and so beautiful a character that it must undoubtedly rank
among the very greatest discoveries which have yet been
made in practical astronomy.

The earth is moving at the rate of about eighteen miles
per second, while light is travelling with a velocity about
ten thousand times as great. It follows that while we are
observing the star, the telescope is actually being carried on-
wards with a velocity which is one ten-thousandth part of
the velocity with which the light from the star is coming
towards the earth. Let s (Fig. 91) represent a star to which
a telescope is to be directed. If the telescope is to remain
at rest while the observation is being made, then it is obvious
that the telescope should be directed along the ray x s,
which comes from the star. If, however, the telescope be
in motion, then a little consideration will make it clear that
the telescope must generally not be pointed exactly at the
star, but in a somewhat different direction. Let us suppose
that the relation between the velocity of light and the velocity

of the telescope is such that the telescope will be carried from the position A B to the position X Y in the same time as the light from the star will travel through the distance B X.   Then it is plain that if the light from the star is to reach the eye of the observer at all, the telescope must be pointed not directly towards the star, but in the direction shown by A B in the figure.   It is, indeed, obvious that the star can only be seen when the rays of light therefrom enter the object glass of the telescope, and pass out at the eye piece to the eye of the observer.   The telescope must, therefore, be placed in such a position that this condition can be fulfilled.

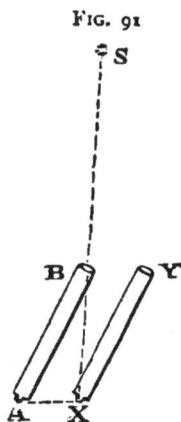

FIG. 91

When the telescope has the position A B the light enters the telescope through the object glass at B ; the motion of the telescope is then sufficient to enable the light to pass down the tube of the telescope without being lost against the sides, and when the telescope reaches the position X Y the light emerges from the eye piece at X and enters the eye of the observer.   The observer being unconscious of his own motion and of that of the telescope, naturally supposes that the light from the star really travels from the direction in which the telescope is pointed ; he, therefore, concludes that the star is in the direction A B, while the star really is in the direction X B.   The observer will, therefore, judge erroneously of the position of the star on the surface of the heavens, and the extent of the error which he makes is represented by the angle A B X, for this is the angle between the direction of the telescope and the direction from which the light of the star is actually coming.

§ 75. *Velocity of Light.*—In the further discussion of the phenomenon of aberration we may proceed in either of two ways.   We may either assume the velocity of light from some of the methods by which that velocity is known, and then deduce the amount of the aberration, or we may determine the quantity of aberration from observation, and then deduce the velocity of light. We shall adopt the former method, though it should be noted, as a striking confirmation of Bradley's theory, that the value of the velocity of light which can be deduced from the observed coefficient of aberration, agrees in a most remarkable manner with the velocity of light as it has been determined by the other methods which are available for that purpose.

.. It has been already stated that the velocity of light is about ten thousand times as great as the velocity of the earth in its orbit. It follows that in the triangle B A x the line B x must be about ten thousand times as long as the line A x. The angle B A x can, therefore, under the most favourable circumstances, never exceed the angle whose circular measure is one ten-thousandth, i.e. about 20″. It, therefore, appears that the greatest effect which aberration is capable of producing is to cause a star to appear in a point on the heaven 20″ distant from the point in which that star would have appeared had the velocity of light been infinitely great with regard to the velocity of the earth, or had the earth been at rest. As, however, the star which is now displaced 20″ in one direction will, in six months, be displaced 20″ in the diametrically opposite direction, the total effect of aberration will amount to a displacement which oscillates within an amplitude of 40″.

The line A x, which indicates the movement of the observer, is, of course, a tangent to the actual orbit of the earth. This tangent is directed to a certain point on the celestial sphere, which is often called *the apex of the earth's way*. The maximum effect of aberration, in deranging the place of a star, will occur when the line s x is perpendicular to A x, the star will then be 90° from the apex of the earth's way, and will, therefore, be situated on that great circle of the celestial sphere, of which the apex is the pole. It is further to be noted that the effect of aberration is to displace

the apparent position of a star in a plane which contains the
line A X, and, therefore, passes through the apex of the
earth's way. The effect of aberration generally is, therefore,
to cause all the stars on the surface of the heavens to appear
to move from their true places slightly towards the apex of
the earth's way. A star actually situated at the apex would
experience no effect from the aberration, while all the stars
on any circle of which the apex was the pole, would aberrate
towards that apex through equal amounts.

FIG. 92.

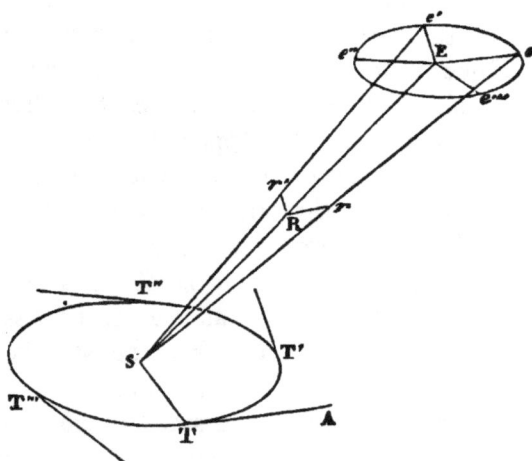

As the apex is daily—indeed, momentarily—changing its
position upon the ecliptic so as to travel completely round
the ecliptic in the course of a year, it appears that the point
towards which a given star is displaced is constantly varying.
It follows from this that the effect of aberration upon the
apparent place of a star is to make that place describe a
small ellipse on the surface of the celestial sphere of which
the centre is the mean place, while the period in which the
ellipse is described is equal to the period of the revolution
of the earth round the sun. Let s (Fig. 92) be the position
of the sun, and let T, T', T", T''' be four positions of the
earth in its orbit, which for the present we may consider to

be circular. Let E be the position of a star as it would be seen from the sun S. When the earth is at T, the line T A drawn touching the circle T T' T'' T''' denotes the direction in which the earth is moving. As the star is at an enormously great distance compared with the distance of the earth from the sun, we may practically consider the line T E as parallel to the line S E, and, therefore, S E indicates the point on the celestial sphere in which the star would be seen if it were not for the effect of aberration. Through any point R on the line S E draw a line R *r* parallel to the tangent T A, and bearing the same ratio to the line S R which the velocity of the earth bears to the velocity of light ; join S *r*, and produce it on to cut the plane drawn through E parallel to the plane of the ecliptic : then the effect of aberration will be to make the star appear to be at *e*, where E *e* is parallel to R *r*. In the same way, when the earth is at T', the star will be seen in a direction which points to the same point on the celestial sphere as is indicated by the line S *e'*. As the lines R *r*, R *r'* are equal, it is obvious that E *e* and E *e'* are also equal ; and hence as the earth assumes the successive positions T, T', T'', T''', &c., the star is found in the corresponding positions *e*, *e'*, *e''*, *e'''*, &c.

The curve *e e' e'' e'''* which the star appears to describe in the course of a year in consequence of aberration must obviously be a circle of which the plane is parallel to the plane of the ecliptic. The effect of this movement upon the apparent place of the star is to be determined by projecting the circle which is actually described upon the tangent to the celestial sphere. Thus the observed path of the star due to aberration will, in general, be an ellipse, of which the axis major is parallel to the ecliptic : the length of the axis major of this ellipse is constant. For a star at the pole of the ecliptic the ellipse reduces to a circle, while for a star actually in the ecliptic the effect of aberration is merely to cause the star to oscillate in the arc of a great circle. A star situated at the pole of the ecliptic will always be seen at the constant distance of about 20" from the position in which it could be seen were the earth stationary.

These results are shown in Fig. 93. K B S A D is the celestial sphere. A B C D is the ecliptic, of which K is the pole. The orbit of the star due to aberration is $e\ e'\ e''\ e'''$, and the observer is at O, the centre of the celestial sphere. The star appears to the observer to describe the ellipse $p\ n\ q\ m$,

FIG. 93.

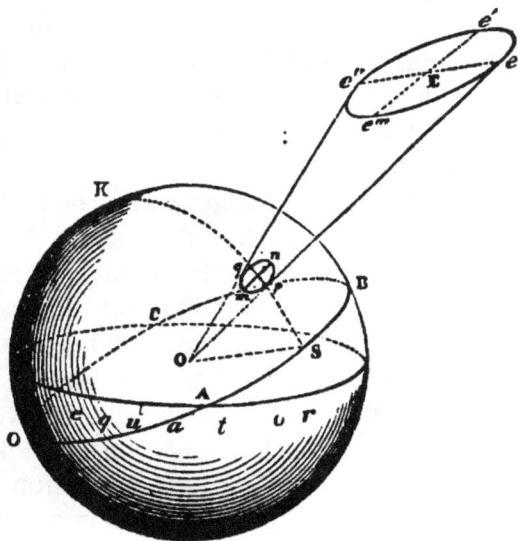

in which the cone drawn from O to $e\ e'\ e''\ e'''$ is cut by the celestial sphere. The major axis $m\ n$ of this ellipse is parallel to the ecliptic, while the minor axis $q\ p$ lies in the circle of latitude K s, which passes through the centre of the ellipse.

§ 76. *Nutation of the Earth's Axis.*—Although the discovery of the aberration afforded a satisfactory explanation of the motion of the stars which Bradley had detected, yet he continued his observations on the zenith distances of stars which culminated near the zenith. It then appeared that there were certain other alterations of the zenith distances, of quite a different character from those of which aberration had already afforded so satisfactory an explanation.

By making allowance for the effect of aberration, we can ascertain the true position in which the star would be seen if the earth was stationary. When this is done, it is found that the places of the stars thus determined are not constant ; but that they indicate a movement very different from the movements of an annual nature which are produced by aberration or by parallax. Thus, for example, Bradley found that from the year 1727 to the year 1736, the star γ Draconis was continually advancing towards the north pole, while, after the latter date, the star began to withdraw from the pole : and similar movements were observed in other stars.

Bradley found that the changes which he had observed, in the polar distances of certain stars, could be explained by the supposition that the pole of the celestial sphere oscillated, through narrow limits, about its mean place. This phenomenon he termed *nutation.* As it appeared from the observations that the period of this change was about eighteen years, it was surmised, by Bradley, that the cause of the phenomenon was to be sought in the motion of the moon's nodes, which complete their revolution in a little more than eighteen years. The explanation of this depends upon the theory of gravitation, but for the present we shall only attempt to describe the actual character of the phenomenon now called nutation.

The axis of the earth points to the celestial pole, and any movement in the celestial pole is, therefore, to be attributed to a change in the position of the earth's axis of rotation. It is not to be supposed that any change in the actual position of the earth's axis *in the earth itself* is indicated either in the phenomenon of precession or in that of nutation. So far as observation has hitherto gone, it would appear as if the axis, about which the earth rotates, were actually fixed in the earth ; and that in the

phenomenon of precession and nutation the movements of
the earth are similar to what they would be, if a fixed axis
were driven through the earth, and this axis were made to
describe the movements which observation has shown in
the motion of the axis of the celestial sphere.

The nature of the movement of the axis which produces
the phenomenon of nutation is shown in Fig. 94: T is the
position of the earth in its orbit, the direction of the revolution round the sun being shown by the arrow; T K is the
perpendicular to the plane of the ecliptic, and K is the pole
of the ecliptic. Let T *o*
be the axis of rotation of
the earth, then, accord-
ing to the phenomenon of
precession, the line T *o*
revolves around T K on
the surface of the cone
made by a line which
makes a constant angle
of 23° 27' with T K.
Owing, however, to the
superposition of the
small effects of nutation
upon those of preces-
sion, the actual move-

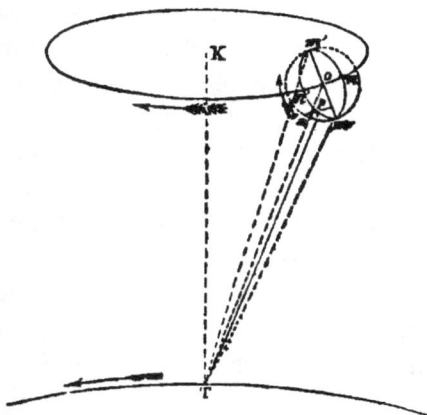

FIG. 94.

ment of the earth's axis is on the surface of a small cone,
T *m n m' n'*; while at the same time the axis T *o* of this
cone moves round on the surface of the cone of revolution, of which T K is the axis. The movement of the axis
of the small cone around the large one constitutes the
phenomenon of precession, while the motion of the axis
of the earth on the surface of the small cone constitutes
the phenomenon of nutation.

The major axis *m m'* of the ellipse, which forms the
base of the cone of nutation, lies in the plane which passes

through the axis of the earth T o and the perpendicular to the plane of the ecliptic T K.

The relative sizes of the cones is necessarily very much exaggerated in the figure, for while the semi-angle, at the vertex of the large cone, is 23° 27′, the axis major $m\,m'$ of the ellipse at the base of the small cone only subtends an angle of 18″·45 at T. The minor axis $n\,n'$ subtends an angle of 13″·73. The pole describes the circumference of the small ellipse in a period of 18⅔ years, but its motion during that time is not uniform. In order to ascertain the actual position of the pole at any given epoch, describe a circle $m\,z\,m'$, of which the axis major of the ellipse $m\,m'\,n\,n'$ is a diameter. If the point z move uniformly round the circle so found, and if it start from the point m at the same epoch as the pole is coincident with m, then the position of the pole at any other epoch will be found by letting fall a perpendicular from the corresponding position of z upon the axis major of the ellipse. The point p in which this perpendicular intersects the ellipse is the position of the pole.

A remarkable consequence of the nutation of the earth's axis consists in the variations which it produces in the obliquity of the ecliptic. The obliquity is equal to the angle between T K and T P, and the greatest and least values of the obliquity correspond to the epochs at which p is at one or other extremity of the major axis of the ellipse of nutation. These fluctuations are limited to 9″·22 on either side of its mean value.

§ 77. *Annual Parallax of Stars.*—We have already mentioned that the discovery of aberration was made by Bradley, while in the course of a series of observations undertaken with the object of discovering some variations in the positions of the stars arising directly from the annual movement of the earth. The reason that Bradley did not find the movements of which he was in search, was that the effects of annual parallax on the stars which he observed were too small to be detected by the method of observing which he adopted. Any annual parallax which has even up to the present been detected is an exceedingly small quantity compared with the coefficient of aberration which Bradley's observations revealed.

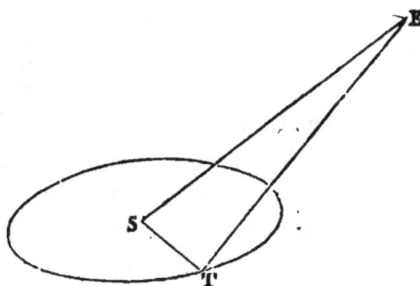

Let s (Fig. 95) represent the sun, and let T be the earth, while E is a star. An observer situated at the sun will see the star in the direction s E, while an observer on the earth will see the star in the direction T E. The position on the celestial sphere to which the star will be referred will therefore be different, according as the position of the observer changes by the annual revolution of the earth around the sun.

FIG. 95.

It is easy to show that the effect thus produced on the apparent place of a star referred to the celestial sphere is to make the star describe a certain small ellipse; for the observer will always see the star in the direction T E, and as the earth moves round the sun the line T E will describe a cone, of which E is the vertex, while the orbit of the earth is the base. The star will, therefore, always appear to be on the surface of the cone. But the star must also be referred to the surface of the celestial sphere, and hence the place of the star will be limited to the line in which the cone just referred to cuts the celestial sphere. As the portion of the celestial sphere included in the cone is extremely small, the portion of the sphere may be represented by its tangent plane, and then we see that the orbit of the star is the section of the cone by a certain plane; but the section of a cone is in this case an ellipse, and hence it appears that the effect of annual parallax upon a star is to cause that star to appear to describe an ellipse on the celestial sphere of which the mean place of the star is the centre. It is easy to see that if the star happened to be situated at the pole of the ecliptic the tangent plane would cut the cone in a circle (assuming the earth's orbit to be circular), and that therefore the apparent orbit of a star so situated would be a circle. In other situations the ellipse would have an eccentricity which gradually increases as we consider stars nearer to the ecliptic, while for a star actually situated in the plane of the ecliptic the minor axis of the ellipse would be zero, and the changes in the position of the star would merely cause it to oscillate to and fro in the ecliptic.

Under all circumstances the major axis of the parallactic ellipse (as it is called) must be parallel to the ecliptic, and the dimensions of the major axis will depend upon the actual distance at which the star is separated from the earth.

The star when situated at one of the extremities of the major axis of the parallactic ellipse is at the greatest distance from its mean place. The angular magnitude of the semi-axis major of the ellipse is equal to the greatest angle between the lines S E and S T. Since S T, being the radius of the earth's orbit, is always small compared with S E, being the distance of the star, the greatest angle between S E and T E is obviously the angle whose circular measure is S T ÷ S E. This angle is therefore the axis major of the parallactic ellipse. The dimensions of the ellipse are thus dependent upon the distance of the star. If the star were so exceedingly remote that the radius of the earth's orbit was absolutely inappreciable when compared with the distance, then the angular value of the semi-axis major of the parallactic ellipse would be inappreciable, and consequently no effect of parallax could be observed. This appears to be the case with the majority of stars so far as our present knowledge extends. The parallactic ellipse is so minute that in the majority of cases our ordinary methods of measurement fail to detect any considerable alterations in the positions of the stars which can be ascribed to this cause. Under these circumstances we infer that the stars in question are so exceedingly distant that the radius of the earth's orbit, great, no doubt, as that really is, is an inappreciable magnitude compared with their distance from us.

By adopting, however, very special modes of measurement, certain minute changes have been noted in the positions of certain stars which are unquestionably the effect of annual parallax. The nature of these observations must now be detailed.

By the observations of right ascension and declination made in the manner already described, it is possible to determine the position of a star on the surface of the celestial sphere. By repeating these observations at intervals throughout the year, the corresponding positions of the star at the different dates are ascertained, and we may conceive that the positions are marked upon the globe, the effect of aberration, nutation, and precession being allowed for. If the star was sufficiently near the earth to have a large parallax, these positions would undoubtedly indicate with more or less accuracy the form of the parallactic ellipse. The irregularities of the observations which are unavoidable, even with the best instruments, would prevent the observed positions from conforming exactly to the form of the ellipse, but a sort of general arrangement of the positions in an elliptic form would, no doubt, be traced. Owing, however, to the extremely great distances of the stars, the parallactic ellipse is so very minute, that its form is entirely lost sight of in the irregularities of the observations which, though still small, yet bear such a large proportion to the dimensions of the ellipse that the latter cannot be discerned. In particular we may mention that, so far as observations hitherto published go, there appears to be no star in the northern half of the celestial sphere which has an annual parallax exceeding about half a second of arc. Even with this parallax the angular value of the axis major of the parallactic ellipse is only a single second of arc, and, unless the star were situated at the pole of the ecliptic, the minor axis of the ellipse must be less than that quantity. The question then arises as to whether our present means of determining the absolute places of stars are sufficiently exact to enable us to detect the existence of an ellipse of such minute dimensions by a series of observations extended over a year. There is no prospect that the meridional determinations of the positions of the stars can ever attain such accuracy as to enable the form and dimensions of an ellipse of this order of magnitude to be accurately constructed.

Not to mention other sources of irregularity, it is sufficient to remember that the apparent declination of a star is affected by atmospheric refraction. The larger part of this refraction can be determined by calculation, and its effect can be allowed for ; but even after this has been done with every attention, there are still irregularities outstanding, which, although too small to be material for many purposes, are still too large to permit of the determination of the ellipse which is the consequence of parallax. It is true that by choosing, as Bradley did, a star which culminates near the zenith, the irregularities of refraction are of comparatively small importance, but there are still so many sources of small irregularity, that even independently of refraction the method indicated would be inapplicable.

The star a Centauri in the southern hemisphere appears to have a parallax much larger than any star which has yet been investigated with this object in the northern hemisphere. In this case the parallax appears to amount to about a second, and consequently the total displacement of the star, from one extremity of the major axis of the ellipse to the other, amounts to about two seconds. A displacement so considerable as this is no doubt capable of being detected, and even measured with some pretensions to accuracy by meridian observations. It was, in fact, by observations of this kind, that the remarkable parallax of this star was originally discovered.

- It is, however, possible to arrange a method of observing by which much smaller parallaxes can be detected and measured than can be done in the case of meridian observations. This method is founded upon the principle that, whatever be the irregularities which have affected the observations, we may presume that two stars which are close together on the surface of the heavens will be affected to nearly the same extent. Suppose, then, that there are two stars, of which one is considerably nearer to us than the other, while the two stars are so nearly in the same visual ray that they appear to be close together on the heavens. Each of the stars will, in consequence of annual parallax, appear to describe a small ellipse on the heavens, but the ellipse formed by the nearer of these two stars will be larger than that described by the more distant star. The apparent distance measured as an arc of the celestial sphere between the two stars will therefore change, and the direction of the arc joining the two stars will also change, and the period of these changes will be a year. If the two stars subtend an angle of not more than two or three minutes, the distance between them and the position angle can be measured by means of the micrometer. All irregularities arising from the instrument may be presumed to affect the two stars equally, and though the refraction will slightly alter the distance and position, yet its effects can be accurately computed and allowed for. By micrometric observations we have, therefore, the means of determining the changes in the position and the distance which occur in the course of a year, and thus of determining the difference between the annual parallaxes of the two stars.

An annual parallax of even one second would require that the star should be distant from us by 206,265 times the radius of the earth's orbit. The actual distance of the star from the earth would therefore amount to about twenty billions of miles (20,000,000,000,000). So far as we know at present it appears that a sphere of which the earth was the centre, and of which the radius was 20,000,000,000,000 miles, would hardly include a single star.

The vast distances here referred to can be illustrated by a comparison derived from the velocity of light. Light travels at the rate of about 186,700 miles per second, and with this velocity it would require eight minutes to come from the sun to the earth. To travel from the nearest star to the earth, even with this prodigious velocity, it would seem that at least three years would be required.

There are several stars of which the parallax has been determined (see Appendix) : thus, for example, the bright star Vega has a parallax of only 0″·18, and the light from this star takes a period of about 18 years to travel to the earth.

Viewed from the distances at which the fixed stars are situated, the earth itself would have become invisible to the most powerful telescopes, the vast orbit in which the earth revolves around the sun would have shrunk to a tiny object, while, if the sun himself were visible, it would only be as a star of no very special brilliancy or importance.

CHAPTER VI

THE MOON.

§ 78. *Apparent Movements of the Moon.*—Next to the sun, the moon naturally demands our attention as the most conspicuous of the remaining celestial objects. It is easy to observe that the place of the moon among the stars is continually changing. We are not now referring to the apparent diurnal motion of the heavens in which the moon participates in common with all the other celestial bodies. If the place of the moon be compared with the stars in its neighbourhood whose brilliancy is such that they are still visible notwithstanding the superior splendour of the moon, it will be found that even in the course of a few hours the position of the moon will have perceptibly changed. This is shown in Fig. 96, which indicates the displacement of the moon with respect to the fixed stars. The movement of the moon is about thirteen times more rapid than that of the sun.

FIG. 96.

By continued observation of the moon it will be seen to move completely around the celestial sphere, and if its path be laid down on a globe or map, it is found that the path is a great circle which lies close to, though it is still distinct from, the ecliptic in which the sun moves. The moon is sometimes a little to the north and sometimes a little to the south of the ecliptic, but it never deviates much therefrom. The direction in which the moon moves round on the surface of the heavens also agrees with that of the sun.

§ **79.** *Phases of the Moon.*—The sun always presents to us a complete circular disk. The moon, on the contrary, only shows us a complete circular disk for a few hours in each revolution. The figure of the moon changes with great rapidity, and in the period of twenty-nine or thirty days, during which it completes the circuit of the heavens and rejoins the sun, it presents to us every possible gradation between a circular disk perfectly illuminated and perfectly dark.

The period of revolution of the moon relatively to that of the sun probably suggested to the ancients the idea of dividing time into months. Perhaps also the idea of the week had its origin in the circumstance that a period of seven days was nearly the interval from new moon to first quarter, or from first quarter to full moon.

It is not unlikely that the ancients had made use of the phases of the moon as a means of measuring time before they were actually acquainted with the cause of the phases. A little reflection will, however, render the cause of the moon's phases sufficiently obvious.

To follow the phenomena presented by the moon in a methodical manner, let us commence by selecting an evening in which, just after the sun has set, the moon appears low down in the west. The moon then appears to us in the form of a narrow crescent, of which the exterior circumference is a semi-circle, and of which the interior circumference is a semi-ellipse of but little eccentricity, and the major axis of which coincides with a diameter of the semi-circle. The moon sets a little after the sun, and we remark that the luminous segment is turned towards the sun—that is to say, the line of greatest width of the crescent points towards the sun. The points of the crescent are generally called the cusps, and it will be noticed that the line joining the cusps is oblique to the horizon, and that each of the cusps is equidistant from the sun.

Fig. 97.

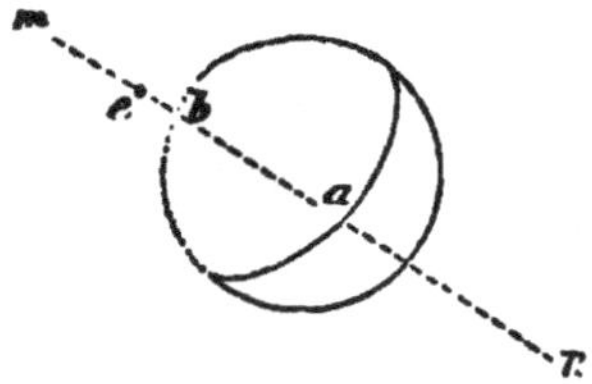

From day to day the crescent enlarges, though its cusps still remain at the extremities of a diameter, the interior curve becomes a semi-ellipse of increasing eccentricity, the moon sets later and illuminates a considerable part of the night, but the line of cusps is still always inclined to the horizon at setting, and the greatest width of the crescent still points towards the sun.

On the seventh day the moon appears like a semi-circle and it is visible during a great part of the night. On the following days the luminous portion continues to increase, the interior elliptic curve takes an opposite position, but the semi-ellipse and the semi-circle have always a common dia-meter.

On the 14th or 15th day the disk of the moon is entirely luminous. The intensity of the light is not, how-ever, quite uniform ; there are some places especially dark, to which the term ' seas ' has been applied. The telescope, however, shows that this term is incorrect, because we can perceive in the so-called ' seas ' numerous features showing an irregular surface.

On the following day the western edge of the moon is seen to be somewhat less sharply defined ; gradually this part becomes more and more obscured, and the curved line which bounds the luminous region becomes a semi-ellipse of more and more eccentricity. On the 22nd day the moon is again a semi-circle, and all the phenomena previously described reappear in an inverse manner ; about the 28th day the moon has drawn very near the sun, after which it becomes invisible for a few days, again to reappear in the west in the form of a narrow crescent, and to repeat the cycle already described.

During the entire revolution of the moon, the illuminated portion is always turned towards the sun, and the dark part is turned away from the sun. Now it might be supposed that half the moon was self-luminous and the other half was dark, and that the phases in the moon were produced by a rotation of the moon on its axis, which at new moon turned towards the earth the dark side, at full moon the bright side, while for the intermediate phases the hemisphere turned to-wards the earth is partly bright and partly dark. If this were the case, then the ' seas ' and other markings upon the disk of the moon would not always occupy the same portion of the visible disk of the moon, but the illuminated hemi-sphere would always contain the same objects, and the boundary between light and darkness would always occupy the same position with reference to the lunar markings.

Observation, however, shows that this is not the case. Among the many interesting features of the moon, there is perhaps none more remarkable than that the moon always turns the same face to the earth. It follows that the markings on the moon always occupy the same position on the visible disk, and the boundary between light and darkness, so far from occupying a constant position with respect to the lunar objects, may actually be observed creeping along from night to night, and illuminating one lunar mountain or valley after another.

From these considerations it follows that the moon cannot be self-luminous, and that therefore the light which we receive from it must be reflected by the moon from some other source.

As the moon is naturally dark, from what source is its light derived? There can be but one answer to this question. The sun is the only body in the universe sufficiently brilliant to explain the brightness of the moon. This idea is confirmed by the following simple observations. When the moon is full it comes on the meridian at midnight ; but at midnight the sun is at its lower culmination, hence the sun and the moon are then separated by about 180°, and the sun must therefore occupy a point of the celestial sphere nearly opposite to that occupied by the moon. We say nearly opposite, for if the sun and the moon were upon the same diameter of the celestial sphere of which the earth occupies the centre, the shadow of the earth would be thrown upon the moon, and we would have the phenomenon of an eclipse of the moon. As the great majority of full moons pass without an eclipse, we infer that the declinations of the sun and moon at full moon are not generally equal in magnitude and opposite in sign, as they would be if the two bodies were situated upon the same diameter of the celestial sphere, and as they are on the days when the moon is eclipsed.

An observer who was stationed upon the moon would

perceive phases on the earth analogous to those which we
see on the moon. When the crescent of the moon is very
narrow, either shortly before or shortly after new moon, the
earth as seen from the moon would be nearly full, and ought,
therefore, to illuminate the moon as the full moon illuminates the earth. Indeed, the efficiency of the earth in this
respect considerably exceeds that of the moon, for the area
of the earth as seen from the moon is more than thirteen
times the area of the moon as seen from the earth.

It follows that the light which the moon receives from
the earth is more than thirteen times greater than the light
which the earth receives from the moon.

It is this light reflected from the earth which often
enables us to see that part of the disk of the moon which
is not illuminated by the sun.

§ 80. *The Eccentricity of the Moon's Orbit.*—One of the
most simple problems in connexion with the moon is the
determination of the eccentricity of the ellipse which we
shall now assume to be the orbit of the moon. It is easily
found by observation that the angular diameter of the moon
varies considerably, and the eccentricity may be taken to be
about 0·055.

Thus the eccentricity of the orbit of the moon is much
greater than that of the orbit of the earth around the sun,
which is 0·0168.

§ 81. *Motion of the Perigee.*—To determine the position of
the moon's perigee (or the point in its orbit in which the moon
is nearest the earth), it is necessary to ascertain the place of
the moon on two different days on which the apparent
diameters of the moon are equal : the place half-way between
these positions is the position of the perigee.

When the position of the moon's perigee is determined at
different periods, it is found to be in rapid motion. In fact,
it makes a complete sidereal revolution in about nine years,
or more accurately in 3232·575 days.

§ 82. *Mean Motion of the Moon.*—It is of primary importance to determine the value of the mean motion of the moon. This can be simply and accurately ascertained by the comparison of two eclipses of the moon separated by a wide interval of time. We may suppose that the opposition of the sun and the moon takes place at the middle of the eclipse; this supposition cannot entail an error of more than a few minutes. If, then, we ascertain by observation the time of the middle of the eclipses, and the number of lunar months or of complete revolutions which have taken place in the interval, we shall have an approximate value of the lunar month by dividing the total time between the two eclipses by the number of full moons which have occurred in the interval. It is obvious that the accuracy of the result will increase with the number of entire revolutions.

By this means it has been ascertained that the Synodic lunar month, or the interval between two consecutive full moons, is 29·530589 days.

§ 83. *Figure of the Moon.*—As the apparent dimensions of the moon are so considerable, it is necessary, as it was in the case of the sun, to select a certain point on the surface of the moon by which its position on the celestial sphere is to be specified. For this it is necessary to examine accurately the shape of the moon's disk as it is presented to us. By careful measurements it is found that the apparent disk of the moon is circular, and therefore it is natural to select the centre of that disk as the point of reference required. Whenever, therefore, the right ascension or the declination of the moon is referred to, we are to understand the right ascension or declination of the centre of the moon's disk. The centre of the moon is not indicated by any mark which we recognise, and therefore it is impossible to observe that point directly ; we can, however, observe the edges of the disk of the moon, and then reduce the observations so as to give the position of the centre. If we were able to observe the transit of the two edges of the circular disk (or limbs), then the mean of these observations would give the transit of the centre. Owing, however, to the fact that only one limb of the moon is generally illuminated, and therefore visible, the method is not generally practicable. We are usually only able to observe the transit of one of the moon's limbs. To determine from a single observation of this kind the position of the moon's centre, we obviously require to know the angular value of the moon's radius. The difficulty is increased by the circumstance that this radius from day to day—nay, even from hour to hour—is constantly changing. The value of the radius may, however, be ascertained at any moment by measuring the distance from the extremity of one cusp of the moon to the other, for the line joining the cusps is always a diameter of the moon, and therefore half their distance is equal to the radius. When allowance has been made for the moon's radius, then the right ascension of the centre is determined from the observation of the limb. The declination may be determined by observing both cusps of the moon with the meridian circle, the mean between the declinations of the two cusps being the declination of the centre. In all that follows we are, therefore, to understand that the centre of the moon is the point to which reference is made.

§ **84**. *Parallax of the Moon.*—The apparent position of the sun on the surface of the celestial sphere depends to a certain extent upon the position of the observer on the surface of the earth ; and in the discussion of observations of the sun it is necessary to reduce the observed positions to those in which the sun would have been seen, had the observer been able to occupy a position at the centre of the earth. The word *parallax* is introduced to signify the dis- tance between the position of the sun as seen by the observer, and as it would be seen from the centre of the earth, while the expression *horizontal parallax* is used to denote the parallax when the sun is rising or setting, or, what comes to the same thing, the angle which the radius of the earth subtends at the sun.

Precisely similar language may also be employed with reference to the moon. The distance between the position of the moon as seen by the observer and the position in which the moon would be seen by an observer at the centre of the earth, is called the *parallax of the moon*, and the angle which the radius of the earth subtends at the moon is the *horizontal parallax* of the moon. There is, however, one great difference between the parallax of the sun and the parallax of the moon. The former is so small a quantity as only to be detected by the most careful observations, while the latter is of very considerable amount, being, in fact, about 400 times greater than that of the sun. It follows that the parallax of the moon requires to be allowed for in nearly all observations of that body.

To determine by observation the parallax of the moon, we proceed as follows :—Let o be the centre of the earth (Fig.

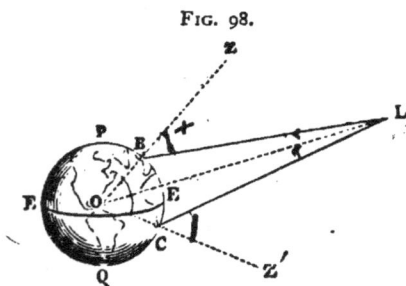

FIG. 98.

98), and let B and C be two points on its surface, which for simplicity we shall suppose to lie on the same meridian, passing through the pole P. Let L be the moon, the position of which is to be observed from the stations B and C. The observer at B determines the angular distance between his zenith z and the moon L—that is, the angle z B L. In a similar way, and at the same moment of time, the observer at C determines the angle L C z', which is the zenith distance of the moon as seen from the position which he occupies. From these observations we can determine the angles B L O and C L O, which are respectively the parallaxes of the moon at B and at C. The geographical latitudes of the observations at B and C are of course known, and therefore also the angle

B O C. This angle is equal to the difference or sum of the latitudes, according as the two stations are on the opposite sides or on the same side of the terrestrial equator. We can then proceed by the following graphical method. Construct an angle equal to the angle B O C, and mark off lengths O B and O C on the legs of this angle, which are equal to each other. Draw through B a line B L, making the angle Z B L equal to the observed zenith distance of the moon from L. Draw through C a line C L, making the angle L C Z' equal to the observed zenith distance of the moon at C. The intersection of these two lines determines the point L, which, being joined to O, determines the angle B L O and C L O, which are the parallaxes at the points B and C.

As a matter of fact, the determination of these angles is really effected by trigonometrical calculation, which gives much more accurate results than could be obtained by any graphical process. The details of the process are very much embarrassed by certain circumstances which we have omitted for the sake of simplicity. It will not, of course, happen that the two observatories at which the observations can be made will lie exactly, or perhaps even nearly, on the same meridian. The effect of this difference between the meridians can, however, be calculated and allowed for in the resulting values of the parallax. So, too. the assumption which we have made that the time at which the moon is observed at one station coincides with the time at which it is observed at the other, cannot generally be realised. The actual motion of the moon in the interval between the observations has therefore to be taken into account, but the nature of that motion is so well known as to enable this to be done with considerable precision.

As the distance of the moon from the earth is comparatively small, we are obliged to take account of the fact that the figure of the earth is not identical with a sphere. In speaking, therefore, of the horizontal parallax of the moon, it becomes necessary to define accurately what radius of the earth is referred to. We may assume that the section of the earth made through the equator is a circle, and therefore that all equatorial radii of the earth are equal, and hence we can give precision to our language by speaking of the *equatorial horizontal parallax* of the moon, meaning thereby the angle which an equatorial radius of the earth subtends at the moon.

From observations made at the observatories of Greenwich and the Cape of Good Hope, the value of the mean equatorial horizontal parallax of the moon has been determined by Mr. Stone (1866) to be

$$57' \; 2'' \cdot 707.$$

From this it follows that the mean distance from the earth to the moon is about 60 times the earth's equatorial radius: but the orbit of the moon is so far from being a circle that the actual distance fluctuates between 56 and 64 times. Remembering that the radius of the sun is about 110 times the radius of the earth, it appears that if the centre of the sun coincided with the centre of the earth, the orbit of the moon, even at its greatest distance from the earth, would be all contained within the volume of the sun.

FIG. 99.

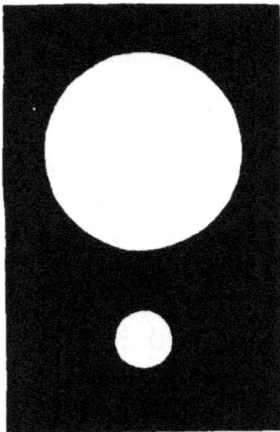

§ **85.** *Dimensions of the Moon.*— When the parallax of the moon is known, and also the angular diameter which the moon subtends, then the actual dimensions of the moon are ascertained. In this way it has been found that the diameter of the moon is about $\frac{3}{11}$ that of the earth. The relative dimensions of the moon and the earth are represented by the white circles in Fig. 99, the larger of these circles being taken to represent the earth ; the smaller of them represents the moon.

§ 86. *Movements of the Moon.*—We have now explained how the position of the moon is to be ascertained by observation, and how this position is to be corrected for the effect of the length of the moon's diameter, and for the effect of parallax. When these corrections have been duly applied, we obtain the position of the centre of the moon on the celestial sphere, as it would be seen from the centre of the earth.

These observations having been repeated, we are enabled to mark the position of the moon on a celestial globe, and then to ascertain the exact nature of the movements which it makes. It will be seen that after the lapse of about twenty-seven days, the moon has accomplished a complete revolution, and has returned to the neighbourhood from whence it started. A great circle can be drawn which coincides very nearly with the several positions of the moon, and this great circle may be regarded as its path. By watching the moon through several revolutions, and drawing the great circle which corresponds to the path of each, it is seen that the path of the moon is not constant. In fact, the inclination of the path to the equator is constantly changing, and in the course of time fluctuates between 18° and 28°. The nature of these changes will be more easily understood by referring the position of the orbit of the moon to the ecliptic. It will then be seen that the inclination of the moon's orbit to the ecliptic is nearly constant, never differing much from 5°, but that the node or point, in which the orbit cuts the ecliptic, is continually changing.

In Fig. 100, E E represents the celestial equator, and A B C D is the ecliptic. The orbit of the moon cuts the ecliptic at the point N at an angle of about 5°, and after completing the circuit of the sphere in the neighbourhood of the ecliptic, cuts the ecliptic again at the point N', which is not coincident with the former point N. The actual path of the moon is thus exceedingly complex, consisting of a series of spirals which are all contained within the zone extending about 5° 9' on each side of the ecliptic.

FIG. 100.

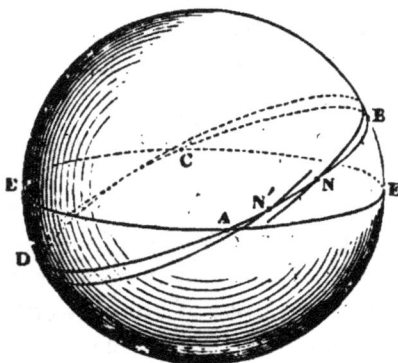

It will simplify the consideration of this somewhat difficult subject, to suppose that the actual path of the moon is in a great circle, but that this great circle is itself changing its position on the sphere. The pole of this great circle makes a nearly constant angle with the pole of the ecliptic, because the inclination of the circle to the ecliptic is nearly constant. In this is easily found the explanation of the changes in the inclination of the apparent orbit of the moon to the celestial equator. The moon's orbit will sometimes be inclined at an angle of 5° 9' less than the obliquity of the ecliptic, and sometimes at an angle of 5° 9' more. Thus the actual inclination of the moon's orbit to the equator varies between 23° 27' − 5° 9', and 23° 27' + 5° 9'.

§ 87. *Motion of the Moon's Nodes.*—The points N, N', in which the orbit of the moon intersects the plane of the ecliptic, are termed the *nodes* of the moon's orbit. The node N, through which the moon passes when entering that hemisphere which contains the north pole, is termed the *ascending node*, while the other node is the *descending node.*

We have already seen that the orbit of the moon is in motion relatively to the ecliptic, and, therefore, the positions of the nodes on the ecliptic are continually changing. Thus the node N moves uniformly round the ecliptic in the opposite direction to that in which the moon is going; the direction and amount of the motion of the node is represented by N N'. The period required by the node to travel com-

pletely round the ecliptic, so as to regain its original position
with respect to the stars, is 6798·279 days. This movement
of the nodes of the moon's orbit is analogous to the movement of the equator of the earth, which forms what we have
described as the precession of the equinoxes. In each case
the obliquity as regards the ecliptic is maintained at a nearly
constant value, while the movements of the intersection with
the ecliptic are nearly uniform. There is, however, one
very conspicuous difference—the revolution of the equinoxes
requires a period of 26,000 years, while that of the moon's
nodes takes but little more than 18 years.

When a close examination of the positions of the moon
at different epochs is made, it is found that the movements
cannot be entirely explained in the manner we have described. It is seen that the inclination of the plane of the
moon's orbit to the ecliptic is not absolutely constant. The
average value of this inclination is 5° 8′ 42″, but the actual
value fluctuates between 5° 17′ 35″ and 5° 0′ 1″. It is also
to be observed that the velocity with which the nodes move
is not absolutely uniform ; this movement is sometimes
accelerated and sometimes retarded. We may, therefore,
regard the actual motion of the moon's node as consisting
of two parts—first, a uniform motion, and then an oscillation
to and fro about the mean position. The normal to the
plane of the moon's orbit thus has a motion round the
normal to the plane of the ecliptic, which may be likened to
the motion of the axis of the earth already explained. In
each case the motion may be considered to take place on
the surface of a small cone, of which the axis describes a
much larger cone around the normal to the plane of the
ecliptic.

§ 88. *Sidereal Revolution of the Moon.*—By comparing
the places of the moon on the surface of the heavens at
remote epochs, it is possible to determine the time occupied
by the moon in making a complete revolution with respect
to the stars. In this way it is found that the sidereal revo-

lution of the moon is 27·321661 days, or very nearly 27 days and a third. It is very remarkable that this period is not constant, but that it is gradually diminishing, so that the time of the sidereal revolution of the moon is less now than it was formerly. The difference is, however, exceedingly small, and can only be detected by comparing the results of many centuries of observation.

§ 89. *Rotation of the Moon.*—The surface of the moon possesses certain well-marked features, visible to the unaided eye, and still more clearly discerned in a telescope. The simplest observations suffice to show that these objects remain constant in position ; in other words, that the face of the moon which is turned towards the earth is always the same. The movement of the moon is, in fact, nearly identical with what it would be could we conceive the moon rigidly attached to a bar, the other extremity of which was capable of turning around the centre of the earth. From this fact that the same face of the moon is always turned towards the earth the very remarkable consequence follows that the period of rotation of the moon on its axis is equal to the period of its revolution in its orbit around the earth.

Fig. 101.

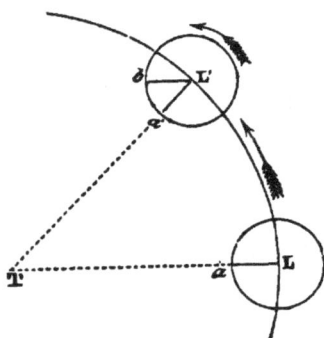

It might at first be supposed that, as we always see the same aspect of the moon, the moon does not rotate upon its axis at all. A little consideration will, however, show that this is a mistake. Let T (Fig. 101) be the earth, and let L and L′ be two positions of the moon. L.*a* is the radius of the moon which is directed towards the earth when the moon is in the position L, and the point *a* indicates a certain locality on the moon's surface. If the moon did not rotate upon its axis, then, when the revolution of the moon had carried it to

L', the line L a would still remain parallel to itself, and be carried into the position L' b.   But observation shows that the objects seen at a when the moon is at L will be seen at a' when the moon is at L'.   It therefore follows that while the moon has moved from L to L' through the angle L T L' the radius L a of the moon must have moved through the angle b L' a'.   But it is plain that the angle b L' a' is equal to the angle L T L', and therefore the moon must be revolving around its axis with an angular velocity equal to the angular velocity of its revolution round the sun.

§ **90.** *Mountains in the Moon.*—As there is no other celestial body so near to us as the moon, so there is none which we can scrutinise so keenly with our telescopes, and with the configuration of whose surface we are so well acquainted.   To see the details on the moon's surface to advantage, the phase of first or last quarter should be chosen, or at all events the time of full moon should be avoided.   The telescopic appearance of the moon shortly before the first quarter is represented in Fig. 102.   The irregularities of the surface are most distinctly seen along the boundaries of light and shade.   The relief there given shows that the surface of the moon is very rugged and irregular—that while there are some regions which appear comparatively level, there are others where mountain ranges and chasms give an appearance of rugged grandeur to which, perhaps, we have no parallel on the surface of the earth.   But the most characteristic objects on the surface of the moon are what appear to be craters of volcanoes, which, though once active, are now nearly, if not quite, extinct.   There are some hundreds of these objects on the surface of the moon which is turned towards us, and many of them are of truly colossal dimensions.

The general type of the lunar objects to which we are now referring is represented in Fig. 103.   A more or less circular plain is surrounded by a high mountainous wall, and in the centre of the plain rises another mountain, or some-

FIG. 102.

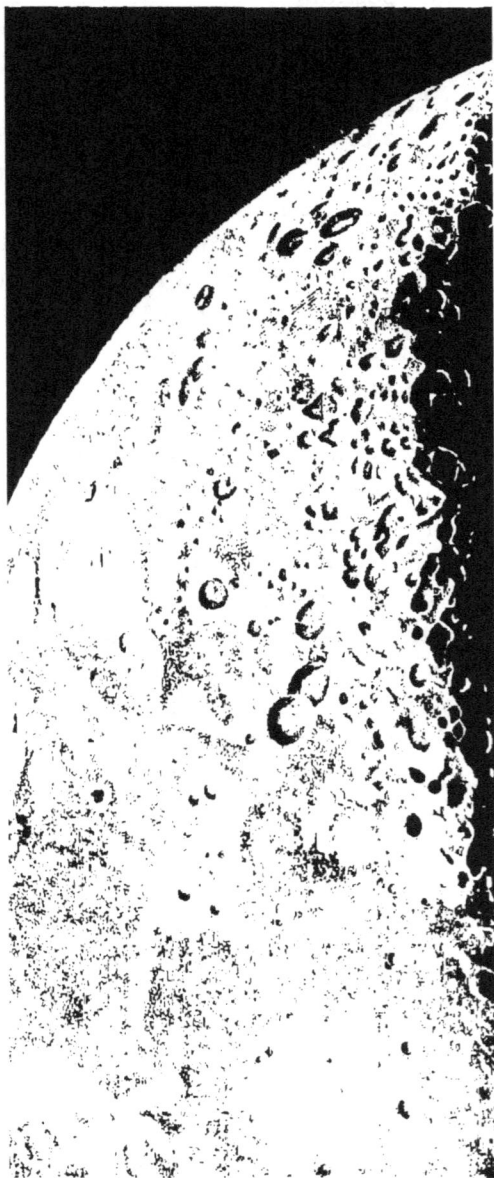

times more than one. This general type is modified in
many instances. The marginal range is not always com-

Fig. 103.

plete, and may indeed be broken into by other craters ; the
central mountain may be absent, and the plain may be more
or less irregular, or marked over with still smaller craters.
Still the general typical crater as here shown is sufficiently
frequent on the moon's surface as to form the most characteristic feature of lunar scenery. The surface of the
moon has been frequently mapped, and the different objects
have been carefully measured and distinguished by names.
Among the most remarkable of these craters we may mention
Tycho and Archimedes, the former of which has a diameter
of fifty-four miles, while that of the latter is sixty miles.

As we never detect any trace of clouds over the surface
of the moon, it has been surmised that the moon is not
surrounded by any atmosphere at all comparable in extent
with that which envelopes the earth. This opinion has been
confirmed by the most careful observations. It may be
remarked in the first place that when the surface of the
moon is only partly illuminated, the boundary between light
and shade is sharply marked. If there were on the moon
any atmosphere approaching in density to that surrounding
the earth, then it would be expected that the phenomenon
of twilight, which is due to the reflection in the upper regions
of the atmosphere, would be manifested. Under these cir-

cumstances a fringe of partial illumination would extend into the dark portion of the moon's surface ; but as such illumination, if it exists at all, is extremely faint, we are justified in concluding that the phenomenon of twilight would not be seen by an observer on the moon's surface, and that consequently, if the moon have any atmosphere, that atmosphere must be of great tenuity.

§ 91. *Periods connected with the Sun and the Moon.*—If the nodes of the moon were fixed, then the period of revolution of the sun with regard to those nodes would be simply equal to the sidereal year. On account, however, of the regression of the moon's nodes, the sun returns to the same node in a period less than a single year. This period amounts to 346·619 days. From a comparison of this period with that of the moon itself a very remarkable result is obtained. If the moon revolved actually in the plane of the ecliptic, then the centre of the moon would in each revolution pass across the centre of the sun, and the moment of this occurrence is called the time of *new moon*. Owing, however, to the circumstance that the orbit of the moon is inclined to the plane of the ecliptic, the moon will not usually pass over the surface of the sun. It is therefore necessary to modify the definition of new moon accordingly. We define the time of new moon to be the moment when the *longitude* of the centre of the moon is equal to the *longitude* of the centre of the sun. The interval between two successive new moons is termed a *lunation*, and it is by this period that the successive phases of the moon are regulated. The length of the lunation is such that 223 lunations make 6585·32 days. Thus 19 periods of the revolution of the sun with respect to the nodes of the moon coincide very nearly with 223 lunations. This remarkable period, amounting to about 18 years 11 days, is of service in the prediction of eclipses. It is known as the Saros.

Another very remarkable period arises from the circumstance that 235 lunations form 6939·69 days, while 19

years of 365·25 days amount to 6939·75 days. We therefore conclude that 19 years are nearly identical with 235
lunations. This is the Cycle of Meton. If the dates of
new moon and full moon are known for a period of 19
years, they can be predicted indefinitely, for in each subsequent 19 years the dates are reproduced in the same
manner. The number which each year bears in the Cycle
of Meton is called the golden number. In 1879 the golden
number is 18. In 1881 the golden number is 1, being the
commencement of a new cycle.

The period called the *Solar Cycle* is founded upon the
recurrence of the day of the week upon the same day of the
month. Owing to the complication produced by leap year,
this period is 28 years. In the year 1879 the Solar Cycle is
said to be 12. This signifies that 1879 is the twelfth of one
of these groups of 28 years. The cycle known as the
Roman Indiction is a period of 15 years. Though this
cycle is not connected with any astronomical phenomenon,
it is still retained. Thus the year 1879 is the seventh year
of the Roman Indiction.

In the almanacs it is usual to find a certain number
stated as the Julian Period. Thus, for example, 1879 is
the 6592nd year of the Julian Period. This cycle arises
from the three numbers 19, 28, 15, which represent the
entire periods of the Cycle of Meton, the Solar Cycle, and
the Roman Indiction respectively. It appears that in a
period of $19 \times 28 \times 15 = 7980$ consecutive years there are
not two years which have the same golden number accompanied with the same solar cycle and the same Roman
Indiction. There is thus a new period, called the Julian
Period, consisting of 7980 years. The first year of this
period is 4713 B.C., which has been adopted because each
of the three other cycles had the value 1 on that year.
This period will continue till the year A.D. 3267.

§ **92.** *Eclipses.*—It occasionally happens that the usually
circular disc of the sun is more or less obscured by the

passage of the moon between the earth and the sun. In some cases the entire light from the sun is cut off, this constituting what is known as a *total eclipse of the sun.* More frequently, however, only a portion of the sun is darkened, and the eclipse is then said to be *partial.* An eclipse visible at some places may not be visible at others, even though the sun be visible ; and an eclipse which is total at some places will only be seen as partial at others. These phenomena of course can only occur at new moon, when the longitude of the moon is identical with that of the sun. It does not, however, follow that there is always, or indeed even usually, an eclipse of the sun at new moon, for as the orbit of the moon is inclined to the ecliptic, the moon will generally at the time of new moon be over or under the sun, and will only be seen against the face of the sun when it happens that the occurrence of new moon is nearly coincident with the time of passage of the moon through one of its nodes.

So too it sometimes happens that the usually brilliant appearance of the moon at the time of full moon is modified by the appearance of a shadow with a circular edge, which passes slowly over the bright disc. It sometimes happens that only a portion of the moon's surface is thus obscured, and the eclipse is then said to be *partial.* Not unfrequently,

FIG. 104.

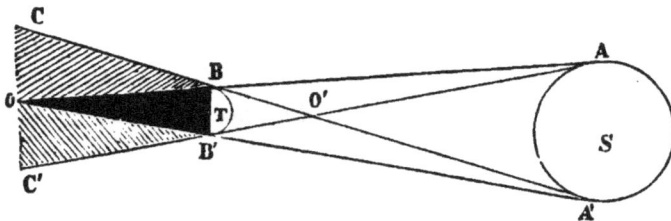

however, the shadow passes completely over the moon, and the eclipse is then said to be *total.*

The circumstances of an eclipse of the moon will be understood from Fig. 104. s represents the sun and T the earth.

A pair of common tangents, A B and A' B', drawn to the earth and the sun, intersect at the point o, while the second pair of common tangents intersect at o'. When the moon, in its revolution around the earth, enters the region c' B' o, it is partially obscured, and a partial eclipse of the moon will be witnessed from the earth when the moon begins to cross B' o. If the moon becomes immersed in the region B O B', then the light from the sun will be altogether intercepted, and a total eclipse of the moon will be seen.

It is a very curious circumstance that even when the moon is plunged entirely in the earth's shadow it still sometimes remains visible, with a peculiar copper-coloured hue, sufficiently bright on some occasions to enable the spots on the surface to be recognised. This is due to the effect of the atmosphere surrounding the earth in refracting the sun's light, and bending it into the cone forming the shadow.

FIG. 105.

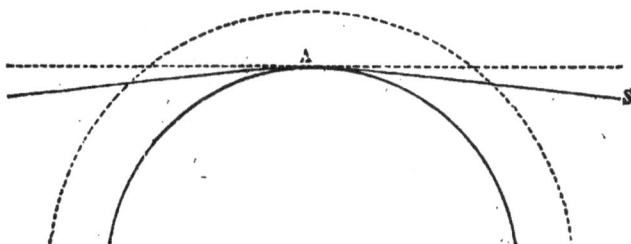

This is shown in Fig. 105, where A represents a point on the surface of the earth, while the dotted circular line is the limit of the atmosphere. If the atmosphere were absent, then a ray of light which just grazed the earth at A would be directed along the straight line which just touches the earth at A. Owing, however, to the presence of the atmosphere, the light is bent so that its direction forms a part of a curve of which the concavity is turned towards the earth. In this way the rays from the sun which pass near enough to the earth to be refracted by its atmosphere are bent into the cone which forms the shadow, and thus illuminate the moon in the manner already described. Nor will it be difficult to explain why the moon under these circumstances exhibits the peculiar copper-coloured light. The rays to which this illumination is due have passed through an enormous thickness of the terrestrial atmosphere, and the absorption of light by the atmosphere tends to render the light which has passed through it of a ruddy colour.

§ 93. *Prediction of Eclipses.*—An eclipse of the moon takes place when the moon, at the time of full moon, is near the node of its orbit on the ecliptic. Let us, for example, suppose that the time of full moon happened to be exactly coincident with the passage of the moon through its node. The position of the node must then be 180° distant from that of the sun. In the period of 18 years 11 days we find that 19 complete revolutions of the sun will have taken place with respect to the moon's nodes. It therefore follows that in 18 years and 11 days after the date of the eclipse the node of the moon will again be 180° distant from the sun. But in very nearly the same period 223 lunations will have been completed. It therefore follows that in 18 years and 11 days after the moon has been full at the node the moon will again be full at the node. An eclipse may occur, even though the time of full moon does not exactly coincide with the time of passing through the node, provided that the moon is sufficiently near the node at the time of full. The same circumstances will recur again at an interval of 18 years and 11 days, and therefore we shall find in general that 18 years and 11 days after the occurrence of a lunar eclipse there will be another lunar eclipse. If, therefore, we know all the eclipses which have occurred in a period of 18 years and 11 days, we are then able to predict future eclipses with considerable accuracy. For example, in the year 1862 a total eclipse of the moon, visible at Greenwich, occurred on June 11. Eighteen years and 11 days from that time bring us to June 22, 1880, and accordingly on that date again there is a total eclipse of the moon. So also the eclipse of the moon on December 5, 1862, is followed at the same interval of 18 years and 11 days by an eclipse which occurs on December 16, 1880.

It should, however, be observed that the numerical relation between the Synodic revolution and the period of the lunation is only approximate. We cannot, therefore, employ this method of prediction with infallible accuracy. It may sometimes happen that a small partial eclipse is not followed by an eclipse in 18 years and 11 days; and also that a partial eclipse may occasionally occur, though no eclipse took place 18 years and 11 days previously. For accurate prediction of the occurrence of eclipses at remote epochs, as well as for an accurate account of the details of eclipses as to the time of commencement, and the duration, with such other particulars as are given in the Nautical Almanac each year, careful calculations have to be made. Such calculations depend upon our knowledge of the motions of the moon, derived from long-continued observations.

§ 94. *Eclipses of the Sun.*—We have already stated that eclipses of the sun are produced at the time of New Moon by the interposition of the dark body of the moon between the earth and the sun. According to the varying circumstances under which the occurrence happens we have a corresponding variety in the character of the eclipse. If the moon should cross the sun in such a manner that its centre passes over or very close to the centre of the sun, then the eclipse assumes one or other of two very remarkable forms. It will sometimes happen that the entire surface of the sun is obscured, in which case the eclipse is said to be total. This phenomenon is, however, a very rare one, and the apparent diameter of the moon is but very little larger than that of the sun, so that the duration of the total eclipse is very short. In fact, the movement of the moon will, in a very few minutes, enable the margin of the sun to be seen again. Sometimes, however, even though the centre

of the moon passes very close to the centre of the sun, the
eclipse will not be total, but a ring of the sun's disk is visible
around the dark edge of the moon. In this case the eclipse
is said to be *annular.* It may seem at first strange that the
moon, passing centrally across the sun, should sometimes
produce a total eclipse and sometimes an annular eclipse.
This arises from the fact that the apparent angular diameter
of the moon is sometimes greater and sometimes less than
the apparent angular diameter of the sun. It will be remembered that the orbits both of the sun and of the moon
are eccentric, and that consequently the distances of these
bodies from the earth are constantly fluctuating within
certain limits. As the angular diameter of an object varies
inversely with its distance, it of course follows that the angular
diameter of the moon and that of the sun are also constantly
fluctuating between certain limits. It happens, curiously
enough, that the mean angular diameter of the moon is nearly
identical with the mean angular diameter of the sun. The
fluctuations arising from the eccentricities of the orbits are,
no doubt, small, but they are sufficiently large to make the
moon sometimes appear larger than the sun, and sometimes
smaller than it. In the former case the central passage of
the moon across the sun produces a total eclipse ; in the
latter case it only produces an annular eclipse.

§ 95. *Frequency of Eclipses.*—Eclipses of the moon are not
so frequent as eclipses of the sun. This will be manifest

FIG. 106.

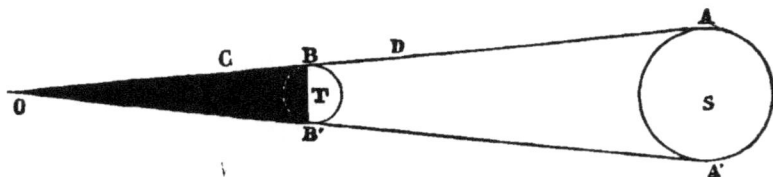

from the consideration of Fig. 106. s represents the sun,
and T is the earth ; the tangents A B and A' B' intersect at o.

An eclipse of the moon will occur whenever the moon enters the cone of the earth's shadow B O B'; an eclipse of the sun will occur whenever the moon enters the portion of the cone which lies between the earth and the sun. The latter portion of the cone has a much larger section than the former : consequently in its revolutions around the earth the moon will more frequently enter the portion of the cone between the earth and the sun than the portion which is formed by the shadow of the earth. For this reason the solar eclipses are more frequent than lunar eclipses. It must not, however, be supposed that at any given place on the earth more solar eclipses will be seen than lunar eclipses. An eclipse of the moon is visible to the inhabitants of a whole hemisphere ; an eclipse of the sun, on the other hand, is only visible to the inhabitants of a portion, and often only a small portion, of a hemisphere. From this it follows that, notwithstanding there are more solar eclipses than lunar eclipses, yet at any given locality more lunar eclipses can actually be seen than solar eclipses. This may be shown by the consideration that the diameter of the earth's shadow at the distance of the moon is greater than the apparent diameter of the sun. *At a given place*, therefore, it will be more usual for the moon to enter into the shadow than for the moon to cross the sun, i.e. it will be more usual for an eclipse of the moon to take place than for an eclipse of the sun.

§ **96**. *Prediction of Solar Eclipses.*—Although it generally happens that a solar eclipse is followed by a solar eclipse in the period of 18 years and 11 days, yet we cannot predict the recurrence of solar eclipses in so simple a manner as we can the recurrence of lunar eclipses. It is not possible by the mere use of this period to say whether the solar eclipse will be visible at a given place, or what the magnitude of the eclipse may be. The sun's eclipses and all their circumstances can, no doubt, be predicted with great accuracy ; but then, such predictions are the result of considerable

calculations of a more intricate character than are required
for predicting the details of the eclipses of the moon.   The
parallax of the moon being so large, the character of the
solar eclipse will depend not merely upon the locality in
which the observer is stationed, but also upon the altitude
of the moon at the time.   As the altitude varies during the
progress of the eclipse, so the parallactic displacement of the
moon changes ; and this displacement must be allowed for,
as well as the actual motion of the moon, when computing the duration and the other circumstances attending
the eclipse.

§ **97.** *Occultations of Fixed Stars.*—As the moon, by passing between the earth and the sun, produces an eclipse of the
sun, so it sometimes happens that the moon passes between
the earth and a star or a planet, and thus temporarily
eclipses them from view.   This phenomenon is known as
an *occultation.*   On account of the parallax of the moon, the
position of the moon on the surface of the heavens, as compared with the fixed stars, varies with the position of the
observer on the earth.   It follows that the occultation of a
star may be seen in one place, while in another the star will
not be occulted, though the moon will pass close thereto.
To predict the occurrence of an occultation the situation of
the observer must be taken into account.   The calculations
necessary are very similar to those which are required for
the prediction of an eclipse of the sun.

The phenomenon of an occultation is peculiarly striking
when the disappearance of the star takes place at the dark
limb of the moon.   In the interval between new moon and
full moon the motion of the moon is such that the dark
limb is on the advancing side.   If, therefore, the path of the
moon happens to cross any star bright enough to be conspicuous, notwithstanding the brilliancy of the moon, then
the star will be instantaneously obscured when the dark edge
of the moon crosses the line joining the eye and the star.
After some time, during which the star remains hidden by

the interposition of the body of the moon, it will reappear
again on the illuminated side.   Between full moon and new
moon the illuminated side of the moon is advancing, and
consequently the disappearance of the star at occultation
takes place at the illuminated limb, while the reappearance
occurs at the dark limb.

As the stars are practically at an infinite distance, we
may consider the rays from them to be parallel.   If, therefore, a circular cylinder be drawn just enveloping the
moon, while the axis of the cylinder is parallel to the rays
from the star, then the star will be occulted by the moon
when the observer is situated in the interior of this cylinder.
The area intercepted on the earth by the cylinder is
therefore the region in which the star will appear to be
occulted.   The actual area is, of course, equal to the
section of the cylinder at the moon itself, i.e. to the section
of the moon by a plane passing through its centre.   As the
area of a section of the moon is but a small fraction of the
entire surface of the earth, it follows that at a given moment
a star is only occulted from observers situated on a comparatively small area on the earth.

§ 98. *Determination of Longitudes by the Moon.*—To determine the difference in longitude between two given stations
on the earth, we require to have the means of comparing
the time at one place with the time shown simultaneously at
the other.   When, for example, it is said that the longitude
of Dunsink is $25^m 21^s$ west of Greenwich, the statement may
be interpreted in two different ways, both of which are
correct.   It may mean that an interval of $25^m 21^s$ of *sidereal*
time will elapse between the transit of a *fixed star* across the
meridian of Greenwich and the transit of the same star
across the meridian of Dunsink ; or it may equally mean
that an interval of $25^m 21^s$ of mean solar time will elapse
between the transit of the mean sun across the meridian of
Greenwich and the transit of the mean sun across the
meridian of Dunsink.   The actual angle formed at the pole

by the meridian passing through Greenwich and Dunsink bears to 360° the same proportion which 25^m 21^s bears to 24 hours. This angle is 6° 20' 15''.

The sidereal clock in the observatory at Dunsink should therefore be always 25^m 21^s slower than the sidereal clock at Greenwich, when the two clocks are correct. So, too, the mean time clock at Dunsink should also be slower than the mean time clock at Greenwich by the same amount. To determine, therefore, the difference between the longitude of Dunsink and that of Greenwich, it is only necessary to have the two clocks at each place correct, and to note their difference. The difficulty in the process is entirely due to the distance between the two stations, which makes the accurate comparison between the two clocks a matter of difficulty. The method formerly adopted was to compare chronometers with the clock at one station and then to carry these chronometers to the second station. This method has been superseded in the case where telegraphic communication exists between the two stations. As the telegraphic signal is practically instantaneous, it can be arranged that at a given signal from one station to the other the time shown by the clocks is to be noted ; and if the clocks are correct, or, what comes to the same thing, it their errors are known, the difference between the two clocks, and, therefore, the difference in longitude, is ascertained. This method is by far the most accurate, and it is always employed when great precision is required, if the telegraphic communication exists.

To determine the longitude at sea or under other circumstances when either of the methods we have suggested are inapplicable, resort may be had to the *method of lunar distances*, now about to be described. It is to be remembered that when we know the time at the station where we are situated, it is only necessary to know the simultaneous time at Greenwich in order to know the longitude. If, therefore, there were a clock in the heavens, the hands of which always indicated true Greenwich time, and if this clock were visible from all parts of the earth, then the determination of longitude would be a comparatively simple matter. The motions of the moon on the surface of the heavens do actually provide us with such a clock. The fixed stars are the numbers on the clock, while the moon is the hand by which the time is indicated. To interpret this clock, and deduce from it the Greenwich time, use is made of the tables of the moon, by which the place of the moon can be predicted with considerable accuracy for many years in the future. From these tables we can deduce the angular distances at which the moon will be separated from the bright stars which lie along its path at stated intervals. These distances are recorded in the Nautical Almanac for intervals of three hours daily. For example, on June 30, 1879, the distance of the moon from the star Regulus at noon, Greenwich time, is 83° 54′ 18″; at III. hours the distance is 85° 43′ 45″; at VI. hours it is 87° 33′ 17″, and at IX. hours 89° 22′ 52″. If, therefore, the actual distance of the moon from the star Regulus be measured, and if the distance coincides with any of those above specified, then the corresponding Greenwich time is at once discovered. It will, of course, generally happen that the observed distance will not coincide with any of the actually recorded distances. A simple interpolation will, however, enable the corresponding mean time to be determined.

In the application of this method it is necessary to allow for the effect of parallax in deranging the position of the moon, while the effect of refraction both on the moon and the star must be taken account of also.

The mode of doing this is shown in the adjoining figure. Let o represent the position of the observer, while z is his zenith, E is the observed position of the star, and L is the observed position of the moon. With the sextant the observer measures the apparent angular distance between E and L—that is, the angle E o L. He also at the same time measures the zenith distances, z E and z L. The effect of refraction upon the star is to make the star appear at E, while, if refraction were absent, the star would really be found at E'. In the same way, if the apparent place of the moon is at L, its true place is at L' so far as refraction is concerned. If, however, the observer were stationed at the centre of the earth, the moon would be thrown up by parallax from L' towards the zenith, and would therefore be found in the position L". The corrected distance, which is to be employed, is therefore E' L". This can be calculated from the spherical triangle Z E' L". For since the distances Z E, Z L, and also the distance E L, are determined by ob-

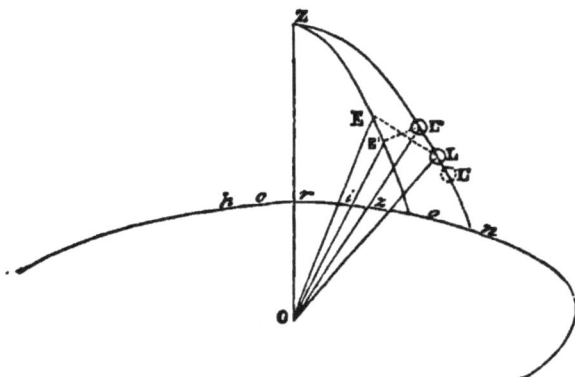

servation, the angle E Z L is known. Also, since the effects of refraction and parallax on the zenith distances can be computed and allowed for, the corrected zenith distances Z E' and Z L" may be regarded as known. In the spherical triangle E' Z L" we therefore know the two sides and the angle included at Z ; the base E' L" is therefore also known. The accuracy of the method of lunar distances as we have here described it is, of course, affected by any irregularities in the tables of the moon's motion, as well as by the actual errors which may arise in the observations.

The occultation of a bright star by the moon affords a method by which the longitude may be determined with

very considerable accuracy. If the time of the occultation be noted at each of two places, and if the suitable allowance be made for the effect of parallax, then a comparison of the observations will show the difference between the local times at the moment of occultation as seen from the earth's centre. This difference is, of course, the longitude. This method possesses the advantage of being but little affected by the errors of the tables of the moon. The occultation of a sufficiently bright star is, however, a comparatively rare occurrence.

CHAPTER VII.

THE PLANETS.

§ 99. *Determination of the Planets.*—The planets of our system which were known to the ancients are Mercury, Venus, Mars, Jupiter, and Saturn. These objects, in a superficial view, resemble somewhat brilliant stars, but a little attention is sufficient to point out that they are of a very different character. If we note the position of one of these bodies with reference to stars which lie in its vicinity, and if we repeat the observation after the interval of a few days, it will be found that, though the stars have retained their relative positions, the position of the planet with respect to the stars has altered. This remarkable feature of the planets attracted attention in the earliest times.

To distinguish a planet we may make use of a celestial globe. If, in comparing the constellations marked on the globe with the constellations in the heavens, a conspicuous starlike object is seen in the heavens which is not to be

found on the maps, we may conclude with considerable
certainty that the object is one of the five planets Mercury,
Venus, Mars, Jupiter, or Saturn. If we wish to solve the
common problem of identifying a given planet in the sky,
we must then make use of the Nautical Almanac, or some
similar work, in which the right ascensions and declinations
of the planets are recorded daily or at frequent intervals
throughout the year. Thus, for example, if we wish to find
out the planet Jupiter on June 23, 1879, we turn to the
Nautical Almanac for 1879, and on p. 253 we see that the
right ascension of Jupiter is 22^h 57^m $33^s \cdot 59$, while its south
declination is $7° 53' 13'' \cdot 1$. With the aid of a telescope
mounted equatorially and provided with graduated circles,
it is only necessary to set the telescope to the required
declination, and then turn the instrument round the polar
axis until the hour angle is equal to the difference between
the sidereal time at the moment and the right ascension of
Jupiter. The planet will then be seen in the field of view.
But a telescope is not necessary when the object is merely
to identify the planet and distinguish it from the stars. By
referring to a celestial map or globe it will be easy, from the
known right ascension and declination, to determine the
position of Jupiter on the celestial sphere. It will be found
to lie in the constellation Aquarius at the date under consideration. If, therefore, the observer looks at that part of
the heavens where Aquarius is situated, he will at once
recognise Jupiter as the very brilliant object there situated.
This he will confirm by noticing that there is no star marked
on the globe in the position in which he sees Jupiter. It
will be of great interest for the beginner, having once recognised Jupiter by either of the methods we have described, to
note the changes in position which it undergoes with refe-·
rence to the stars in its vicinity. If he be provided with a
telescope, then from one night to the next, or even in the
course of a single night, he will be able to detect the movements of Jupiter by carefully comparing his position with

the stars in the neighbourhood. Even without the aid of a telescope careful observation by alignment with the stars will in a few weeks reveal the motion of the planet.

All the movements of the planets which are visible to the unaided eye, including Mercury, Venus, Mars, Jupiter, and Saturn, are performed in orbits which never deviate much from the plane of the ecliptic. It follows that these planets are always seen in a portion of the heavens near the ecliptic. This consideration considerably simplifies the operation of finding out the planets. For example, to take the case of Jupiter in June 1879. Suppose that it is known that this planet crosses the meridian about 23 hours of sidereal time. The place of the planet in the heavens can then be easily found by the globe. On the great circle of the globe which represents the equator the hours of right ascension are marked. Draw from the pole a great circle to the point marked XXIII.; produce this on to cut the ecliptic: then Jupiter must be very close to this intersection, because it is known to be close to the ecliptic, and it is also known to have a right ascension of 23^h. This method will be quite sufficient for the purpose of identifying a planet.

§ 100. *The Zodiac.*—As the planets which are now under consideration never depart very widely from the ecliptic, it is possible to mark out a certain zone in the heavens, on the interior of which the planets are always to be found. This zone extends to 8° on each side of the ecliptic, and it is termed the zodiac. If we divide the ecliptic into twelve equal portions, and then draw through each of the points of division great circles perpendicular to the ecliptic, these great circles will subdivide the zodiac into twelve equal portions. Each of these portions is represented by a certain name, which is also borne by the constellation which each portion contains.

§ 101. *Kepler's Laws.*—By the aid of the meridian circle it is possible to determine a series of positions of a planet at different times, and thus to mark these positions on the celestial sphere with precision. When this is done, it is found that although the *general features* of the motions of the planets are consistent with the supposition that the orbits are perfect circles, yet that when a more minute comparison is instituted between the results of this supposition and the results of actual observation, certain discrepancies are brought to light which are too large and too systematic to admit of being explained away as merely errors of observation.

The first of the three discoveries which bear the name of Kepler may be thus enumerated :—

The path of a planet round the sun is an ellipse, in one focus of which the centre of the sun is situated.

Fig. 108.

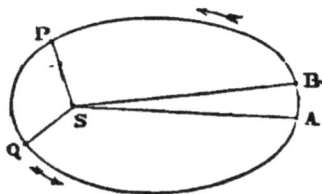

Thus, let s (Fig. 108) represent the centre of the sun ; then the ellipse A B P Q denotes the path of the planet. In none of the principal planets is the deviation from a circle so great as it is represented in the figure, which has been designedly exaggerated.

It is approximately correct to say that the planets move with uniform velocity in these orbits round the sun. When, however, this question comes to be tested by careful observations, it is found that the movements are not quite uniform. The planet is found to be moving more rapidly at some parts of its path than it is at others. The law by which these variations in the velocity of a planet are controlled was also discovered by Kepler, and is expressed by his second law, which is thus stated :—

In the motion of a planet round the sun, the radius vector, drawn from the centre of the sun to the planet, sweeps over equal areas in equal times.

Thus, for example, in Fig 108, when the planet moves from A to B, its radius vector sweeps out the area A S B, and in moving from P to Q the radius vector sweeps out the area P S Q. Kepler's second law asserts that if the area A S B be equal to the area P S Q, then the time taken by the planet in moving from A to B is equal to the time taken by the planet in moving from P to Q.

It is easy to show how, when the planet moves in obedience to this law, the observed changes in the velocity of the planet can be completely accounted for. Since the planet must move from P to Q in the same time as it takes to move from A to B, and since the distance P Q is very much larger than the distance A B, it follows that the velocity of the planet must be greater when it is moving through P Q than when it is moving through A B. We hence see that the planet must be moving with greater velocity according as its distance from the sun is less.

In the two laws of Kepler, which have been already discussed, we have only been considering the motion of *one* planet. We have now to consider the very remarkable law, also discovered by Kepler, which relates to a comparison between *two* planets. This law, known as Kepler's third law, is thus stated :—

The squares of the periodic times of two planets have the same ratio as the cubes of their mean distances from the sun.

To explain this, it must first be observed that by the mean distance of the planet from the sun is to be understood a length which is equal to the semiaxis major of the ellipse, while the periodic time is understood to be the actual number of days and fractions of a day, in which the planet, as seen from the sun, makes a complete revolution round the heavens.

To illustrate the application of Kepler's third law we may take the cases of Venus and the Earth. The periodic times of Venus and the Earth are respectively 365·3 days and 224·7 days, while the mean distance of Venus is 0·7233 if we take the mean distance of the earth as unity. We have for the square of the ratio of the periodic times

$$\left(\frac{365\cdot3}{224\cdot7}\right)^2 = 2\cdot643,$$

while for the cube of the ratio of the mean distances

$$\left(\frac{1\cdot 0000}{0\cdot 7233}\right)^3 = 2\cdot 643,$$

which verifies the law.

Kepler's three laws are found to be borne out completely, even to their minutest details, when proper allowance has been made for every disturbing element.

§ 102. *The Planet Venus.*—With the exception of Mercury, and possibly also of some one or more other planets which may be still closer to the sun, Venus is the nearest of all the planets to the great centre of our system. The actual path of Venus differs but little from a circle of which the sun is the centre, and of which the radius is 67,500,000 miles. Owing, however, to the ellipticity of the orbit, the exact distance fluctuates a little on either side of its average value, but it is never less than 67,000,000 miles, or more than 68,000,000 miles. The planet Venus, therefore, describes around the sun an orbit of about 135,000,000 miles in diameter. This, though no doubt an orbit of majestic proportions, is still considerably smaller than the orbit which is described by the earth, and it is very much smaller than the paths described by the more distant planets. Round her path Venus sweeps in a period of 224 days 15 hours, and though this may seem a considerable time for the journey round the sun to be accomplished, yet the length of that journey is such that the planet has to move at an average rate of twenty-two miles per second, in order to accomplish it in the time. The actual velocity with which Venus moves is not absolutely uniform, but it is never greater than twenty-two and a half miles per second, or less than twenty-one and a half miles. It is instructive to contrast the velocity of Venus with the velocity of the earth, for as the latter is eighteen and a half miles per second, we have an illustration of the general truth that those planets, which are near the sun, move more quickly than those which are farther off.

In order to consider the circumstances under which Venus appears alternately as a morning or evening star, it is necessary to observe that the relative positions of the sun and of Venus, as seen from the earth, are the same as if the earth and the sun remained fixed, and if Venus turned around the sun in a period of 584 days. We have thus a cycle of phenomena which run through their course in a period of 584 days and then begin again. We shall briefly describe the course of the changes in the appearance of Venus in one of these cycles, and we shall take for example the cycle which commenced on May 6, 1877. Venus was then in what is called superior conjunction with the sun ; in other words, Venus was so placed that the sun lay between the planet and the earth. It would not, however, be correct to say that Venus was actually behind the sun, for although this sometimes happens, yet its occurrence is very rare, because of the inclination of the orbit of Venus to the ecliptic. At the date we have mentioned Venus was south of the sun by a distance equal to more than three of the diameters of the sun. (We refer, of course, to the apparent diameters projected on the surface of the celestial sphere.) It is obvious that when Venus is in superior conjunction, very nearly the entire illuminated surface is turned towards the earth, but the distance from the earth to the planet is then so great, that the brilliancy does not amount to the fourth part of what it will ultimately become.

After passing through the position of superior conjunction, Venus begins to move to the east of the sun, and thus to set after the sun. It was thus that late in the autumn of 1877 Venus began to be seen in the west after sunset, gradually increasing in brilliancy as well as in the apparent distance by which it was separated from the sun. The successive changes in the situation of Venus at this time are shown in Fig. 109, which exhibits the actual position of the apparent orbit of Venus about the sun during the cycle referred to. For convenience the sun is represented on the horizon at the moment of setting. The angular distance between the sun and Venus gradually increased until it attained a maximum on December 11, 1877, the planet being then at its greatest elongation east. The angular distance between the sun and Venus was then 47°. We thus come to the first critical epoch of the cycle, 219 days after its commencement. The brilliancy of the planet has also been increasing, and at the time of its greatest elongation the brilliancy has attained three quarters of its greatest value. If looked at through a telescope during its movement, the phases of the planet can be seen exactly resembling in miniature the

FIG. 109.

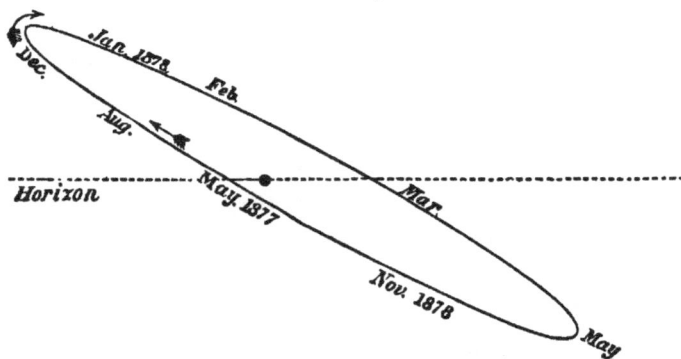

phases of the moon. Thus in superior conjunction the planet appeared full, but gradually changed its form, until at greatest elongation the planet resembled the moon at first quarter. After passing the point of greatest elongation, Venus commences to assume a crescent form, but its brilliancy is still increasing, because the continual diminution of the distance between the earth and the planet more than compensates for the gradual diminution of the portion of the illuminated hemisphere which is turned towards the earth. Each day the shape of the crescent becomes more slender, but each day the apparent diameter of the crescent measured from one horn to the other gradually increases, so that the apparent area of the crescent on which its brightness depends gradually increased until January 16, 1879. On this date, or 256 days after the commencement of the cycle, the brilliancy of the planet attained its maximum. It is thus on one occasion during each cycle, or on an average once every 584 days, that Venus attains its greatest brilliancy as an evening star. The next occasion on which Venus is brightest in the evening occurred on August 18, 1879: the exact interval between these dates is 579 days. The difference of 5 days between this period and that of the average lengths of the cycle, i.e. 584 days, is principally due to the circumstance that the orbit of Venus, though nearly a circle, is still slightly different therefrom. The average of a large number of intervals between two successive recurrences of Venus at its brightest in the evening will be found to equal 584 days, and the differences will never amount to more than a few days on one side or the other of its average value.

After the position of greatest brilliancy has been passed, the crescent formed by Venus becomes more and more slender, though the circle of which the crescent is a part has a diameter which is gradually increasing. The planet is now drawing in again towards the sun. In fact, the elongation or the angular distance between the planet and the sun, which at the time of greatest elongation was equal to 47°, has already diminished to 40° at the time of greatest brilliancy. As the planet approaches the sun, the crescent becomes gradually attenuated to a thin line of light, still a beautiful object in the telescope, though of course, as the planet is now so close to the sun and sets very soon after it, it will become necessary to search for the planet in the day time. On February 20 Venus had approached as nearly as possible between the earth and the sun into the position called inferior conjunction. This occurs 292 days after superior conjunction, and thus half the cycle of 584 days has been accomplished.

S

Owing, however, to the inclination of the orbit of Venus to the plane of the ecliptic, the planet did not actually come between the earth and the sun at this conjunction. After passing through inferior conjunction, Venus moves towards the west of the sun, and, therefore, rises before the sun, and is called a morning star. The actual orbit which Venus describes, relatively to the sun at sunrise, is shown in Fig. 110. From this figure it will be seen that Venus then performed the second half of the cycle of changes in a similar manner to the first half, only with the order reversed.

FIG. 110.

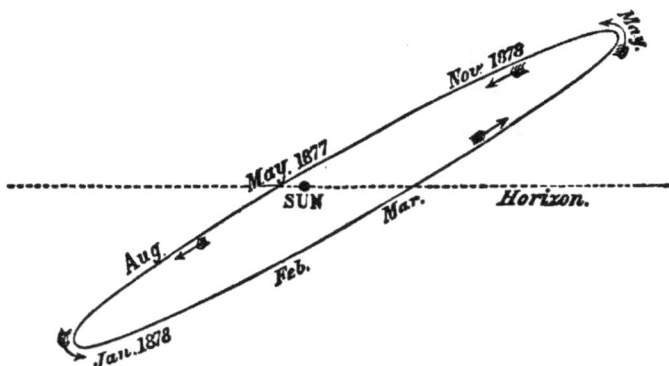

The planet then moved further from the sun, increasing in brilliancy at the same time, until March 29, when, after an interval of 329 days from the commencement of the cycle of changes, the position of greatest brilliancy is again attained. Thus in each cycle of changes we have two epochs of greatest brilliancy, and the interval between these epochs averages 73 days. On the first of these occasions Venus is always an evening star, and on the second a morning star. The angular distance from the sun gradually increases until May 1, when the elongation is 46°. This is the last remarkable phase in the cycle of changes, and occurs on the 365th day. During the remaining 219 days Venus is moving slowly through the remote part of its orbit back to the

position of inferior conjunction, from which we have supposed
it to start.

When Venus is in the neighbourhood of superior conjunction—that is, when it is most remote from the earth—its
apparent diameter is about ten seconds of arc, but when the
planet is in the neighbourhood of inferior conjunction it
becomes a much more imposing object ; for, owing to our
diminished distance, the apparent diameter is augmented to
about sixty seconds. It will thus be readily understood
that the best time to see Venus during each cycle of 584
days is during the period of about four months, when the
planet is passing from the greatest elongation east to the
greatest elongation west.

FIG. III.

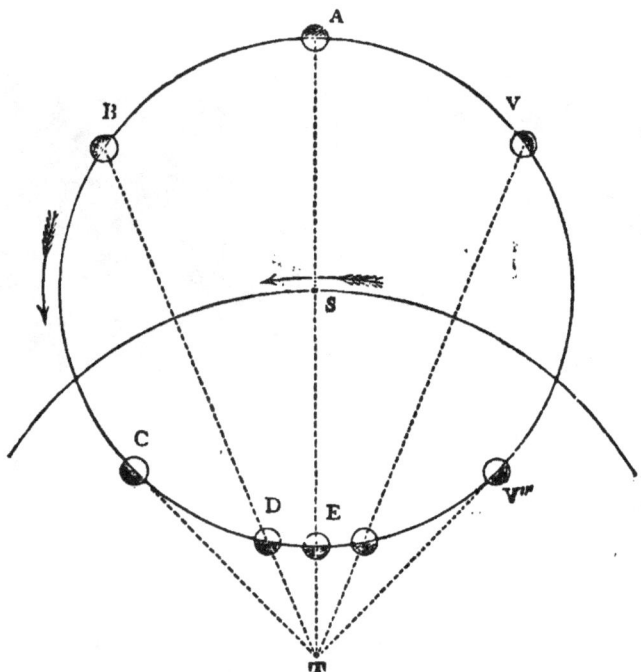

It is easy to show how all these changes in the appearance of Venus can be explained by the actual motion of
Venus around the sun. Let T (Fig. 111) be the earth, and

let s be the sun, while v denotes Venus. Then the circle described about the sun's centre represents the orbit of Venus. When Venus is at the position A, its entire illuminated hemisphere is turned towards the earth, but its distance is so great that the apparent angular diameter is very small, and the actual size of Venus is represented at A (Fig. 112). When Venus moves to B, a portion of the dark hemisphere is directed towards the earth, and, consequently, the planet appears as represented in B (Fig. 112), the apparent diameter of the planet having increased, owing to the diminished distance from the earth. When Venus arrives at c it is in the position of greatest elongation ; the planet then appears with half its disk illuminated, as shown in c (Fig. 112).

FIG. 112.

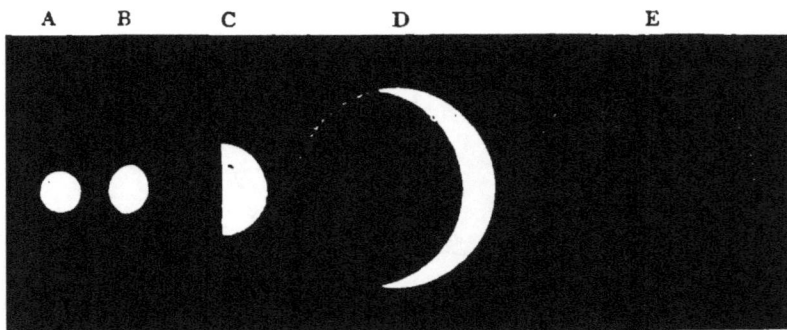

After passing this position, Venus moves on till, in the position D, when it has come comparatively close to the earth, it assumes the crescent form shown in D (Fig. 112). It may be observed that two points, D and B, have been chosen, which lie in the same straight line through T. It follows that the angular distance between the sun and Venus in either of the positions B and D are equal ; yet how different is the actual appearance of the planet in the two cases ; in the one it is shown in B, and in the other in D (Fig. 112). After leaving D the planet moves on to E, where almost the entire dark surface of the planet is directed towards the earth.

The actual size of the disk of Venus in this case is represented at E (Fig. 112).

From the observation of certain markings upon the surface of Venus it has been shown that this planet revolves upon its axis in the same direction as the planet revolves around the sun. Schröter has found that the planet makes a complete revolution in $23^h \, 21^m \, 8^s$, and he has estimated the angle between the plane of its equator and the plane of its orbit at 72°. It appears that the duration of a day on the planet Venus is very nearly equal to a day on the earth, though the year of Venus is but 225 of our days. Venus appears to be surrounded by an atmosphere the existence of which is rendered manifest by a twilight analogous to the same phenomenon on the earth. The hemisphere of the planet which is turned away from the sun is not entirely dark, but it is illuminated for a short distance within the boundary by light which is reflected from the atmosphere. The density of this atmosphere appears to be comparable with that surrounding the earth ; for it would appear from the observations of Mädler that the horizontal refraction at Venus is 44′, while in our atmosphere the horizontal refraction is 33′.

The line of separation between the shadow and the illuminated portion of Venus is irregularly notched and indented, and the cusps or horns of the crescent are frequently seen truncated. This appears to be due to the mountains and asperities upon the surface of Venus. From the measurement of these irregularities it is possible to form some estimate of the altitudes of the mountains of Venus, and it would appear that the mountains there must be considerably more lofty than the terrestrial mountains with which we are acquainted. The highest peaks in the Himalaya only rise above the sea-level to a height which is the 740th part of the radius of the earth, while the mountains on Venus, so far as these observations may be relied upon, ascend to a height which is the 144th part of the radius of the planet.

On those rare occasions in which a transit of Venus takes place, the planet is seen to form a black circular disc projected against the surface of the sun. The measures which have been taken of this disc do not indicate the existence of any ellipticity. It is, however, not impossible that the globe of Venus may have an ellipticity comparable with that of the earth; for even the ellipticity of the earth itself would not be recognisable if it were viewed from the distance at which Venus is separated from the earth.

The apparent diameter of Venus, or rather the angle which its diameter subtends at the eye, varies of course with its distance. According to Tennant (1875) the diameter of Venus, when at a distance equal to that of the earth from the sun, is 16″·9. From this we conclude that the radius of Venus is nearly equal to the radius of the earth.

§ **103.** *Mercury.*—The movements of Mercury are to a certain extent analogous to those of Venus. The planet oscillates from one side of the sun to the other, as Venus does, and is thus occasionally to be seen as a morning star, and occasionally as an evening star. The movements of Mercury are, however, not nearly so regular as those of Venus, owing to the much greater eccentricity of the orbit of Mercury.

Though Mercury is sometimes very easily visible with the unaided eye, it is not very often seen, on account of the planet being always in the proximity of the sun. Mercury is also usually too near the horizon, when visible at all, to be seen to much advantage. It is, therefore, not surprising to find that many persons who are familiar with the four greater planets of our system—viz. Venus, Mars, Jupiter, and Saturn—have never yet seen the planet, which, so far as we certainly know at present, is the nearest of all to the sun.

The planet Mercury has a diameter of about 3,000 miles. It is therefore much smaller than the earth, of which the diameter is about 8,000 miles. The only one of the principal planets which at all approaches Mercury in bulk is Mars, of which the diameter is 4,000 miles. The diameter of Jupiter is more than thirty times as great as the diameter of Mercury, yet the brilliancy of Mercury on some occasions (e.g. February 1868) rivals that of Jupiter in splendour. This arises from two causes. In the first place Mercury is only distant from the sun by about $\frac{1}{14}$th part of the distance by which Jupiter is separated from the sun. It follows that the brilliancy with which the surface of Mercury is illuminated by the sun's rays must be much greater than the brilliancy of the illumination of Jupiter. When this is combined with the circumstance that the earth is much nearer to Mercury than it is to Jupiter, it will be understood how, notwithstanding the small size of Mercury, it may sometimes be nearly, if not quite, as splendid as Jupiter.

The eccentricity of the orbit of Mercury is much more considerable than that of any other of the principal planets of our system. In fact, the greatest and the least distances of Mercury from the sun are respectively 43,000,000 miles and 28,000,000 miles, or very nearly in the ratio of 3 to 2. Mercury accomplishes one complete revolution around the sun in about 88 days. Owing to the revolution of the earth around the sun in 365 days, the apparent time of one revolution of Mercury relatively to the earth is 116 days, and this is the average duration of the cycle, which includes a complete series of the changes in the relative positions of Mercury and the sun. Once in each period of 116 days Mercury appears at its greatest elongation to the east of the sun, and then in about 48 days afterwards the planet has moved to its greatest elongation west of the sun.

The actual angular value of the greatest elongation is not constant, for it depends upon the relative distances of the earth and Mercury from the sun. The distance of the earth is very nearly constant, but the distance of Mercury fluctuates between the wide limits already specified. Under

the most favourable circumstances, at the greatest distance
of Mercury from the sun, combined with the least distance
of the earth from the sun, the elongation might amount
to upwards of 28°. On the other hand, if Mercury happened to be at its least distance from the sun at the time
of greatest elongation, while at the same time the earth was
at its greatest distance, the elongation would only amount to
16° or 17°.

As, however, we pointed out in the case of Venus, it
does not follow that a planet is brightest at the time of
greatest elongation. No doubt, if a planet were situated at
the distance of 41,000,000 miles from the sun, it would
appear brightest at the time of the greatest elongation, and
the appearance of the planet in the telescope would then
resemble that of the moon at first quarter. If the distance
of the planet from the sun be greater or less than 41,000,000
miles, then the maximum brightness will not occur at the
time of greatest elongation. There is, however, an important distinction to be observed. If the distance of the planet
exceed 41,000,000 miles, as is always the case with Venus,
but rarely the case with Mercury, then the greatest brightness occurs after the greatest elongation east, or before the
greatest elongation west; in either case when the planet
appears somewhat crescent-shaped in the telescope. If the
distance of the planet from the sun at the time of greatest
elongation be less than 41,000,000 miles, as is generally the
case with Mercury, then the greatest brightness occurs a few
days before the greatest elongation east, or a few days after
the greatest elongation west ; in either case when the planet
shows more than half its disc illuminated.

Mercury, when visible at all, must be sought for low
down in the west shortly after sunset, or low in the east
shortly before sunrise, according as the planet is at its east
or west elongation. It is not usually a very striking object
in the telescope, but its phases, resembling those of a miniature moon, are readily seen. Many years ago the acute

observer Schröter detected minute irregularities in one of the horns of the crescent of Mercury. This was attributed by him to the presence of a mountainous region on the planet, and the heights of these mountains were roughly estimated to be nearly eleven miles. The same observer detected traces of an atmosphere surrounding the planet, and from various appearances it was inferred that the planet rotated on its axis in a period of about 23^h 0^m 53^s.

§ 104. *The Superior Planets.*—The *inferior* planets, Mercury and Venus, are, as we have seen, always to be found in the neighbourhood of the sun. The case is very different with the planets Mars, Jupiter, and Saturn, now to be considered. These planets are, it is true, always found in or near the ecliptic, but so far from being limited to that part of the ecliptic which is near the sun, they may occupy any position with respect to the sun, and are sometimes even 180° distant from the sun on the opposite side of the circle.

§ 105. *The Planet Mars.*—The nearest of the superior planets to the earth is Mars, and we shall therefore commence the study of the exterior planets by an examination of his movements. Let us imagine for a moment that the earth and Mars are each revolving around the sun in circles of which the sun occupies the centre. The motion of the earth is of course completed in 365 days, but the motion of Mars requires the longer period of 687 days. It follows that, as the earth moves more rapidly than Mars, it must occasionally overtake him, and pass between him and the sun. This phenomenon is known as an opposition of Mars, and it occurs once in every 780 days. When this is the case, Mars is 180° distant from the sun, and he consequently comes on the meridian at midnight. A moment's consideration will also show that when Mars is in opposition he is nearer to the earth than he is when he occupies any other position. If we take the mean distance of the earth from the sun to be 92,000,000 miles, then the mean distance of

Mars from the sun is 140,000,000 miles. It is therefore plain that when Mars is in opposition, the distance by which he is separated from the earth is 48,000,000 miles. This, therefore, is the distance at which we would see Mars at every opposition ; and supposing the orbits of the earth and Mars were exact circles, then this distance would be the same at every opposition.

But the orbits are not exact circles. The orbit of the earth is an ellipse, which, indeed, has such a small eccentricity that it differs but little from a circle. But the orbit of Mars differs a good deal from a circle, and hence it follows that at some oppositions Mars is nearer to the earth than he is at others. To state the matter accurately, we find that though the average distance between the earth and the sun is 92,000,000 miles, yet that the real distance is sometimes as great as 93,500,000 miles, and sometimes as small as 90.500,000. Similarly it is shown by observations of Mars that the distance at which he is separated from the sun ranges between 153,000,000 miles and 127,000,000 miles. If it should happen that at the time of opposition Mars was at its greatest distance from the sun, and the earth at its least distance, then the actual distance between the earth and Mars at such an opposition could not be less than 62,500,000 miles. But if at the time of opposition Mars was found at its least distance from the sun, while the earth happened to be at its greatest distance, then the actual distance between the earth and Mars would be only 33,500,000 miles. We learn from these considerations that at the time of opposition the distance between Mars and the earth cannot possibly be greater than 62,500,000 miles, or less than 33,500,000 miles.

Mars is of course most favourably situated for observation when he is in opposition. He then comes on the meridian at midnight, and his distance from the earth is at its minimum value. The figures we have just given show, however, that some oppositions may be much more favourable than others. The distance at the time of opposition, when the planet is most favourably situated, is scarcely more than half what that distance is under the least favourable circumstances. The advantage with which a distant object may be scrutinised varies inversely as the square of the distance. We therefore find that some oppositions may be nearly four times as favourable as others for the purposes of the astronomer.

At the great majority of oppositions, of which one occurs about every two years, the distance between the earth and Mars lies well between the limits we have mentioned. But in the opposition of September 1877 a combination of favourable circumstances brought the distance of Mars from the earth to very nearly as low a point as that distance is capable of attaining. On September 2, 1877, Mars approached the earth to a distance of 34,700,000 miles, which is very close to the lowest limit. It was on August 17 in that year that the inner of the two satellites of Mars was discovered by Professor Hall, at Washington, the distance of the planet from the earth being then only 36,300,000 miles. It is necessary to go back to the year 1845 in order to find an opposition at which Mars approached the earth so closely as it did in September 1877. It appears that on August 18, 1845, the distance between the earth and Mars was 34,300,000 miles. On this occasion, therefore, Mars was nearly half a million miles nearer the earth than it was in 1877. The next nearest approach was in the year 1860, when on July 2 the distance of Mars was 36,000,000 miles. It may be observed that although this approach is not so close as that witnessed in the opposition of 1877, yet Mars was actually closer to the earth then than he was at the time when the two satellites were first discovered.

Another past opposition may be noted for the greatness of the distance between the earth and Mars, even when they were closest. On February 13, 1869, the distance from the earth to the planet then in opposition was 62,300,000 miles.

This, it will be observed, approaches very closely to the maximum value of the distance at opposition between the earth and the planet.

Assuming, as we may for our present purpose, that the orbit of the earth is circular, there is another mode of viewing the subject which may perhaps tend to make it clearer. It has been found that Mars completes his revolution round the sun in a period of 687 days. In the course of this period the distance of Mars from the sun goes through all its changes, increasing from 127,000,000 miles up to 153,000,000, and then again decreasing to the same limit. If we conceive an observer stationed at the sun, then the direction in which he will see Mars when that planet is closest to the sun is the same for every revolution. This, in fact, amounts to the statement that the direction of the major axis of the orbit of Mars is practically constant, for we may overlook for the present purpose the small and slow changes of that direction which arise from the action of the other planets. This line produced backwards will point to that part of the heavens in which Mars is always to be found when he is at the greatest distance from the sun. Twice every year the earth in its annual motion crosses or passes extremely close to this line. On August 26 each year the earth passes between the sun and that point of the heavens in which Mars must be situated when he is nearest to the sun. Similarly on February 22 in each year the earth passes between the sun and that point of the heavens in which Mars must be situated when he is at the greatest distance from the sun.

We can now understand the circumstances under which an opposition is favourable for the purpose of observing Mars. If it so happened that an opposition occurred on August 26 in any year, Mars would be as near as possible to the sun, and therefore as near as possible to the earth, because the earth is then situated exactly between Mars and the sun. If, on the other hand, an opposition occurred on

February 22, then matters would be as unfavourable as possible, because for an opposition at any other date the earth and the planet would be nearer together. We have, therefore, in the date alone a very simple test as to the excellence of an opposition of Mars for astronomical purposes. The nearer that date is to August 26, the better it is, and the nearer that date is to February 22, the worse it is. These considerations are illustrated by the opposition of 1877, which occurred on September 5, or only ten days after the best date possible. In 1845 the opposition took place eight days before the critical date, and therefore, as we have already mentioned, it was somewhat more favourable than the opposition of 1877. On the other hand, a very unfavourable opposition occurred on February 13, 1869, the day of the year being not far from that on which an opposition would be least favourable.

It is worth noticing that the revolution of Mars round the sun, and the revolution of the earth around the sun, are so related that Mars accomplishes 17 revolutions in very nearly 32 years, or, with still greater accuracy, 25 revolutions in 47 years. From the last-mentioned numbers it appears that if the earth and Mars occupy a certain position at a given date, they will in 47 years be again found in the same relative position, Mars having in the interval performed 25 revolutions. It therefore appears that a favourable opposition of Mars will be followed in 32 years, and also in 47 years, by favourable oppositions. For example, we have mentioned that 32 years before the opposition in 1877 a favourable opposition occurred in 1845. We therefore infer that 47 years after this date, i.e. in 1892, a favourable opposition may be looked for.

The accompanying diagram (Fig. 113), which is drawn accurately to scale, exhibits the orbit of the earth and the orbit of Mars. E E′ E″ represents the orbit of the earth, and M M′ M″ the orbit of Mars. The axis major of the earth's orbit is M L, that of Mars's orbit is P Q. A favourable opposition obviously occurs whenever Mars is near P while the earth is near A, while the most unfavourable opposition takes place whenever Mars is near Q while the earth is near B. It was on August 11, 1877, when the earth and Mars had the position represented by E and M, that the first satellite

FIG. 113.

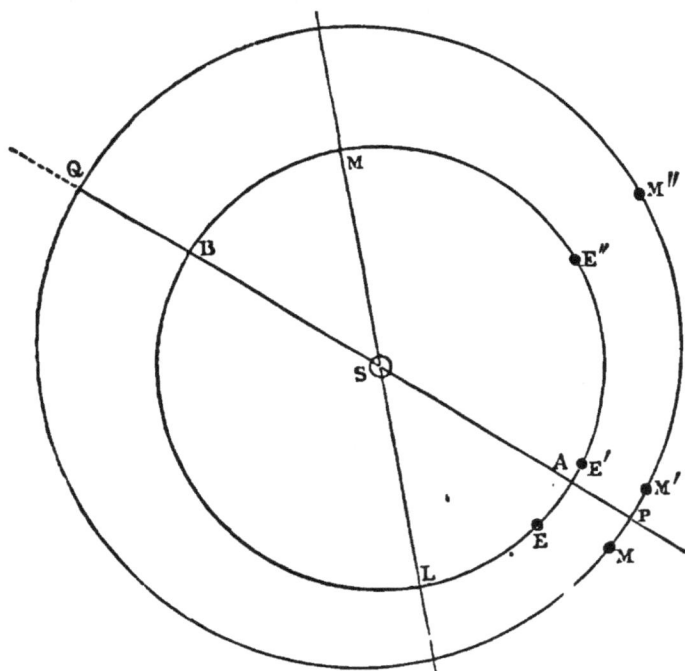

of Mars was discovered. The actual closest approach of the planet to the earth took place on the following September 2, when the relative positions of the two bodies are shown by E and M'. At the opposition which occurred in November 1879 the closest approach of the planet to the earth took place on the 4th, when the planet and the earth had the position represented by E'' and M''.

The movement of Mars among the stars appears sometimes to be direct, i.e. in the same direction as that in which the sun or the moon move, and sometimes to be re-

trograde, i.e. in the opposite direction. It will be easy to
render an explanation of these features of the movements
from what we have already seen.

Let the outer of the two circles (Fig. 114) represent the
orbit of Mars, while the inner arc is the orbit of the earth, s
being the sun in the centre. As we are
only considering the apparent movements of Mars, we may suppose that
Mars is at rest at M, while the earth is
moving with a corresponding slower
motion. Let us first suppose the
earth to be at x ; then Mars, which is
situated at M, will be seen in the direction x M, and will be referred to the
stars in the position P. As the earth
moves on in the direction towards
c, Mars appears to move towards
H, but this motion will gradually be-

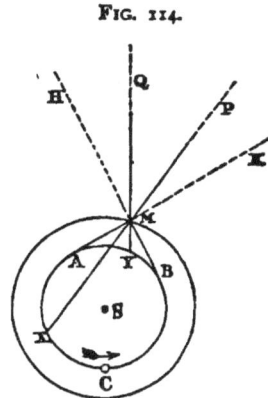

FIG. 114.

come more slow, until at the moment when the earth
reaches B, which is the point of contact of the tangent
drawn from M, the motion of Mars will apparently cease.
As the earth passes from B towards Y Mars will begin to
move backwards, so that when the earth arrives at Y Mars
will have returned to Q. As the earth approaches A Mars
will move gradually towards K, so that when the earth
reaches A Mars will have reached K, after which it will
gradually return again towards H. It will thus be seen that
the retrograde movement of Mars occurs about the time of
opposition.

§ 106. *Appearance of Mars.*—From observations of
certain spots or markings upon the surface of Mars it has
been concluded that that planet, like Venus and the earth,
revolves on its axis. The actual duration of this period has
been estimated by various astronomers. By comparing the
observations of Hooke in 1666 with those of Sir W. Herschel
in 1783 and of Dawes in 1856–1867, Proctor has found that

the duration of one revolution of Mars on its axis is 24^h 37^m 22^s·735. The equator of Mars is inclined to the plane of its orbit at an angle of about 28°. The angular diameter of Mars when viewed from the mean distance of the earth is about 9$''$·4 (Engelmann, 1873). The ellipticity of its disc is practically insensible.

§ 107. *Satellites of Mars.*—Although many astronomers had carefully examined Mars with the view of finding whether he was attended with any satellites, yet it was not until the opposition of 1877 that the two satellites which attend Mars were really found. This very interesting discovery was made by Mr. Asaph Hall, with the great object-glass of the observatory at Washington. The following is the account of the discovery in his own words :—

'My search for a satellite was begun early in August, as soon as the geocentric motion of the planet made the detection of a satellite easy. At first my attention was directed to faint objects at some distance from the planet, but all these proving to be fixed stars, on August 10 I began to examine the region close to the planet, and within the glare of light that surrounded it. This was done by sliding the eye piece so as to keep the planet just outside the field of view, and then turning the eye piece in order to pass completely around the planet. On this night I found nothing. The image of the planet was very blazing and unsteady, and the satellites being at that time near the planet, I did not see them. The sweep around the planet was repeated several times on the night of the 11th, and at half-past two o'clock I found a faint object on the following side and a little north of the planet, which afterwards proved to be the outer satellite. I had hardly time to secure an observation of its position when fog from the Potomac river stopped the work. Cloudy weather intervened for several days. On the night of August 15 the sky cleared up at eleven o'clock, and the search was resumed ; but the atmosphere was in a very bad condition, and nothing was seen of the object, which we now know was at that time so near the planet as to be invisible. On August 16 the object was found again on the following side of the planet, and the observations of that night showed that it was moving with the planet, and, if a satellite, was near one of its elongations. On August 17, while waiting and watching for the outer satellite, I discovered the inner one. The observations of the 17th and 18th put beyond doubt the character of these objects, and the discovery was publicly announced by Admiral Rodgers. Still, for several days the inner moon was a puzzle. It would appear on different sides of the planet in the same night, and at first I thought there were two or three inner moons, since it seemed to me at that time very improbable that a satellite should revolve around its primary in less time than that in which the primary rotates. To decide this point I watched this moon throughout the nights of August 20 and 21, and saw that there was in fact but one inner moon, which made its revolution around the primary in less than one-third the time of the primary's rotation, a case unique in our solar system.'

The names [1] which have been chosen for the satellites are—

Deimos for the outer satellite,
Phobos for the inner satellite.

From the observations made in 1877 the following values of the times of revolution have been found by the discoverer :—

Deimos 30^h 17^m $53^s \cdot 86$;
Phobos   7   39 $13 \cdot 996$.

The planes of the orbits of both the satellites are very

---

[1] Δεῖμος and Φόβος are the sons of Mars whom he summons to yoke his steeds. See *Il.*, xv. 119. I am indebted to Professor Tyrrell for the following version of the lines referred to :—

'He spake ; and called Dismay and Rout to yoke
His steeds ; and he did on his harness sheen.'

T

nearly coincident with the plane of the equator of Mars. The orbit of Deimos is practically circular, but that of Phobos has a certain small eccentricity.

As the hourly motion of Phobos is about 47° round the centre of Mars, and as it is so near to the surface of the planet, this satellite will present a very singular appearance to an observer situated on Mars. Phobos will rise in the west and set in the east, and will meet and pass Deimos, whose hourly motion is only 11°·882. The distances of the satellites from the centre of the planet are for Deimos 14,500 miles and for Phobos 5,800 miles. The semi-diameter of Mars being 2,100 miles, the horizontal parallaxes of the satellites are very large, amounting to 21° for Phobos.

The sizes of the satellites have not been satisfactorily ascertained. All that is certainly known is that they must be very small, not having a diameter of more than a few miles.[1]

§ 108. *Jupiter.*—Although Jupiter is much more dis-

[1] In connection with this discovery it is curious to note the following letter of Kepler, written to one of his friends soon after the discovery by Galileo in 1610 of the four satellites of Jupiter, and when doubts had been expressed as to the reality of this discovery. The news of the discovery was communicated to him by his friend Wachenfels, and Kepler says, 'Such a fit of wonder seized me at a report which seemed to be so very absurd, and I was thrown into such agitation at seeing an old dispute between us decided in this way, that between his joy, my colouring, and the laughter of both, confounded as we were by such a novelty, we were hardly capable he of speaking or I of listening. On our parting I immediately began to think how there could be any addition to the number of the planets without overturning my "cosmographic mystery," according to which Euclid's five regular solids do not allow more than six planets round the sun. . . . I am so far from disbelieving the existence of the four circumjovial planets that I long for a telescope to anticipate you, if possible, in discovering *two* round Mars, as the proportion seems to require, *six* or *eight* round Saturn, and perhaps *one* each round Mercury and Venus.' The anticipations of Kepler with respect to Mercury and Venus have not yet been realised.

tant from the earth than Mars, yet the magnitude of the disc
of Jupiter is so great, that it is a very conspicuous and remarkable object in the telescope. The most striking feature
of the disc is the presence of bands parallel to Jupiter's
equator. It seems certain that the surface we really see
is not the actual solid surface of the planet, but merely a
stratum of clouds or vapours in which Jupiter is enveloped.

The ellipticity of Jupiter is very well marked, amounting
to about one-seventeenth. The equatorial and polar radii
are respectively equal to 11·16 times and 10·51 times the
radius of the earth.

To the unaided eye Jupiter appears as one of the most
brilliant stars : its brightness is about equal to that of Venus.
Jupiter rotates on his axis in a period of $9^h 55^m 25^s \cdot 70$
(Schmidt). When Jupiter is viewed through a telescope
even of quite moderate power, certain small points of light
are seen in his neighbourhood, which are constantly changing
their relative positions, though always accompanying Jupiter
in his movement round the heavens. These small objects
are the four satellites, or moons, of Jupiter, which were discovered by Galileo and were one of the firstfruits of the
invention of the telescope.

Observation shows that the movements of the satellites
of Jupiter around the planet obey Kepler's laws, just as the
planets do around the sun. The following table gives the
mean distances of the satellites from Jupiter in terms of the
radii of Jupiter (Laplace) :—

	Mean Distances.	Periodic Times.
1st satellite	5·698	1·77 days.
2nd ,,	9·067	3·55 ,,
3rd ,,	14·462	7·15 ,,
4th ,,	25·436	16·69 ,,

The eccentricities of the orbits of the first and second
satellites are insensible, and those of the third and fourth are

very small : the planes of the orbits of the satellites are very close to the plane of Jupiter's equator.

The orbits of the four satellites of Jupiter are represented in Fig. 115, in which the relative proportions of the orbits, as well to each other as to the bulk of Jupiter, are shown. Jupiter projects from the side which is turned away from the sun a cone of shadow, into which the satellites every

FIG. 115.

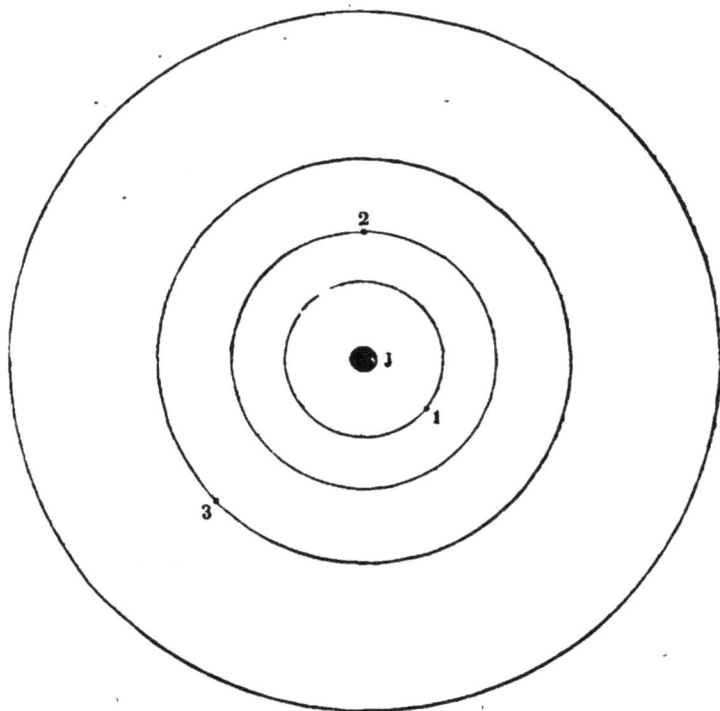

now and then enter, and undergo an eclipse, which resembles an eclipse of our moon. As Jupiter is much larger than the earth, and as it is much farther away from the sun, it is easy to see that the length of the cone which forms the shadow of Jupiter greatly exceeds the length of the cone which is formed by the shadow of the earth. It follows that the dimensions of the section of the cone of Jupiter's shadow, even at the distance of the outermost satellite, is very nearly as large as the section of Jupiter himself. The eclipses of the satellites of Jupiter are consequently much more frequent than the eclipses of the moon. The three first satellites enter the cone of Jupiter's shadow once during each revolution. The fourth satellite alone sometimes passes above or below the shadow, so as to escape being eclipsed

As the eclipses of the satellites are capable of being observed simultaneously at different places, these phenomena are sometimes employed for the purpose of determining longitudes. The method, however, is not a very accurate one. Independently of other difficulties, the eclipse of a satellite is not an instantaneous phenomenon ; the illumination of the satellite is at first enfeebled by being plunged in the penumbra before entering the cone of total darkness, and the disc of the satellite, being of appreciable dimensions, does not cross the boundary instantaneously. For both these reasons, the decrease in the light of the satellite is not instantaneous, and consequently the time when the satellite ceases to be visible will depend upon the power of the telescope which is employed. The eclipse, therefore, wants the precision and definiteness which render it a suitable signal for observers at the different stations to note their local times, and then by comparison find out the difference between their longitudes.

The eclipse of a satellite of Jupiter must be carefully distinguished from an analogous phenomenon which is often witnessed, i.e. the occultation of a satellite by the disc of the planet. In the case of an eclipse the satellite ceases to be visible because the mass of Jupiter is interposed between the satellite and the sun, and thus cuts off the sun's light from the satellite, which ceases to be luminous and is therefore invisible. But in the case of the occultation of a satellite, what really happens is that the satellite actually goes into such a position that the body of Jupiter is interposed between the satellite and the earth, thus rendering the satellite invisible.

It will often happen that a satellite is seen in transit across the face of Jupiter. The shadow thrown by the satellite then forms a circular dark spot on the face of Jupiter, which can be seen in the telescope. An observer who was stationed on the interior of that spot would find the satellite interposed between himself and the sun, and that consequently the sun was in a state of total eclipse.

§ 109. *Saturn.*—When Galileo directed his newly-dis-covered telescope to Saturn, he saw at once that the planet

FIG. 116.

was not a simple disc like Mars, or like Jupiter, but that it had appendages which caused Saturn to present a peculiar appearance, the nature of which he was not able to under-stand. By careful observation it was shown by Huyghens that the appearance which had puzzled Galileo was really due to a remarkable circular ring, or rather series of rings, which completely surround Saturn without touching it. The appearance of Saturn is shown in Fig. 116, in which we see a portion of the ring in front, while another portion is hid-den behind the globe of the planet. The projecting portions on each side of the globe are often called the *ansæ*. In the revolution of the planet round the sun, the plane of the ring remains constantly parallel to itself. This is shown in Fig. 117, where the point s represents the sun, and where T is the position of the earth in its orbit round the sun. It is obvious

FIG. 117.

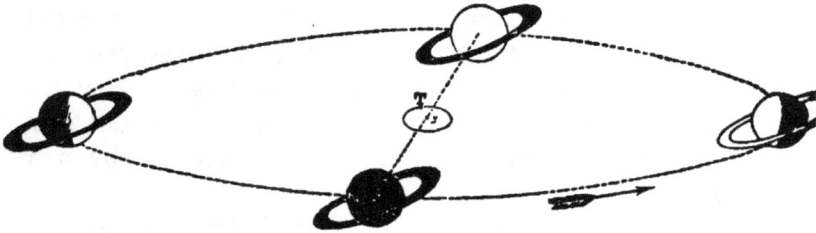

that an observer stationed on T can only see the ring obliquely. Thus, though the actual contour of the ring is a circle, yet as we only see the projection of that circle, it appears to us like an ellipse.

§ 110. *Law of Bode.*—There exists a very remarkable law which connects together the distances of the planets from the sun. This law is generally known by the name of *Bode's Law.* Attention was drawn to it in 1778 by the astronomer Bode, but he was not really the author of the law.

To express this law we write the following numbers :—

0, 3, 6, 12, 24, 48, 96.

This series is easily remembered by the circumstance that, with the exception of the first, each number is double the one which precedes it. If we add 4 to each of these numbers, the series becomes—

4, 7, 10, 16, 28, 52, 100,

which series was known to Kepler. These numbers are, with the exception of 28, sensibly proportional to the distances at which the principal planets are separated from the sun. In fact, the actual distances are represented as follows :

Mercury.	Venus.	Earth.	Mars.	Jupiter.	Saturn.
3·9	7·2	10	15·2	52·9	95·4

These numbers agree tolerably well with those indicated by Bode's law. It is to be observed that this law does not appear to have any theoretical foundation, and for that reason stands in quite a different category from the laws of Kepler.

§ 111. *Discovery of New Planets.*—Including the earth itself, the number of planets known to the ancients was but six. The invention of the telescope has, however, revealed to us the existence of numerous other planets ; so that the number of these bodies at present recognised is about 200, and frequent additions are being made to the list by new discoveries.

On March 13, 1781, William Herschel was examining the stars in the constellation Gemini with a telescope of considerable magnifying power, when he perceived an object which was not a luminous point, as ordinary stars are, but which had a diameter of appreciable dimensions. By continued observation of the same object he noticed that it also differed from the stars in another important respect, for while the stars remained fixed in their relative positions, this new object was in motion. He soon after recognised that this new object was really a planet moving like the other planets in a nearly circular orbit round the sun, and in a plane but little inclined to the plane of the ecliptic. This planet has received the name of *Uranus*, and it is attended by four satellites.

Adopting the distance of the earth from the sun as unity, the distance of Uranus from the sun is 19·18. Thus the orbit of Uranus lies considerably outside that of Saturn, and its radius is very nearly double that of Saturn. The law of Bode is also fulfilled approximately in the case of Uranus. The distance which that law assigns for the planet immediately outside Saturn is 196, which does not differ much from 191·8, which is found by multiplying the distance of Uranus, expressed in terms of the mean radius of the earth's orbit, from the sun by 10.

The astronomer Bouvard, having published in 1821 the

tables of Uranus, by which the position of the planet at different epochs could be found, discovered that the calculated positions of the planet could not be reconciled with the observed positions. The discrepancies could, he suggested, be reconciled by supposing that they arose from some extraneous and unknown influence disturbing the movements of Uranus. Bouvard even suggested that this extraneous influence might be due to a planet which circulated in an orbit exterior to that of Uranus.

The research for this unknown planet was taken up independently by M. Leverrier in France, and by Mr. Adams in England. It was shown that when every allowance was made for the action of Jupiter and Saturn upon Uranus, there were still outstanding discrepancies. Leverrier showed that these discrepancies could be reconciled by the existence of an exterior planet, and he predicted the situation of the planet so accurately that on the night of September 23, 1846, Neptune was actually found by Dr. Galle, at Berlin, in the place indicated. Professor Challis, at Cambridge, had previously commenced a search in accordance with the indications of Adams, and he had actually observed the planet on August 4 and 12, 1846. There can be no doubt that when these observations were compared he would have recognised the planetary nature of the object. The merit of this brilliant discovery must therefore be shared between Leverrier and Adams.

Neptune is not visible to the unaided eye. Viewed in a telescope of inconsiderable magnifying power, this planet seems like a star of the eighth magnitude. When the magnifying power of the instrument is increased, the dimensions of the planet become more considerable, and its circular disc can be perceived. The apparent angular diameter of Neptune is 2″·o. The actual diameter of Neptune is about 4 times the diameter of the earth. Neptune is accompanied by a single satellite.

Neptune is, so far as we know at present, the outermost

planet of the solar system. The law of Bode is, in the case of this planet, very much astray. It would indicate 388 as the distance of the planet immediately outside Uranus, while the actual distance of Neptune is only 300·4.

§ 112. *The Minor Planets.*—According to the indications of Bode's law there ought to be a planet situated in the interval between Mars and Jupiter at a distance represented by 28. This has been abundantly confirmed by the discovery within the present century of a great number of planets which circulate round the sun at about the distance which was indicated by Bode's law. These planets are all small objects, requiring telescopic power; they now number more than 200. The following are a few of the more remarkable of these bodies, with the names and dates of their discoveries.

1. *Ceres*, discovered by Piazzi at Palermo on January 1, 1801. The distance of this planet from the sun is 2·77. Multiply this by 10 ; we obtain 27·7 instead of 28, as indicated by Bode's law.

2. *Pallas*, discovered by Olbers at Bremen on March 28, 1802. Its distance from the sun is 2·77.

3. *Juno*, discovered by Harding at Gottingen September 1, 1804. Its distance from the sun is 2·67.

4. *Vesta*, discovered by Olbers at Bremen March 29, 1807. Its distance from the sun is 2·36.

The search for minor planets is a work now carried on in several observatories. For this purpose maps have been prepared, on which the stars are recorded with great accuracy and minuteness. These maps need not extend to the whole surface of the heavens, as the planets are generally found in or near the ecliptic, and therefore it is only the regions of the sky in or near the ecliptic which require to be mapped for this purpose. The astronomer compares the stars which he sees through his telescope with the stars which he sees recorded on the map. If he detects an object in the heavens which is not represented on the map, then the question arises as to what the object may be. If it be very small, it may possibly be a star which has escaped the attention of the astronomer by whose observations the maps were constructed. It may even be a star which, by reason of variability in its brightness, was not noticed when the maps were made, but which has become conspicuous when the comparison of the maps with the heavens is made in the manner we are now supposing. The first test to apply to an object of this kind is to ascertain whether its position is constant or whether it has a movement relatively to the stars. If in the course of a few nights the object remains sensibly in the same position as compared with the stars in its vicinity, we are then justified in supposing that the object is really a star which from one cause or another previously escaped notice. If, however, the star really has changed its place in the course of a night or two, or if, as sometimes happens, the observations of a single night are sufficien to show marked changes in the position, then we may infer that the unknown object is really a planet or a comet. If it be a minute body resembling a star, we may assume that the object is a planet ; but if it be a diffused mass of luminous matter, we may generally assume that it is a comet. The question will then remain as to whether the object thus revealed to us is new or whether it has been previously observed. In the case of the majority of the comets their orbits are such that they only visit the neighbourhood of the earth so as to become visible once for a few weeks or months, and then again retreat to the depths of space from which they have come. The case of the minor planets is, however, different. They revolve in nearly circular orbits around the sun, the periods of such revolutions being only a few years. It is, therefore, quite possible that the planet which has been observed has been already seen on previous occasions. To ascertain this point, reference must be made to an ephemeris in which the orbits of all the planets which are known are given, and in which will be found predictions of the places of the planets. The ephemeris for this purpose is contained in the 'Berliner Jahrbuch.' Here will be found the position of each minor planet at intervals of 20 days throughout the year, while the place is given for each day in the case of the passage of a planet through opposition. If, therefore, the observer detects a planet, he first refers to the 'Berliner Jahrbuch,' to find whether any of the planets already discovered may not be situated in the position in which the new object appears.

§ 113. *Elements of the Movements of a Planet.*—Observations of the minor planets show that they also obey the three laws of Kepler. Each of the planets revolves in an ellipse of which the sun occupies one of the foci; each of the planets sweeps over equal areas in equal times, and the squares of the periodic times are proportional to the cubes of the mean distances.

In order to define completely the nature of the orbit of a planet, it is necessary to specify certain quantities with regard to that orbit which are called its *elements*. In the first place we must have the shape and size of the ellipse specified. This will involve two quantities—namely, the length of the semi-axis major and the eccentricity of the ellipse. From Kepler's third law it appears that the period of revolution is known when the length of the semi-axis major of the ellipse is known, and therefore this element is really included in the two already specified. But besides knowing the shape and size of the ellipse, certain other elements are necessary in order to specify the orbit. We require to know the plane in which the ellipse lies, as well as the actual position of the ellipse in its plane. The position of the plane of the orbit is specified by its line of *nodes*—that is, by the line in which it intersects the plane of the ecliptic and by the *inclination* at which the plane of the orbit is inclined to the plane of the ecliptic. Still one more quantity is required to determine the ellipse. We know that the focus of the ellipse is situated at the sun, but the direction of the axis ' major of the ellipse is still undetermined. This may be defined by the *longitude of the perihelion*, which is the sum of the longitudes of the ascending node and the angular distance (measured in the direction of the motion) of the perihelion from the node. Finally, we require to know the date at which the planet occupied a certain specific point of its orbit. When this and the other elements of the orbit are known, the place of the planet at any time can be predicted.

To determine the six elements of the orbit of a planet, it is necessary to have three observations of the planet both in right ascension and declination. The computation of the actual shape and position of the orbit from these observations is a matter of no little complexity. It is, however, easy to render a general account of the method so far as to show that three observations will be adequate for the purpose.

When the right ascension and the declination of a celestial body are known, all that we really infer is that at the time when the observation was made the body was seen in a certain direction. The position of the observer has to be taken account of, because that position is constantly changing in consequence of the annual motion of the earth around the sun (we may for the present overlook the diurnal rotation of the earth). The date of the observation being known, the precise position of the observer on the ecliptic at that date is also known ; and as the direction of the planet is determined by the observation, it follows that at the moment of the observation the planet must have been situated upon a straight line of which the position in · space was actually known. The same takes place at the two other observations, and consequently the result of the three observations is, that, on dates which are given, the planet was situated respectively on three lines of which the positions in space are given.

The problem which has now to be solved may be thus stated. To describe an ellipse of which the sun is at the focus, which shall intersect three given straight lines in space. Draw any plane through the sun; this will cut the three lines in three points, and a conic section can be drawn to pass through these points, and so that the sun shall be situated in the focus. Unless, however, the plane be properly chosen this conic will not be the ellipse which the planet is describing. In accordance with Kepler's laws the area swept over by the radius vector to a planet is proportional to the time, and, consequently, as we know the dates of the observations, the area swept over by the radius vector between the first and second observations, and also between the first and the third, are also known. The plane must therefore be so drawn through the sun, that the conic passing through the three points shall be an ellipse, and that the areas subtended by the two arcs of the ellipse at the focus shall be given. We thus have two conditions imposed on the plane, and these two conditions will be sufficient to determine the plane, for it will be remembered that, when a plane is drawn through a given point, two other elements are sufficient to fix the position of the plane. These two elements may conveniently be taken as two of the direction angles of the normal to the plane. In this way we have shown that three complete observations of a planet are adequate for the determination of its orbit.

§ 114. *The Parallax of the Sun.*—There are several distinct methods by which the distance of the sun from the earth may be ascertained. The best known of these methods, even if not the most trustworthy, is that which is afforded by the rare occurrence of the phenomenon which is known as the transit of Venus.

Owing to the inclination of the orbit of Venus to the orbit of the earth, it usually happens that when in conjunction Venus passes over the sun or under the sun. If, however, the conjunction occur when Venus is near the node, then Venus will pass actually between the earth and the sun, and will be seen like a dark spot upon the face of the sun. Thus, suppose the circle c A B (Fig. 118) to repre-

sent the apparent disc of the sun, the planet Venus appears
to enter on the disc of the sun at A, and then moving across
the sun in a direction which is indicated
by the dotted line A B, leaves the sun
at B.

FIG. 118.

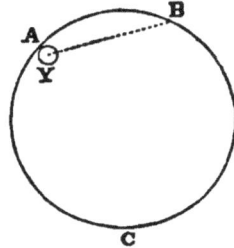

In discussing this question no little
complexity arises from the inclination
of the orbit of Venus to that of the
earth. We shall therefore simplify the
subject by supposing that the orbit of
Venus coincides with the ecliptic, as the
principle of the method will not be affected. We may also
suppose that the orbits of the earth and of Venus are circles.
Under the circumstances we have supposed the path of
Venus across the disc of the sun would pass very near the
centre of the sun when seen from any point on the earth.
When the planet has just entered completely upon the disc
of the sun, as at A (Fig. 118), the phenomenon is said to be
first internal contact. After passing across the face of the
sun the planet reaches B, and when the circular edge of the
planet just touches the circular edge of the sun, the phase
called *last internal contact* is reached. Observations of the
transit of Venus are mainly devoted to the accurate determination of the time at which first internal contact and last
internal contact occur. It is from these observations, made
at different parts of the earth, that the parallax of the sun is
to be concluded. We shall not enter into any detail in the
matter, but merely point out how such observations will
enable the parallax of the sun to be ascertained. Let P Q T
(Fig. 119) denote the sun, of which s is the centre. Let A B
represent the earth. Draw A T a common tangent to the
sun and the earth, and also the common tangent B T, which
touches the earth at B and the sun in a point so close to T
as to be undistinguishable therefrom. The circle through
A B denotes the orbit of the earth, and the circle through X Y
denotes the orbit of Venus, and the arrows indicate the

direction in which the earth and Venus are moving. As
Venus is moving faster than the earth, it follows that Venus
will overtake the earth, and thus at a certain time will arrive

FIG. 119.

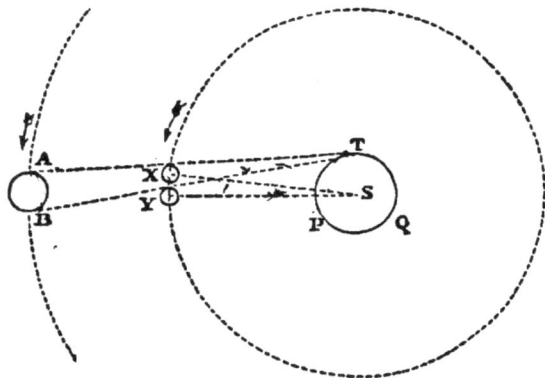

at the position x. Remembering that we have assumed
that the orbit of Venus lies in the same plane as that of the
earth, it follows that an observer stationed at A will just see
Venus at first internal contact, and, having previously regulated his clock accurately, he will be able to note the moment
when Venus occupies the position x.

Venus, however, is moving around the sun with a greater
angular velocity than that which the earth has round the
sun : it follows that after a certain interval of time Venus will
have attained the position represented by Y, where it touches
the second common tangent drawn to the earth and the sun.
An observer at the point on the earth's surface denoted by B
will therefore now see Venus at first internal contact. It is
therefore clear that A is the spot on the earth at which the
internal contact is first seen, and that B is the spot where the
internal contact is last seen. A correction will have to be
made here on account of the rotation of the earth on its axis,
for during the time occupied by Venus in passing from x to
Y the earth will have turned through an angle which is quite
appreciable. Consequently the real point B on the earth's

surface, where the internal contact is last seen, is different from what it would have been had the earth not rotated on its axis after the first internal contact was observed from A. Astronomers, however, know how to allow for this difference, and we shall not here consider it further.

We shall, therefore, suppose that expeditions are sent to the two stations A and B (of course, as a matter of fact, they can only be sent to the places nearest A and B which are suitable from geographical considerations), and that at each of these two stations the moment of first internal contact is observed. We may also suppose that a telegraph wire is laid from A to B, so that at the instant of contact as seen from A a telegraphic signal is despatched to B. The observer at B then notes the arrival of this signal by his clock, and when he himself sees the internal contact at a time also marked by his own clock he is able with the greatest precision to determine the interval of time between the two contacts. We thus learn from these observations the actual time which Venus takes to pass from the position X to the position Y. But we also know that the entire time which Venus requires for performing a revolution around the sun *relatively to the earth* is 584 days. Hence, if we assume that Venus moves uniformly, we can, by a simple proportion, find the angle X S Y. The radius T S is small compared with the distance S X ; we may therefore, with sufficient accuracy for our present purpose, suppose that the angle X T Y is equal to the angle X S Y. We can now consider the problem to be solved, for the distance A B, being very nearly equal to the diameter of the earth, is therefore known, also the angle A T B is known, and therefore the distance A T from the sun to the earth is known.

From the transit of Venus which occurred in 1874, Sir George Airy, after a discussion of the results obtained by the expeditions sent by the British Government, concludes that the equatorial or horizontal parallax of the sun is 8″·760. From this it appears that the mean distance of the sun from the earth is 93,300,000.[1] The next transit of Venus will occur in the year 1882.

The astronomical importance of an accurate knowledge of the sun's distance can hardly be over-estimated. With a single exception it is the unit in which the distances of the heavenly bodies are invariably expressed. The exception to which we refer is the moon, the distance of which is directly determined by observations of the zenith distance of the moon in the way already described. Once, however, the distance of the sun from the earth is found, then from Kepler's third law the distances of all the planets are found. From this will follow the dimensions of the planets and of the orbits of their satellites. Thus, to determine the scale of the whole solar system, it is only necessary to determine the distance of the sun from the earth. So also when we seek to determine the distances of the fixed stars from the earth, the quantity which the observations give is the angle which the radius of the earth's orbit subtends at the star. The determination of the distance of the star is therefore expressed in terms of the radius of the earth's orbit.

There are many other methods by which the distance of the sun from the earth can be found, besides that which is afforded by the transit of Venus. One of the best of these methods is by the observations of the parallax of a planet as seen from different places on the earth's surface. In the application of this method, either Mars or a small planet is chosen, and its apparent position relatively to the fixed stars in its neighbourhood is measured from two stations on the earth. If the planet were as far off as the stars, then no apparent difference would be seen in the places afforded by the two different sets of observations ; but as the planet is comparatively near the earth, the distance between the two stations is sufficiently great to cause an appreciable change in the place of the planet when viewed from the two places.

[1] Report ordered to be printed by the House of Commons, July 6, 1877.

In this way the distance of the planet can be found, and thus the scale of the solar system, and the distance of the sun itself, is ascertained.

A still more simple method, and one which has much to recommend it, is a modification of that which has just been described. Owing to the rotation of the earth, the position of the observer himself is in constant motion. If, therefore, the planet be observed with reference to the stars in its neighbourhood, and if the observer repeat these observations some hours later, he will find that the apparent place of the planet has changed. The alterations in the place of the planet are due to three causes—1st, to the actual motion of the planet itself; 2nd, to the displacement of the observer, due to the revolution of the earth around the sun; 3rd, to the displacement of the observer, due to the rotation of the earth on its axis. By suitable discussion of the observations it is possible to distinguish the amount of displacement arising from the last-mentioned cause, and thence to determine the ratio of the distance of the planet to the diameter of the earth. For the application of this method, it is obviously desirable to have the planet as close to the earth as possible.

CHAPTER VIII.

COMETS AND METEORS.

§ 115. *Comets.*—Besides the planets, there are certain other bodies belonging to the solar system which are sometimes seen moving among the constellations. These bodies are called *comets*. A comet usually consists of a more or less brilliant point or nucleus surrounded with a nebulosity which is often extended in one direction so as to form a tail

(Fig. 120). Frequently, however, comets are seen which do not appear to have any conspicuous tail, and sometimes they have more than one. Sometimes also the nucleus is very minute, or entirely wanting. The actual dimension to which a comet extends is often enormously great. Thus, in the month of March 1843, a comet appeared of which the tail extended to a length of 40°. The tail of the comet of 1680 had a length of 90°, and that of 1618 was stretched across the sky, through an arc not less than 104°. The comet of 1811 had a tail 23° long, and in 1744 a comet appeared with six tails. The most recent great comet which was well seen is that of Donati in 1858. The examples which we have alluded to are those of comets which, from their splendour, were exceedingly conspicuous objects. The ordinary comets are much smaller, and indeed the great majority of these bodies are faint telescopic objects, of which a few are usually found every year.

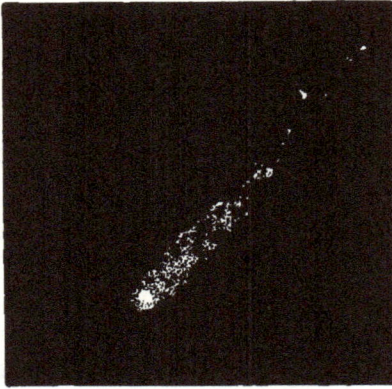

Fig. 120.

A comet is only visible in the heavens for a limited period. It first appears in some region of the heavens where nothing of the kind had been seen a few days previously; from day to day the comet changes its position and its brightness, and even its shape. It will remain visible for some weeks, or perhaps months. Occasionally it becomes lost to view from having gone too close to the sun, and then again the comet is seen on the other side of the sun, when, after remaining visible for some time, it gradually recedes, becomes fainter, and is finally lost to sight.

It was Newton who first explained the real motion of

comets. He discovered that comets really move in conic sections with the sun in one of the foci Many of these orbits appear to be parabolic, but the majority are perhaps very elongated ellipses, and the path of the comet in the vicinity of the sun, when alone we see it, is in this case hardly to be distinguished from a parabola.

When a comet appears, its right ascension and declination are determined, and from three such observations the orbit of the comet can be completely ascertained. To accurately describe this orbit its *elements* must be ascertained. The elements of the orbit of a comet are— 1st, the *inclination* of the plane of the orbit to the plane of the ecliptic ; 2nd, the *longitude of the ascending node* of the orbit, that is, the angle which the line of intersection of the comet's orbit on the ecliptic makes with a line drawn through the sun parallel to the line of the equinoxes; 3rd, the *longitude of the perihelion,* which is in the case of direct motion the sum of the longitude of the ascending node and the angular distance (measured in the direction of the motion) of the perihelion from this node ; 4th, the *perihelion distance,* that is to say, the shortest distance of the parabola from the centre of the sun, the radius of the earth's orbit being taken as unity ; 5th, *the epoch,* or the time at which the comet has passed through its perihelion ; 6th, the direction of the motion which is *direct* from west to east like the sun, or *retrograde* from east to west.

§ **116**. *Periodic Comets.*—If the orbit of a comet were really parabolic, then, after once passing round the sun, the comet would retreat to the depths of space from whence it came, and would never again become visible. If, however, the orbit be elliptic, then after a greater or less time, depending upon the length of the axis major of the ellipse, the comet would return, and thus form what is known as a periodic comet.

When a comet appears, the question arises as to whether this comet is to be regarded as parabolic, or whether it is not one of the periodic comets. This question cannot be decided by the appearance of the comet. The actual shape and dimensions of a comet vary so much, even in the course of a single apparition, that it is quite hopeless to expect that the identity of the comet at future returns can be detected from its resemblance to comets previously seen. The only method of determining the identity of a comet is, to ascertain the elements of its orbit, and then seek in the record of previous comets for a comet which has an orbit of a similar character.

§ 117. *Halley's Comet.*—To illustrate this subject, we shall take the case of the comet which appeared in 1682, and which is generally known as Halley's comet. From the observations made by Lahire, Picard, Hévélius, and Flamsteed, the following elements were found for the orbit of the comet during the apparition of 1682 :—

Inclination.	Longitude of Node.	Longitude of Perihelion.	Perihelion Distance.	Direction of Motion.
17° 44' 45"	51° 11' 18"	301° 55' 37"	0·58	Retrograde.

But from the observations of Kepler and Longomontanus it appeared that a great comet, observed in 1607, had the elements—

Inclination.	Longitude of Node.	Longitude of Perihelion.	Perihelion Distance.	Direction of Motion.
17° 6' 17"	48° 14 9 '	300° 46 59"	0·58	Retrograde.

From the close resemblance between these elements and those of the comet of 1682, Halley concluded that the comet of 1682 and that of 1607 were really the same object, and that the period of its revolution about the sun was 75 years. It further appeared that the comet observed by Apian in 1531, i.e. 76 years before 1607, had an orbit with the following elements :—

Inclination.	Longitude of Node.	Longitude of Perihelion.	Perihelion Distance.	Direction of Motion.
17° 0'	45° 30'	301° 12'	0·58	Retrograde.

The identity of these elements with those found for the comets of 1607 and 1682 left no doubt as to the identity of the objects. If this were true, then Halley predicted that the comet would reappear in 1758. The influence of the planets in perturbing the motion of this comet is very considerable. Clairaut computed that by the action of Jupiter the comet would be retarded 528 days, and 200 days by that of Saturn. He therefore predicted that the comet would return to perihelion about the middle of April 1759, but that, owing to the calculations being only approximate, an error of 30 days on one side or the other was possible. The comet did actually return very close to the predicted time, and passed the perihelion on March 12, 1759. The following elements were deduced from the observations then made :—

Inclination.	Longitude of Node.	Longitude of Perihelion.	Perihelion Distance.	Direction of Motion.
17° 36′ 52″	53° 50′ 27″	303° 10′ 28″	0·58	Retrograde.

Observing that the period of the comet is about 75 years, another return was expected in the year 1835, and, by accurate calculation of the perturbation, the date of the perihelion passage was fixed for November 13. The perihelion passage actually did take place on November 16.

As we know the periodic time of the comet, we are enabled, by means of Kepler's third law, to find the mean distance of the comet from the sun. Representing the mean distance of the earth by unity, the mean distance of Halley's comet is 35·9. The actual shape and dimension of the orbit of Halley's comet, as compared with the planets, is shown in Fig. 121, where it will be seen that the orbit of the comet passes a little outside that of Neptune.

§ 118. *Distinction between Planets and Comets.*—As both planets and comets are members of the solar system revolving around the sun in orbits which are described in accordance with the laws of Kepler, it is necessary to specify the wide differences in their movements which separate the class of bodies we call planets from the class of bodies we call comets.

FIG. 121.

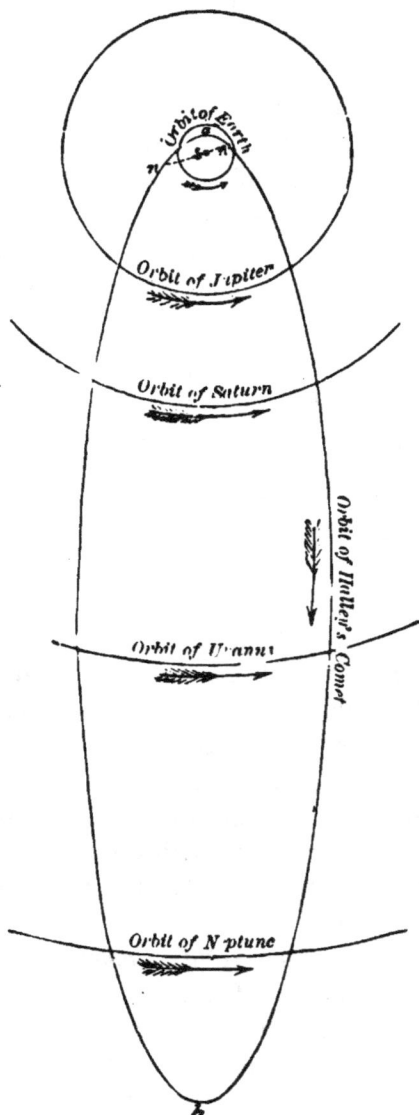

The planets all move round the sun in the same direction, and their orbits are, generally speaking, but very little inclined to each other. The eccentricities of the orbits of planets are but small, so that the ellipses which they describe differ but little from circles. Comets, on the other hand, move in orbits whose planes are inclined at all possible angles to the plane of the ecliptic; some have a direct motion, while that of others is retrograde. Most of them describe orbits of such great eccentricity that, while they are visible, the orbit cannot be distinguished from a parabola.

§ **119**. *Shooting Stars.* —In addition to the planets and their satellites, and the vast host of comets, there are numerous other small bodies generally recognised as belonging to the solar system. The study of these small bodies is a branch of astronomy in which the telescope is of but little use. We do not see these bodies under ordinary circumstances with the telescope, but we become aware of their existence in other ways. We sometimes find them actually tumbling down upon the earth: we can then take them in our hands, weigh them, and analyse them ; they are the bodies which we call *Meteorites.* Collections of these objects are to be seen in the British Museum and elsewhere.

But there is another very different way by which we become aware of the existence of these small bodies. The descent of a *meteorite* is a comparatively rare. and at all times very noteworthy, phenomenon ; yet everyone is familiar with what are called 'shooting stars,' while occasionally the somewhat more imposing phenomena which are known as *fire-balls* are witnessed. We have the best reasons for knowing that what we call shooting stars are really small objects which dash into our atmosphere from external regions. Thus meteorites and shooting stars are the evidences which we possess of the existence of comparatively minute bodies in space, which are external to the earth and to our atmosphere.

As an illustration on a somewhat imposing scale of the circumstances under which these external bodies (often called *meteors*) enter our atmosphere and thus become manifest, we may take a very remarkable fire-ball which occurred on November 6, 1869. This fire-ball was very extensively seen from different parts of England, and by combining and comparing these observations very accurate information has been obtained as to the height of this object and the velocity with which it travelled.

Let us suppose that an object many miles up in the air is seen simultaneously by two observers stationed at a considerable distance apart, and that each observer has noted carefully the direction in which he saw the object ; then the height of the object can be determined. For, take a map on any convenient scale, and at the two stations of the observers insert straight wires projecting from the map in the

in which the object was seen from each of the
These two wires must, of course, intersect at the

FIG. 122.

in order to represent a set of parallel lines in the object
which is being copied. When we are looking at the shooting stars we really only see the projections of their paths
upon the surface of the heavens. But we have shown that
all these projections pass nearly through the same point, and
therefore we infer that the shooting stars belonging to the
same shower are moving in nearly parallel lines. It is not
of course meant that throughout all space the paths of the
meteors producing the shooting stars are a group of parallel
straight lines. We refer only at present to the very short
portions of those paths which are described by the meteors
when we see them as shooting stars just at the moment of
their dissolution.

We are now able to ascertain the actual direction in
which the shooting stars of a shower are moving ; in fact, a
line drawn from the eye of the observer to the radiant must
be parallel to the straight lines along which the shooting
stars move relatively to the earth.

It will next be necessary to consider the true significance
of the special day of the year on which the shower of the
Andromedes was witnessed. The following table gives the
names of the three principal showers of shooting stars, as
well as the days of the year on which they may be looked
for : —

Name of shower.				Date of appearance.	
Perseids	.	.	.	August	9 to 11
Leonids	.	.	.	November	12 to 14
Andromedes	.	.	.	November	27 to 29

As the earth moves round the sun once in a year, the
earth is on each day of the year in the same part of its orbit
as it was on the same day of the preceding year. We are
therefore led to connect the annual recurrence of the shooting star showers with the position of the earth in its orbit
round the sun. Thus, when the earth is in one part of its
orbit it is likely to meet the Leonids, when it is in another

it is likely to meet the Perseids, and so on. We therefore infer with certainty that the stream of meteors which form the Andromedes, when they enter our atmosphere as shooting stars, must pass through that part of the earth's orbit in which the earth is situated at the end of November.

　　The question has now assumed a geometrical form of considerable interest. Let E (Fig. 123) denote the position

FIG. 123.

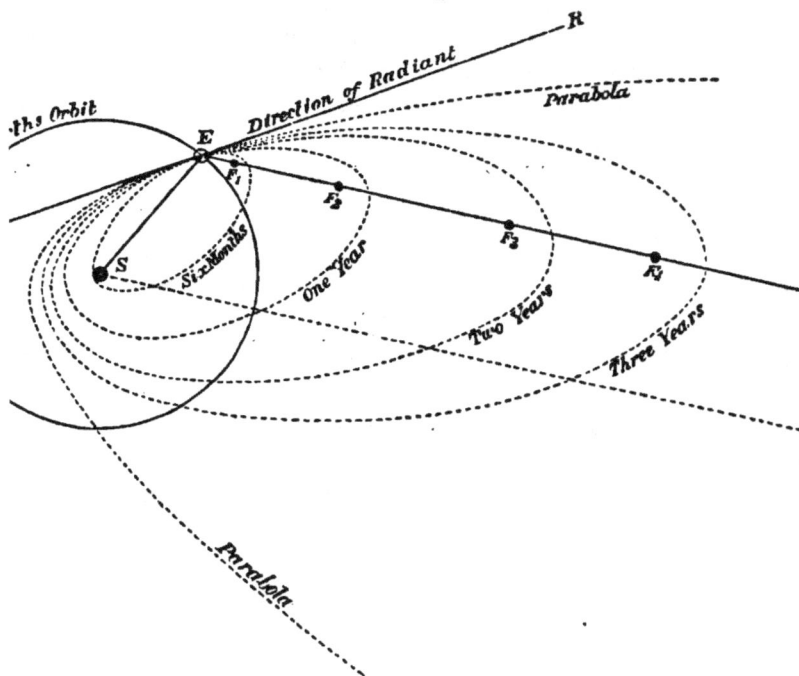

of the earth at the time of the occurrence of a shower, and let the line E R, drawn through E, denote the direction of the radiant, after allowing for the motion of the earth. Then the line E R must be a tangent to the orbit of the meteors, and of course the sun is a focus of that orbit. The position of the other focus is, however, indeterminate. All that we know at present is that, from a property of the ellipse, the second focus must lie on the line E F₁, which is drawn through

E so as to make with the tangent the same angle as E S makes with the tangent. To determine the position of this second focus some additional datum is indispensable ; the most suitable datum is the time of revolution of the meteors. In the figure we have shown orbits corresponding to periods of six months, one year, two years, and three years, all drawn to scale. As a matter of fact, it would be impossible to draw on this scale the actual orbits of any of the meteor showers, as the periods are all much longer than those here referred to. When, however, the periodic time is known, then, from Kepler's third law, the axis major of the ellipse is known, and therefore the position of the focus is determined. All, therefore, that is necessary to determine the orbit of a swarm of meteors is given, when we know the day on which the shower takes place, the position of the radiant, and the periodic time of the swarm.

Pausing for a little in our consideration of the Andromedes, let us turn to the periodic comet which was discovered more than a century ago, and which is now known by the name of Biela's comet. It must be regarded as a very remarkable circumstance that the path of Biela's comet actually crosses the path of the earth. This is of importance in connection with the Andromedes, because the point of the earth's orbit, where it is crossed by the orbit of Biela's comet, is precisely that spot which is occupied by the earth on November 27 in each year. It surely is a remarkable coincidence that the earth should encounter the Andromedes at the very moment when it is crossing the track of Biela's comet. We are at once tempted to make the inference that the comet and the meteors are in some way connected, and the justice of this inference is corroborated in the most astonishing manner by three additional circumstances :—

I. We have already explained how the direction from which the Andromedes come is to be found, and we can also find the direction from which Biela's comet comes. These two directions are identical.

2. Biela's comet sweeps around the sun in an elliptic orbit, with a period a little under seven years. Thus, after coming into the vicinity of the earth and being occasionally visible, the comet again withdraws to a vast distance, to return again in about seven years more. Now it so happens that at the end of 1872 the time had arrived for the return of Biela's comet, and thus the occurrence of the great shower of the Andromedes took place at the time when we knew that Biela's comet must, comparatively speaking, have been in the vicinity of the earth, though it had not yet been observed.

3. Professor Klinkerfues ingeniously argued that if a comet, coming from the radiant of the Andromedes, actually brushed past the earth on the night of November 27, 1872, the comet ought to be found immediately afterwards in that region of the heavens which is diametrically opposite to the radiant of the Andromedes. He therefore telegraphed to Mr. Pogson at Madras, requesting him to search in the region thus indicated. The search was made, and the comet (or a comet) was found. Unfortunately, bad weather prevented sufficient observations of the comet being made, but there can be little or no doubt that it was connected with the Andromedes.

The connection thus established between comets and shooting stars has been verified in several other cases, especially in the great showers of the Leonids and the Perseids. We are thus led to regard the association with comets as a characteristic feature of the periodic showers of shooting stars.

Out of the thousands of comets there are but comparatively few of which the orbits cross the earth's orbit ; consequently, supposing that comets are often accompanied by meteor streams, we are led to the conclusion that there are many comet meteor streams in the solar system which we never see. It therefore appears that the streams of cometary meteors may have an existence quite independent of the

earth, and it may be regarded as a merely accidental circumstance that the earth's orbit intersects a few of the comet-
meteor orbits, and thus enables us to become aware of the
existence of the meteors when they enter our atmosphere as
shooting stars.

Although the great showers only recur at distant
intervals of time, yet it frequently happens that, even in
ordinary years when no large shower is seen, a number of
shooting stars are observed when the earth crosses the track
of one of the great meteor currents. Thus, for example,
though the great showers of the Leonids only recur at
intervals of 33¼ years, yet it frequently happens that from
November 12 to November 14 meteors are seen which
diverge from the well-known region in the constellation Leo,
and are undoubtedly moving in the orbit of the Leonids.
The cause of this may be explained by Fig. 124.

FIG. 124.

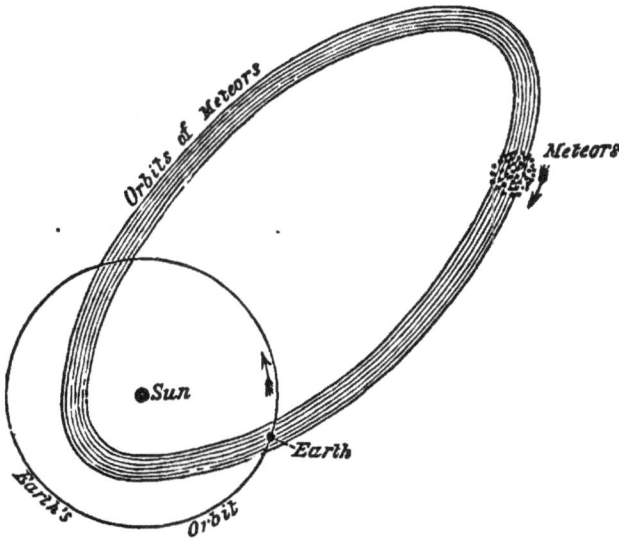

If we suppose that the several meteors of the shoal
describe similar ellipses, then those which are moving on

the larger ellipses will have a larger periodic time than those which are moving in the smaller ellipses. The consequence is that the shoal gradually lengthens out until the more erratic members become dispersed round the entire orbit. It is these meteors, thus distributed, which the earth meets in ordinary years, while when the earth encounters the main shoal, a grand display is witnessed.

CHAPTER IX.

UNIVERSAL GRAVITATION.

§ **120.** *The Law of Gravitation.*—When Kepler's laws of the motion of the planets were announced, they were merely empirical results deduced from actual observations. It was Newton who showed that Kepler's laws are really only the consequences of the grand law of nature which is called the *law of gravitation.*

We know from the *first law of motion,* that when a body is not acted upon by any force it moves with uniform velocity in a straight line, but that if the body be moving in a curved path, or if its velocity be not uniform, then force must be acting upon that body. Kepler's laws state that planets move in ellipses and not in straight lines, and also that the velocities of the planets do not remain constant. We are, therefore, forced to admit that some force must constantly act upon the planets. It remains to discover at each moment the direction and intensity of this force.

We can first prove that the direction of the force which acts upon the planet passes constantly through the sun. This was shown by Newton to be a consequence of the law discovered by Kepler that equal areas are described in equal times by the radius vector, drawn from the sun to the planet. The orbit of the planet, although really a curve, may be considered to form a polygon of an indefinitely large number of sides. We may further suppose that the length of each side of the polygon is proportional to the velocity with which the planet is moving on that side, so that equal time is occupied in describing each of the sides. The actual force which acts upon the planet may be conceived to be an instantaneous impulse which the planet receives at each of the corners of the polygon, while during the passage from one corner to the next no force is in action. As the time taken to describe a side of the polygon is constant, the property of equal areas in equal times amounts to the assertion that the areas subtended by the sides of the polygon at the centre of the sun are all equal.

From the second law of motion it appears that if a body

FIG. 125. FIG. 126.

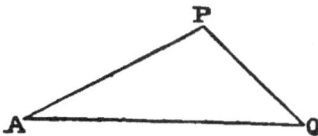
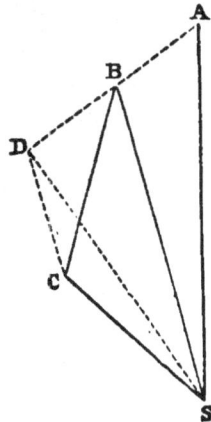

have a velocity which is represented both in magnitude and direction by the line A P (Fig. 125), and if by the sudden application of a force the direction and magnitude of the velocity are changed into A Q, then the force which has produced this effect will be parallel to P Q. Let A B C (Fig. 126) be three consecutive corners of the polygon described by the planet. Produce A B until B D is equal to A B. Then B D will represent the velocity both in direction and magnitude which the planet has while describing A B, and B C represents both in magnitude and direction the velocity which the planet has along the side B C. From the principle just explained, the instantaneous force which acted upon the planet at B, must have been parallel to the

line D C. But as A B is equal to D B, the area of the triangle
S B A is equal to the area of the triangle S B D, and consequently the area of S B D must be equal to that of S B C. It
hence follows that the line D C is parallel to B S, and therefore the impulsive force which acted upon the planet at B
must be parallel to the line B S.

We therefore see that the direction of the impulsive force
which acts at each corner of the polygon points towards the
centre of the sun. Let the number of sides of the polygon
be increased indefinitely, and we see that the planet will at
each position be acted upon by a force which is directed
towards the centre of the sun.

Thus Kepler's discovery of the law of equal areas in equal
times shows that the force which acts upon the planets must
be directed towards the sun. It can also be shown con-
versely that, when the motion of a body takes place under
the action of a force directed towards a fixed point, then
the body describes equal areas in equal times.

All that the second law of Kepler has shown us, is the
direction of the force which acts upon the planets: it has told
us nothing of the law according to which the *intensity* of
that force varies. To solve this question we must have re-
course to the two remaining laws.

The third law asserts that the squares of the periodic
times are in the same ratio as the cubes of the mean
distances, and as there is no reference made to the eccentricities, we may, for the purpose of considering this law,
assume the hypothetical case of planets which revolve in
circular orbits around the sun in the centre. This supposition is not excluded by the first law, which states that the
orbit of a planet is an ellipse ; for there are ellipses of every
degree of eccentricity, including the circle as an extreme
case. In the case of a planet revolving in a circular orbit,
the second law of motion, which asserts the description of
equal areas in equal times, requires that the velocity shall
be uniform.

Let us, then, suppose the case of two planets moving in circular orbits and with uniform velocities around the sun. Let the radii of the circles be R_1 and R_2, and the times of revolution be T_1 and T_2; then Kepler's third law asserts the proportion—

$$T_1^2 : T_2^2 :: R_1^3 : R_2^3.$$

But we know from the principles of mechanics that when a body is revolving in a circle with uniform velocity, the body must be constantly acted upon by a force which is directed towards the centre of the circle, and that the magnitude of that force is proportional to

$$\frac{4\pi^2 R_1}{T_1^2},$$

where, as before, R_1 is the radius of the circle, and T_1 is the periodic time.

If, therefore, we denote by F_1 and F_2 the intensities of the two forces, we must have the proportion:—

$$F_1 : F_2 :: \frac{R_1}{T_1^2} : \frac{R_2}{T_2^2};$$

but, from what we have already seen—

$$\frac{R_1}{T_1^2} : \frac{R_2}{T_2^2} :: \frac{1}{R_1^2} : \frac{1}{R_2^2},$$

and hence we have—

$$F_1 : F_2 :: \frac{1}{R_1^2} : \frac{1}{R_2^2}.$$

From this we deduce the very important result that in the case of two planets revolving around the sun in circles, the intensities of the forces vary *inversely as the square of the distances.* We may state this result with increased generality as follows :—

Each planet is acted upon by a force directed towards the

sun, and varying inversely as the square of the distance from the sun.

Assuming this law to be true, Newton discovered that the orbit of a planet would be a conic section of which the sun was situated in one of the foci, thus explaining in the most simple manner the first of Kepler's laws.

In the same way as the planets describe ellipses in consequence of their attraction to the sun, so the moon describes an ellipse around the earth, and the satellites describe ellipses around their primaries. More generally it is believed that every body in the universe attracts every other body, and this is termed the law of universal gravitation.

It remains to explain how the magnitude of the gravitation between two bodies is affected by the masses of those bodies. If the mass of either body be doubled, the intensity of the attraction is doubled. It therefore appears that the intensity of gravitation must vary proportionally to the product of the masses. The exact expression for the gravitation of two masses m and m', separated by a distance r, is

$$c\, m\, m' \div r^2,$$

where c is a certain constant depending upon the nature of gravitation itself, and is equal to the number of units of force in the gravitation between two units of mass when placed at the unit of distance apart.

A very remarkable circumstance connected with the force of gravitation must be here adverted to. We have stated that the intensity of the force is proportional to the product of the masses, but it appears to be quite independent of the nature of those masses. For example, two masses of lead placed at a certain distance are attracted by the same force, as two equal masses of iron would be when separated by the same distance. The attraction of gravitation is therefore a very different force from that kind of attraction called magnetic attraction, where the character of the masses is of the utmost importance.

It is not difficult to show that the force by which the moon is retained in its orbit around the earth is really produced by the same attraction of the earth which causes a body to fall at the earth's surface.

A body falling freely near the surface of the earth will in one second move over a distance of 16·1 feet. Remembering that the distance of the moon from the centre of the earth is about 60 radii of the earth, and that the intensity of gravity varies inversely as the square of the distance, it appears that, at the distance of the moon, a body let fall would in one second move towards the earth through a distance 16·1÷3600 feet=0·053 inch very nearly. The moon is moving in an orbit which for our present purpose we may regard as a circle, of which the earth is the centre (Fig. 127). If the

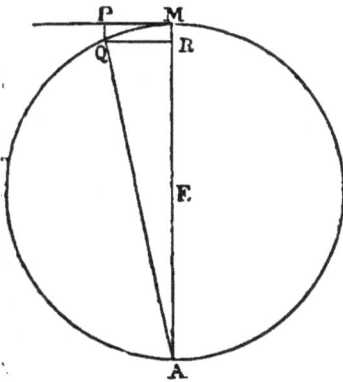

Fig. 127.

attraction of the earth were suspended, then the moon would move in a straight line, and we shall suppose that in one second the moon at M would move along the tangent to its orbit at M to a distance M P. Owing, however, to the attraction of the earth, the moon, instead of being found at P, is really at Q, and as P is very close to M, we may consider the line P Q as parallel to M E. The moon, therefore, has in one second fallen in towards the earth through a distance P Q. It remains to calculate the length of this line P Q. Produce M E to intersect the circle at A, and let fall Q R perpendicular on A M, then it is easy to show that the triangles M Q R and M A Q are similar, and that consequently

$$M R : M Q :: M Q : M A,$$

whence

$$\dot{P}\,Q = M\,R = M\,Q^2 \div M\,A.$$

As the sidereal revolution of the moon occupies 27·32 days, and as the radius of the moon's orbit is about sixty times the radius of the earth, it is easy to compute that the distance M Q through which the moon moves in one second is about 40,120 inches. The diameter of the moon's orbit is approximately 30,130,000,000 inches.

Whence we have

$$P\,Q = 40,120^2 \div 30,130,000,000 = 0\cdot053.$$

The identity between the actual value thus found for P Q and the distance through which a body ought to fall in one second when at a distance from the earth equal to the radius of the earth's orbit, places it beyond any doubt that the motion of the moon is controlled by precisely the same force as that which causes a stone to fall at the surface of the earth.

§ 121. *Perturbations of the Planets.*—The movements of the planets around the sun are not quite so simple as would be the case if the laws of Kepler were accurately fulfilled. The vast mass of the sun, which so enormously exceeds that of all the planets taken together, no doubt subordinates the planets, so that, as a first approximation, we may suppose that the movements of the planets are entirely and solely controlled by the sun. When, however, a nicer calculation is made, it is found that while Kepler's laws are very nearly fulfilled, there are still well-marked divergencies therefrom. The orbits do not constantly remain in the same plane; they are not accurately ellipses, nor is the law of the description of equal areas in equal times accurately fulfilled. The theory of universal gravitation has, however, rendered a most satisfactory account of these irregularities. According to that theory, not only must the sun attract the planets, but the planets must attract the sun, and also each other, and it

is in the mutual attractions of the planets, that the explanation is found of the irregularities to which we have referred.
The complete account which theory thus gives of these
irregularities is a most wonderful confirmation of the truth
of the law of universal gravitation. Conceive a fictitious
planet moving exactly according to Kepler's laws in an orbit
of which the elements are changing slowly and continuously.
The motion of this fictitious planet can be so chosen, that it
shall represent closely the actual motion of the real planet.
The position of the real planet will sometimes be in advance
of, and sometimes behind, the fictitious planet, and these
oscillations are usually called the *periodic variations.* The
changes in the elements of the orbit of the fictitious planet,
which are found necessary to keep that planet in the vicinity
of the real planet, are termed the *secular variations* of the
orbits.

By the theoretical study of the secular variations, certain
very remarkable propositions have been discovered, some of
which shall be here considered. For simplicity we shall
suppose that only two planets are under consideration, and
we shall consider the derangement produced in the orbit of
each planet by the action of the other planet. It has been
demonstrated that, notwithstanding the changes which all the
other elements of each orbit undergo, the length of the axis
major of each ellipse, or the mean distance of each planet
from the sun, remains constant. The importance of this
conclusion will be manifest, when it is remembered that the
time of revolution of a planet round the sun is, by Kepler's
third law, solely dependent upon the axis major of the
elliptic orbit. From the constancy of the length of this axis
major, we can therefore infer that, notwithstanding the
perturbations, the periodic time of each planet remains
constant.

The eccentricities of the orbits of the disturbed planets
undergo certain secular changes. Yet these changes are
confined within such narrow limits, that no considerable

alteration in the configuration of the solar system can arise from this cause. When two planets moving in the same direction are mutually disturbed, the sum of the squares of their eccentricities, each multiplied by a certain numerical factor, remains constant. As the eccentricities are at present small, the constant is small, but if the sum of two positive quantities be small, then each of the quantities must be small: hence it follows that the eccentricities of the disturbed orbits must for ever remain small. A similar proposition holds good for the inclinations of the orbits to the plane of the ecliptic, so that, though the inclination may change, yet it never can change to any large extent.

The propositions we have just referred to, relative to the major axis, the eccentricities, and the inclinations of the orbits of the planets, indicate the stability of the planetary system, as at present constituted. We thus see that the orbits of the planets must preserve very nearly the same relations as they have at present.

§ 122. *Masses of the Planets.*—By the aid of the theory of gravitation, we are enabled to solve the very important problem of determining the masses of the different bodies in the solar system. In the first place, we shall show how the relative values of the masses of the earth and the sun are to be obtained. If we had the means of determining the attraction exerted by the sun and the attraction exerted by the earth on an object at an equal distance from the two bodies, then, since the ratio of these attractions would be equal to the ratio of the masses of the sun and the earth, we could determine the latter ratio. Now we have by the motion of the earth itself a means of making this calculation. The attraction of the earth would make a body at its surface fall 16·1 feet in one second. As the distance of the sun is about 23,300 radii of the earth, the attraction of the earth on a body situated at the distance of the sun is such as to make the body fall in one second through a distance $16·1 \div 23,300 \times 23,300$. From the motion of the earth around the sun we can find the distance through which the earth falls in one second towards the sun (see § 120). We hence deduce that the mass of the sun is 324,000 times that of the earth, or, taking the mass of the sun as unity, we have for the mass of the earth $1 \div 324,000$.

There is but little difficulty in finding the mass of a planet which, like Mars, Jupiter, Saturn, Uranus, or Neptune, is attended by one or more satellites. From the observations of the motions of the satellite, the distance a body falls through in one second towards the planet can be computed, and, from the motion of the planet round the sun, the distance the planet falls in towards the sun in one second is found, whence we can deduce the ratio of the mass of the planet to the mass of the sun.

In the case of planets which, like Venus and Mercury, are not attended by visible satellites, the determination of the masses constitutes a more difficult problem. We are then obliged to resort to the perturbations which those planets produce in the movements of other bodies belonging to the solar system. The theoretical expressions of these perturbations involve the mass of the disturbing body, and as the actual amounts of the perturbations are determined from the observations, it is possible by a comparison between theory and observation to obtain the value of the unknown mass. Although this method may seem very recondite, yet such is the perfection of the theory of perturbation, that it is possible by these means to obtain the masses of certain planets with considerable accuracy. This was very remarkably confirmed in the case of the planet Mars, for, until the discovery of the satellites in 1877 placed a more accurate means of finding the mass of the planet within our reach, our knowledge of the mass of Mars was solely deduced from the perturbations which it produces. Yet the value of the mass which was arrived at from the observations of the satellites merely confirmed, with a slight correction, the value of the mass which had been determined from the perturbations.

The masses of comets have also been estimated.

The orbits of the comets are so irregularly placed with regard to the planets, that it has frequently happened that a comet passes so close to a planet as to be very greatly deranged by the attraction of the planet. Yet in such cases it has appeared that the action of the comet upon the planet is inappreciable, and thus it follows that the mass of the comet must be very small.

§ 123. *Gravitation at the Surface of the Celestial Bodies.* —The gravitation at the surface of a celestial body, which may be regarded as spherical, depends partly upon the mass of the body and partly upon its radius. When we know the mass and also the radius we are enabled to compute the gravitation. Thus in the case of the sun, it appears that the actual intensity of gravitation at his surface is about twenty-seven times greater than the intensity of gravitation at the surface of the earth. By this it is meant that a mass of one pound would require as much to support it at the surface of the sun, as would be adequate to support a mass of twenty-seven pounds at the surface of the earth. In the case of Jupiter the gravitation at the surface is about two and a half times as great as the gravitation at the earth. On the other hand, owing principally to the small mass of the moon, the gravitation at the surface of the moon is about one-sixth part of what it is on the earth. A man on the surface of the moon would be able to raise a load containing six times as much matter as he would be able to lift on the earth.

§ 124. *Perturbations of the Moon.*—According to Kepler's laws, the moon should accurately describe an ellipse about the centre of the earth as one of the foci. It is, however, very easily shown by observation that the movements of the moon are not by any means of so simple a character as those which Kepler's laws would prescribe. The orbit of the moon is not exactly an ellipse, nor does the plane of the orbit remain constant. The moon is, in fact, perturbed just as we have already seen that the planets are perturbed by their mutual actions.

The perturbations of the moon arise, however, from a very different cause from those of the planets. The moon is an appendage to the earth, and it is by the gravitation of the moon towards the earth that the motion of the moon is mainly controlled. The disturbing body in the case of the moon is really the sun. The earth and the moon are of course both attracted by the sun, but the moon is sometimes nearer to the sun (for example, at or near new moon) than it is on other occasions (for example, at or near full moon). Under these circumstances the moon is therefore more powerfully attracted by the sun at some parts of its path than it is at others, and this irregularity in the intensity of the sun's attraction is the cause of the lunar perturbations.

At the time of new moon, the moon, being nearer the sun, is more powerfully attracted by the sun than the earth is attracted by the sun, and consequently the distance from the earth to the moon is augmented. At full moon the attraction of the sun is more powerful on the earth than it is on the moon, and consequently the earth is more drawn in to the sun than the moon, the effect of which is also to increase the distance between the earth and the moon. At the time of first and last quarter, the earth and the moon are practically at the same distance from the sun, and as the distance of the sun is so great, the earth and moon may be considered to be displaced along parallel lines through equal distances, and therefore the distance of the earth and the moon is unaltered. As, however, the distance of the earth from the moon is generally increased by the disturbing effect of the sun, we may assert, as one consequence of the disturbance of the sun, that the orbit of the moon is larger than it would be were that source of disturbance absent. The efficiency of the sun in producing this disturbance increases when the earth is near perihelion ; for then, the distance of the sun from the earth and moon being diminished, the difference of its effects upon the earth and moon becomes more manifest. The orbit of the moon, therefore, varies in size with the different positions of the earth in its annual revolution : according to Kepler's third law the time of revolution of the moon increases when the size of the orbit increases ; and hence it appears that when the earth is in perihelion the periodic time of the moon is a maximum, and when the earth is in aphelion, the periodic time of the moon is a minimum. The effect of this upon the apparent place of the moon is to produce a certain derangement called the *annual equation.* This inequality of the moon was discovered by Tycho Brahe b yobservation, long before the explanation of it was known.

§ 125. *The Secular Acceleration.*—Were the eccentricity of the earth's orbit constant, then the motion of the moon at the end of the year would be the same as it was at the commencement, so far as the annual equation is concerned. Owing, however, to the secular alterations of the earth's orbit, which are produced by the perturbations of the planets, the eccentricity is changing, and consequently there is a gradual alteration in the orbit of the moon. As the axis major of the earth's orbit remains constant, it can be shown that the sun causes the orbit of the moon to be larger, the greater is the eccentricity of the orbit. At present the eccentricity of the earth's orbit from one century to another is gradually decreasing, and consequently there is a gradual decrease in the size of the moon's orbit, and therefore a gradual decrease in the periodic time of the revolution of the moon. About one half of the observed value of the acceleration of the moon's motion can be explained in this manner.

The gradual diminution of the eccentricity of the earth's orbit will not continue indefinitely. All the secular inequalities of the planet's orbits are really periodic, though usually requiring vast durations of time to run through their changes. The time will, however, come, when the diminution of the eccentricity of the earth's orbit will be turned into an increase. The acceleration of the moon now in progress will then, so far as it is due to this cause, be turned into a retardation. This again after the lapse of ages will be turned into an acceleration, and so on indefinitely.

§ **126.** *Cause of the Precession of the Equinoxes.*—The diurnal rotation of the earth about its polar axis is subject to a very remarkable disturbance which gives rise to the phenomena of the precession of the equinoxes and nutation. The disturbance is due to the attraction of the sun and the moon upon the protuberant portions at the earth's equator.

To explain this we have to make use of a theorem in dynamics which we cannot demonstrate in this volume. If the earth be rotating around its polar axis, then that rotation will not be disturbed by any force which passes through the centre of gravity of the earth. In fact, so far as the mere rotation of the earth upon its axis is concerned, we might regard the centre of gravity as a fixed point, and then the force which passed through the centre of gravity could be neutralised by the reaction of the fixed point.

If the earth were a perfect homogeneous sphere, the attraction of the sun or the moon would be a force passing through the centre of the sphere, and would leave the rotation unaffected. Even though the earth were not a perfect sphere, or not homogeneous, still if the attracting body were so far off that all points of the earth might be considered practically at the same distance from the attracting body, the attraction would be a force passing through the centre of gravity of the earth. The sun and the moon, however, are both so comparatively near the earth that we are not entitled to make this supposition, and consequently neither the attraction of the sun or of the moon passes through the earth's centre. To this is due the phenomenon of the precession of the equinoxes.

Let P Q (Fig. 128) represent the axis of the earth, and let s be the position of the attracting body ; then, since the attraction varies inversely as the square of the distance, it follows that the portion of the earth turned towards that

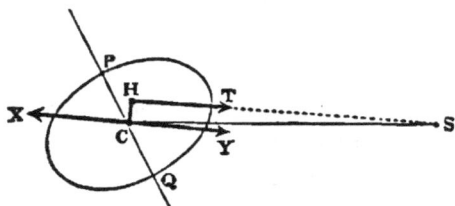

attracting body will be acted upon by a greater force than
the portion towards the remote side, and consequently the
total attraction will be
directed along the line
H S, which passes above
the centre of gravity of
the earth C. Let H T
represent the magnitude
of this force both in
intensity and direction.

FIG. 128.

Through the centre C draw a line X Y parallel to
H T, and let us suppose that equal and opposite forces are
applied at the centre C, each of these forces being equal to
H T. The force C Y may now be left out of view, for as it
acts through the centre of gravity it can have no effect upon
the rotation of the earth around its axis. Thus the effect of
the attracting body upon the earth may, so far as the rotation
of the earth is concerned, be represented by the pair of equal
parallel and opposite forces H T and C X. Such a pair form
what is known in mechanics as a *couple.*

It would seem as if the immediate effect of this couple
would be to turn the earth so as to bring its polar axis C P
perpendicular to the line C S, or (supposing the sun to be
the attracting body under consideration) to bring the plane
of the equator to coincide with the plane of the ecliptic.
The effect of the couple is, however, so entirely modified by
the fact that the earth is in a state of rapid rotation, that,
paradoxical as it may appear, the real effect of the couple
is not to move C P in the plane of the paper, but to make
C P move perpendicular to the plane of the paper.

In explanation of this apparent paradox, we may remark
that, in a miniature form, every schoolboy is already
acquainted with a precisely analogous phenomenon in the
motion of a common peg-top. In Fig. 129 the line P Z is
vertical, P C is the axis of the peg-top; and C is the centre of
gravity of the peg-top. If the peg-top, when not spinning,

were placed in the position represented in the figure, the force of gravity acting along c H would immediately cause it to tumble over, the line c P moving in the plane of the paper. But when the peg-top is in a state of very rapid rotation, the circumstances are entirely different. Everyone has observed that the axis c P, so far from falling in the plane of the paper, commences to move perpendicularly to the plane of the paper, and will, in fact, describe a right circular cone around P z as an axis. It is undoubtedly true that after a time the angle z P c begins to increase, and that before long

Fig. 129.

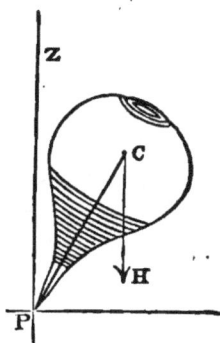

the peg-top really does tumble down, but this is solely due to the influence of disturbing forces—viz. friction at the point, and the resistance of the air—and if these forces could be evaded the speed with which the peg-top spins would be undiminished, and so long as that speed remained unaltered, so long would the axis of the peg-top continue to describe the right circular cone around the line P z.

Assuming that what holds good in the case of the peg-top holds good in the colossal case of the earth itself, we should expect to find that the axis P c (Fig. 128), instead of moving towards s, and thus diminishing the obliquity of the ecliptic, would commence to move perpendicularly to the plane of the paper and thus not alter the obliquity of the ecliptic at all. The axis of the earth would thus describe a right circular cone, of which the axis is perpendicular to the plane of the ecliptic, and this is actually the motion which the precession of the equinoxes requires.

The precession of the equinoxes is due to the action of both the sun and the moon. Owing, however, to the proximity of the moon, its effect is greater than that of the sun. In fact, of the total amount, about one-third is due to the sun and the remainder to the moon.

§ **127.** *Nutation.*—The efficiency of the moon in the production of the phenomenon just considered depends upon the position of the orbit of the moon with respect to the ecliptic, and as this is changing, so the efficiency of the moon will undergo certain changes, the period of which is equal to that of the revolution of the moon's nodes. The effect of this is that the phenomenon of precession is not so simple as it would be were the moon actually in the plane of the ecliptic. Thus the true position of the pole oscillates about its mean place as determined by precession, and this oscillation is the *nutation.*

CHAPTER X.

STARS AND NEBULÆ.

§ **128.** *Star Clusters.*—The stars are very irregularly distributed over the surface of the heavens. This is, indeed, sufficiently obvious to the unaided eye, and it is confirmed by the telescope. In certain places we have a dense aggregation of stars of so marked a character as to make it almost certain that the group must be in some way connected together, and that, consequently, the aggregation is real, and not only apparent, as it might be if the stars were really only accidentally near to the same line of sight, and, consequently, appeared to be densely crowded together, when, in reality, they might be at vast distances apart.

Of such a group we have a very well known example in the group called the Pleiades, which we have already mentioned (§ 25). Most persons can see six stars in the Pleiades without difficulty, but with unusually acute vision more can be detected. With the slightest instrumental aid, however, the number is very greatly increased, and the group is seen to consist of perhaps 100 stars.

Another illustration of such a group is an object in the constellation Cancer known as the Præsepe, or the Beehive. To the unaided eye this is merely a dullish spot on the sky, not well seen unless the night is very clear. A telescope shows, however, that this dullish spot is really an aggregation of perhaps 60 small stars.

By far the most splendid object of this kind in the northern hemisphere is the cluster in the Sword-handle of Perseus. We have here two groups of stars close together, and, when seen in a good telescope, the multitudes of these stars, and their intrinsic brightness, form a most superb spectacle.

§ 129. *Globular Clusters.*—The objects known as *star clusters* are exceedingly numerous. Among them are several which are remarkable telescopic objects, not for the brightness (even in the telescope) of the individual stars composing the star cluster, but for the vast numbers in which the stars are present, and for the closeness in which they lie together. These objects are often known as *globular clusters*, because the stars forming them seem to lie within a globular portion of space, and they frequently appear to be much more densely compacted together towards the centre of the globe. In fact, at the centre of one of these splendid objects it is in some cases almost impossible to discriminate the individual stars, so closely is their light blended. As Sir John Herschel says, 'it would be a vain task to attempt to count the stars in one of these globular clusters. They are not to be reckoned by hundreds, and on a rough calculation grounded on the apparent intervals between them at the borders and the angular diameter of the whole group, it would appear that many clusters of this description must contain at least five thousand stars compacted and wedged together in a round space whose angular diameter does not exceed eight or ten minutes—that is to say, in an area not more than a tenth part of that covered by the moon.'

The most remarkable of these objects in the northern hemisphere is the globular cluster in Hercules (right ascension, $16^h 37^m$; declination, $+36° 43'$). To the unaided eye or in a small telescope this looks like a dull nebulous spot, and it requires a good telescope to exhibit it adequately.

§ 130. *Telescopic Appearance of a Star.*—The appearance of a star in a telescope differs in a most marked manner from the appearance of one of the larger planets. In the case of the planet we can see what is called the 'disk:' we can actually observe that the planet appears circular, and that it is presumably a globe with an appreciable diameter. In most cases too we can discern markings upon the globe of the planet of which drawings may be made. Indeed, as we have already mentioned (§ 102), we can see in the planet Venus changes precisely analogous to the phases of the moon, thus proving, of course, that the planet possesses an appreciable disk. By increasing the magnifying power of the telescope the size of the disk can be increased, though, of course, at the expense of its intrinsic brightness.

Widely different, however, is the telescopic appearance of a fixed star. Even the most powerful telescope only shows a star as a little point of light. By increasing the optical power of the telescope, the brilliancy of the radiation from this point can be increased, but no augmentation of the magnifying power has hitherto sufficed to show any appreciable 'disk' in the great majority of the fixed stars which have been examined. How is this to be explained? The answer is to be sought not in the real minuteness of the stars, but in the vast distances at which they are situated. In order to form some estimate of the real diameter which the stars do subtend at the eye, let us suppose that our sun were to be moved away from us to a distance comparable with that by which we are separated from those stars which are nearest to us.

For this purpose the sun would have to be transferred to a distance not less than 200,000 times as far as his present distance from the earth. Let A B (Fig. 130) denote the diameter of the sun, and let E be the position of the earth. Then, as we have already seen (§ 11), the circular measure of the angle which the sun subtends at the earth is practically equal to A B ÷ E B. Now suppose the sun be transferred to the position indicated by A′ B′, then the angle which he would subtend in the new position is A′ B′ ÷ E B′.

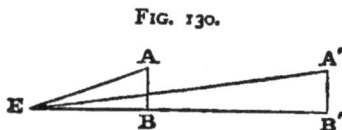

FIG. 130.

Hence the ratio of the angles which the apparent diameter of the sun subtends at the eye at the two different distances is

$$\frac{A\,B}{E\,B} \div \frac{A'\,B'}{E\,B'};$$

but as the *real* diameter of the sun is the same in both cases, we must have

$$A\,B = A'\,B',$$

and hence the ratio just written becomes

$$\frac{E\,B'}{E\,B}$$

Hence we infer that the angle which the sun's diameter subtends at the eye varies inversely as his distance from the observer.

If, therefore, the sun were to be carried away from us to a distance 200,000 times greater than his present distance, the angle which his diameter at present subtends would be diminished to the 200,000th part of what it is at present. Assuming, as we may do for rough purposes, that the sun's apparent diameter is half a degree, it follows that the apparent diameter when translated to the distance of a star would be expressed in seconds by the fraction

$$\frac{1800}{200000} = 0''·009.$$

In other words, the sun's diameter would then subtend an angle less than the hundredth part of a single second.

It is at present, at all events, quite out of the question to suppose that a quantity so minute as this could be detected by our instruments. Even were it ten times as great it would be barely appreciable, nor unless it were at least fifty times as great would we be able to measure it with any approach to precision.

It is, therefore, clear that we cannot infer the actual dimensions of the stars from their minute apparent size in the telescope.

§ 131. *Variable Stars.*—We have mentioned the mode of classifying stars by their magnitudes; we have now to add that there are some stars to which this method cannot be applied. These are called *variable* stars, inasmuch as their brightness is not constant, as that of the majority of stars appears to be. There are some hundreds of stars in the heavens the brightness of which is now known to change. It would be difficult here to describe in detail the different classes of the variable stars, so we shall merely give a brief account of one or two of the most remarkable.

In the constellation Perseus is a bright star, Algol (right ascension, $3^h 0^m$; declination, $+40°27'$). Owing to the convenient situation of this star, it may be seen every night in the northern hemisphere. Algol is usually of the second magnitude, but in a period of between two and three days, or more accurately in a period of $2^d 20^h 48^m 55^s$, it goes through a most remarkable cycle of changes. These changes commence by a gradual diminution of the brightness of the star from the second magnitude down to the fourth in a period of three or four hours. At the fourth magnitude the star remains for twenty minutes, and then begins to increase in brightness again until, after another interval of three or four hours, it regains the second magnitude. Algol continues at the second magnitude for about $2^d 13^h$, when the same series of changes commences anew.

Another very remarkable star belonging to the class of variables is o Ceti or Mira (right ascension, $2^h 13^m$; declination, $-3° 34'$). The period of the changes of this star is $331^d 8^h$. For about five months of this time the star is quite invisible; it then gradually increases in brightness until it becomes nearly of the second magnitude. After remaining at its greatest brightness for some time it again gradually sinks down to invisibility.

§ 132. *Proper Motion of Stars.*—We have hitnerto frequently used the expression ' fixed stars ; ' we have now to introduce a qualification which must be made as to the use of the word *fixed* with reference to the stars. Compared with the planets, the places of which are continually changing upon the surface of the celestial sphere, the stars may, no doubt, be termed fixed ; but when accurate observations of the places of the stars made at widely distant intervals of time are compared together it is found that to some of the stars the adjective *fixed* cannot be literally applied, as it is undoubtedly true that they are moving. It is true that the great majority of what are called fixed stars do not appear to have any discernible motion, and, even those which move most rapidly, when viewed from the vast distances by which they are separated from the earth, only traverse but a very minute arc of the heavens in the course of a year. The most rapidly moving star hardly moves over an arc on the celestial sphere of $10''$ per annum. A motion so slow as this is inappreciable without very refined observations. The moon has a diameter which subtends at the eye an angle which we may roughly estimate at half a degree, and to move over a space equal to the diameter of the moon on the surface of the heavens would require a couple of centuries even for the most rapidly moving star.

We have already had frequent occasion to discriminate between real motion and apparent motion, and are therefore naturally tempted to enquire whether the motions of the stars which we have been considering are real, or whether they can be explained as merely apparent motions. Now where must we look for the cause of the apparent motion? It is manifest that the annual motion of the earth around the sun could not possibly explain the appearances which have been observed. The annual motion of the earth around the sun would have an effect which must be clearly periodic in its nature. In fact, it would be merely the annual parallax which we have already considered (§ 77). The motions which we have to explain are not (so far as we know at present) of a periodical character; for the stars which possess this motion usually appear to move continually along great circles.

§ **133.** *Motion of the Sun through Space.*—It was therefore suggested by Sir W. Herschel that possibly a portion of the proper motions of the stars could be explained by the supposition that the sun, carrying also the retinue of planets, and all the other bodies forming the solar system, was actually moving in space. On this supposition, it is clear that those stars which were sufficiently near to us must have an apparent proper motion. If the motion of the sun were directed along a straight line towards a certain point of the heavens, then the apparent place of a star at that point would be unaffected by the motion of the sun; but all other stars would spread away from that point, just as when you are travelling along a straight road the objects on each side of the road appear to spread away, as it were, from the point towards which your journey was directed.

It was found by Sir W. Herschel that a considerable portion of the observed proper motions of the stars could be explained by the supposition that the sun was moving towards a point in the heavens near to the star λ Herculis. The investigations of other astronomers have tended to confirm this very remarkable deduction as to the sun's motion in space, and have led them to conclude that the sun is moving towards a point in the heavens which (considering the difficulty of the investigation) is exceedingly close to the point determined by Sir W. Herschel. The right ascension of the point thus determined is 263° 43′ ·9, and its declination is 25° 0′·5 (Dunkin).

Not only has the direction in which the sun moves been determined, but the observations also determine the velocity of the motion. It is found that in one year the sun probably moves through an arc of 0″·4103 as seen from the distance of a star of the first magnitude.

We thus see that the real motion of the earth in space is of a very complicated character; for though it describes an ellipse about the sun in the focus, yet the sun is itself in constant motion, and consequently the real motion of the earth is a composite movement, partly arising from its own proper motion around the sun, and partly arising from the fact that, as a member of the solar system, the earth partakes of the motion of the solar system in space.

§ 134. *Real Proper Motion of the Stars.*—It should, however, be observed that after every possible allowance has been made for the effect of the motion of the solar system, there remain still outstanding certain portions of the proper motions of the stars, which are only to be explained by the fact that the stars in question really are in actual movement.

Nor, if we reflect for a moment, is there much in the last conclusion to cause surprise. The first law of motion combined with the most elementary notions of probabilities will show us how exceedingly improbable *rest* really is. Among all the possible kinds of motion, infinitely various both in regard to velocity and in regard to direction, there is no one which is not *à priori* just as probable as another; there is no one which is not *à priori* just as probable as *rest*. Hence even if there were no causes tending to produce change from an initial state of things, it would be infinitely improbable that any body in the universe was absolutely at rest. But even if a body were originally at rest it could not remain so. Distant as the stars are from the sun, and from each other, they must still, so far as we know at present, act upon each other. It is true that these forces acting across such vast distances may be slender, but, great or small, they are incompatible with rest, and hence we may be assured that every particle in the universe (with, it is conceivable, one exception) is in motion.

We are thus led to believe that the fact that proper motion has only been detected in comparatively few stars is to be attributed, not to the actual absence of proper motion, but rather to the circumstance that the stars are so exceedingly far off that, viewed from this distance, the motions appear so small that they have not hitherto been detected. It can hardly be doubted that, could we compare the places of stars now with the places of the same stars 1,000 years ago, most of them would be found to have changed. Unfortunately, however, the birth of accurate astronomical observation is so recent that we have only imperfect means of making this comparison, for the ancient observations which have been handed down to us are not sufficiently accurate to afford trustworthy results.

§ 135. *Double Stars.*—We have already alluded to the occasional close proximity in which stars are found on the celestial sphere. In many cases we have the phenomenon which is known as a *double star*. Two stars are frequently found which appear to be so exceedingly close together, that the angular distance by which the stars are separated is less than one second of arc, and an exceedingly good telescope is required to 'divide' such an object, which, when viewed in an inferior instrument, would appear to consist only of a single star. The great majority of double stars known at present are, however, not nearly so close together. About 10,000 objects have now been discovered which are included under the term double stars, though it must be added that the components of many of these are at a considerable distance apart.

We shall here describe briefly a few of the most remarkable of these very interesting objects.

§ 136. *The Double Star Castor.*—One of the finest double stars in the heavens is Castor (α Geminorum). (Right ascension, 7^h 26^m; declination, $+32°$ $17'$.) Viewed by the unaided eye, the two stars together resemble but a single star, but in a moderately good telescope it is seen that what appears like one star is really two separate stars. The angular distance at which these two stars are separated is about five seconds. One of the stars is of the third magnitude, and the other is somewhat less. The reason why the unaided eye cannot distinguish the separate components is their great proximity. The angular distance of the components is the same angle as that which is subtended by a length of one inch at a distance of 1,146 yards, and is therefore quite inappreciable without instrumental aid. The question now arises whether the propinquity of the two stars forming Castor is apparent or real. This propinquity might be explained by the supposition that the two stars were really close together compared with the distance by which they are separated from us. Or it could equally be explained by supposing that the two stars, though really far apart, yet appeared so nearly in the same line of vision that, when projected on the surface of the heavens, they seemed to be close together. It cannot be doubted that in the case of many of the double stars, especially those in which the components appear tolerably distant, the propinquity is only apparent and arises from the two stars being near the same line of vision. But it is also undoubtedly true that, in the case of very many of the double stars, especially among those belonging to the class which includes Castor, the two stars are really at about the same distance from us, and therefore, as compared with that distance, they are really close together.

Many double stars of this description exhibit a phenomenon of the greatest possible interest. If we imagine a
great circle to be drawn from one of the two component stars
to the north pole of the celestial sphere, then the angle
between this great circle and the great circle which joins the
two stars is termed the *position angle* of the double star. By
the ingenious instrument called *the micrometer* (§ 48), which is
attached to the eye end of a telescope mounted equatorially,
it is possible to measure both the position angle of the two
components of a double star, and also the *distance* of the
two stars expressed in seconds of arc. When observations
made in this way are compared with similar observations of
the same double star, made after an interval of some years,
it is found in many cases that there is a decided change both
in the distance and in the position angle. In the case of
the double star Castor, at present under consideration, it is
true that the movement is very slow. It is, however, un-
doubted that in the course of some centuries [1] one of the
components will revolve completely around the other.

§ **137.** *Motion of a Binary Star.*—The theory of gravitation affords us the explanation of these changes. We have
seen how in the case of the sun and the planets, each
planet describes around the sun an orbit of which the figure is
an ellipse, with the sun in one focus, while the law according
to which the velocity changes is defined by the fact that
equal areas must be swept out in equal times. The circumstances presented by the sun and a planet (the earth, for
example) are somewhat peculiar, and Kepler's laws must be
stated somewhat differently before they can be applied with
strict generality to the motion of a *binary star* (as one of the
revolving double stars is termed). In the case of the sun and
the earth we have a comparatively minute body moving
around a very large body. In fact, as the mass of the sun
is more than 300,000 times greater than the mass of the
earth, we may neglect the mass of the earth in comparison

---

[1] The periods assigned for the time of revolution of Castor vary from
232 years (Mädler) to 996 years (Thiele).

with the mass of the sun. Thus, in speaking of Kepler's laws as applied to the motion of a planet around the sun, we often regard the centre of the sun as a fixed point, and attribute all the motion which is observed to the planet.

It is manifest, however, that some modification of Kepler's laws is necessary before we can apply them to the case of most of the binary stars. In the case of Castor, though the two components are not exactly equal, yet they are so nearly so that it would obviously be absurd to regard even the larger of them as a fixed point while the whole orbital motion was performed round it by the other. The fact of the matter is, that *both the components* are in motion, each under the influence of the attraction of the other, and that what we actually observe and measure is only the relative motion of the components.

It would lead us beyond the limits of this book to endeavour to prove the more generalised conception of Kepler's laws which we shall now enunciate. Let us suppose the case of a binary star so far removed from the influence of other stars or celestial bodies that their attraction may be regarded as insensible. Then each of the two components of the binary star is acted upon by the attraction of the other component, but by no other force. We suppose a straight line A B to be drawn connecting the centres of the stars, and we divide this line into two parts, A G and B G, in the proportions of the masses of the two stars, so that the point of division G lies between the two stars and nearer to that star, A, which has the greater mass. The point G thus determined is the *centre of inertia* of the two stars. It can be proved that, however the stars A and B may move in consequence of their mutual attractions, the point G will either remain at rest or will move uniformly in a straight line. It can be shown that each of the stars A and B will move in an elliptic orbit around the point G as the focus, and that each star will describe equal areas in equal times.

It can also be shown that, although both of the stars are in motion, yet the relative motion of one star about the other—i.e. the motion of the star B about the star A as it would be seen by an observer who was stationed on A—is precisely the same as if the mass of the star A were augmented by the mass of the star B, and as if A were then at rest and B moved round it just as a planet does around the sun.   To this apparent motion of B around A, Kepler's laws will strictly apply.   The orbit of B is an ellipse of which A is one of the foci, and the radius vector drawn from A to B will sweep out equal areas in equal times.

It is natural to enquire whether these theoretical anticipations with respect to the motions of the binary stars are borne out by observation.   We have no reason to expect that we shall actually see motions of the simple character which we have described.   It is to be recollected that the plane in which the orbit is described may be inclined *in any way* to the surface of the celestial sphere.   Consequently the orbit which we shall see may only be the projection of the real orbit upon a plane which is perpendicular to the line joining the binary star to the eye.   We have therefore to consider what modifications the orbit may undergo by projection.   It can be shown that if the original orbit be elliptic, the projected orbit will be elliptic also; but it also appears that though the star A was the focus of the original orbit, it would, generally, not be the focus of the projected orbit.   The law of the description of equal areas in equal times would hold equally true both in the original orbit and in the projected orbit.

By a comparison of observations made at different times it is possible to plot down the actual position of the star B with respect to the star A, at the corresponding dates.   It is found that in the case of several binary stars the orbit thus formed is elliptic, and it is possible, by a consideration of the position of the point A in this ellipse, to determine the position of the true orbit with reference to the celestial sphere and the various circumstances connected with the motion.

In this way the true orbits of several of the most re-markable among the binary stars have been determined. One of the most rapidly revolving double stars appears to be 42 Comæ Berenices, which accomplishes its revolution in a period of 25·7 years. The two components of this star are exceedingly close together, the greatest distance being about one second of arc. There is very great difficulty in making accurate measurements of a double star so close as this one. Consequently more reliance may be placed upon the determination of the orbits of other binary stars, the components of which are farther apart than those of 42 Comæ Berenices. Among these we may mention a very remarkable binary star, ξ Ursæ Majoris. The distance of the two components of this star varies from one second of arc to three seconds. The first recorded observation of the distance and position angle of this star was by Sir W. Herschel in 1781, and since that date it has been repeatedly observed. From a comparison of all the measurements which have been made it appears that the periodic time of the revolution of one component of ξ Ursæ about the other is 60 years, and it is exceedingly improbable that this could be erroneous to the extent of a single year. Thus this star has been observed through more than one entire revolution.

§ **138.** *Dimensions of the Orbit of a Binary Star.*—In the determination of the size of the orbit of a binary star all we can generally ascertain is, of course, the diameter of the orbit measured in seconds of arc. Actually to determine the number of miles in the diameter of the orbit, it would be further necessary for us to know the distance at which the binary star is separated from the earth. This distance is in the great majority of cases entirely unknown to us at present. There are, however, one or two exceptions.. Of these we shall mention Sirius. Early in the present century the proper motion of this star was found to be affected by an irregularity which showed that an unseen body must be moving around it and disturbing its motion by its attraction.

After a hundred years of observation the orbit of this body was calculated, and it was shown that the irregular motion of Sirius could be accounted for by supposing that the disturbing satellite had a period of about fifty years. The satellite was actually discovered by Alvan Clark in 1862, and was found to be moving around Sirius at a mean angular distance of about seven seconds. The annual parallax of Sirius is found to be 0″·25—that is to say, the radius of the earth's orbit, viewed from the distance of Sirius, subtends an angle of 0″·25. It therefore follows that the real distance of the companion of Sirius must exceed the distance of the earth from the sun in the ratio that 7″ exceeds 0″·25—that is, it is 28 times as great.

§ **139.** *Determination of the Mass of a Binary Star.*— When we know the diameter of the orbit of a binary star and its periodic time, we are able to compute the sum of the masses of the two component stars. This is an exceedingly interesting subject, inasmuch as it affords us a method of comparing the importance of our sun to the other stars as far as *mass* is concerned.

Let us first consider what the periodic time of a planet would be if it revolved round the sun in an orbit of which the radius were twenty-eight times that of the earth's orbit. According to Kepler's third law, the square of the periodic time is proportional to the cube of the distance. Consequently, since the earth revolves around the sun in one year, it follows that a planet such as we have supposed would revolve around the sun in a period of time which was equal to the square root of the cube of 28, i.e. to 148 years very nearly. According to the latest results it would appear that the periodic time of the revolution of the satellite of Sirius is 49·3 years, i.e. the velocity with which the motion takes place is greater than it would be if the mass of Sirius only equalled the mass of the sun.

The ratio which the mass of the sun (augmented, it should in strictness be said, by the mass of the earth) bears to that of Sirius and its satellite taken together can be ascertained. For this we require the following principle, which for the present we shall take for granted.

If two bodies, A and B, are revolving in consequence of their mutual attraction, then the sum of the masses is inversely proportional to the square of the periodic time, supposing the mean distance of A and B to remain unaltered. It therefore appears that the following proportion is true:—

$$\frac{\text{Mass of Sun and Earth}}{\text{Mass of Sirius and Satellite}} = \left(\frac{49\cdot3}{148}\right)^2$$

It follows from this that the mass of Sirius and its satellite taken together are about nine times the mass of the sun.

Now though it is true that subsequent observations may necessitate corrections in these results, yet we may be pretty confident that the mass of Sirius is several times as great as that of our sun. The most uncertain part of the data is the annual parallax of Sirius, which has not yet been certainly determined, and may deviate from $0''\cdot25$ by a considerable fraction of its total amount.

§ 140. *Colours of Double Stars.*—Among the most pleasing and remarkable phenomena presented by double stars are the beautiful colours which they often present. The effect is occasionally heightened by the circumstance that the colours of the two components are frequently not only different, but are contrasted in a marked manner. Conspicuous among these objects is a very beautiful double star, γ Andromedæ. The two components of this star are orange and greenish blue. Attentive examination with a powerful telescope shows also that the greenish blue component consists of two exceedingly small stars close together. While considering this subject, it should be remarked that isolated stars of a more or less reddish hue are tolerably common in the heavens, the catalogues containing some four or five hundred stars of this character. Among those visible to the naked eye perhaps the most conspicuous is the bright star α Orionis. Stars of a greenish or bluish hue are much less common, and it is very remarkable that, with very few exceptions, a star of this colour is not found isolated, but always occurs as one of the two components of a 'double star.'

§ 141. *Nebulæ.*—There are a great number and variety of objects in the heavens which are known under the general term of 'Nebulæ.' The great majority of these objects are invisible to the naked eye, but with the aid of powerful telescopes some thousands of such objects have been already discovered. Of these objects, which for convenience are grouped together, many are undoubtedly mere clusters of stars such as those of which we have already given some account. It is, nevertheless, tolerably certain that many of the objects termed nebulæ are not to be considered as mere clusters of stars, though their real nature has, as yet, been only partially determined.

§ 142. *Classification of Nebulæ.*—The following analysis of the different objects, which are generally classed under the name of nebulæ, has been made by Sir William Herschel, to whom the discovery of a vast number of nebulæ is due :—

1. Clusters of stars, in which the stars are clearly distinguishable ; these are again divided into globular and irregular clusters.

2. Resolvable nebulæ, or such as excite a suspicion that they consist of stars, and which any increase of the optical power of the telescope may be expected to resolve into distinct stars.

3. Nebulæ, properly so called, in which there is no appearance whatever of stars, which again have been subdivided into subordinate ones, according to their brightness and size.

4. Planetary nebulæ.

5. Stellar nebulæ.

6. Nebulous stars.

The first of these classes is that which we have already described (§§ 128, 129). The resolvable nebulæ, which form the second class, are to be regarded as clusters of stars, which are either too remote from us, or the individual stars of which are too faint to enable them to be distinguished. Among the most remarkable objects at present under consideration are the oval nebulæ. They are of all degrees of eccentricity, some being nearly circular, while others are so elongated as to form what have been called 'rays.' The finest object of this class is the well-known nebula in the girdle of Andromeda. This object is just visible to the naked eye as a dullish spot on the heavens. Viewed in a powerful telescope it is seen to be a nebula about $2\frac{1}{2}°$ in length and $1°$ in breadth. It thus occupies a region on the heavens five times as long and twice as broad as the diameter of the full moon. The marginal portions are faint, but the brightness gradually increases towards the centre, which consists of a bright nucleus. This nebula has never actually been resolved, though it is seen to contain such a multitude of minute stars that there can be little doubt that, with suitable instrumental power, it would be completely resolved.

Among the rarest, and indeed the most remarkable, nebulæ are those which are known under the name of the 'Annular Nebulæ.' The most conspicuous of these is to be found in the constellation Lyra; it consists of a luminous ring; but the central vacuity is not quite dark, but is filled in with faint nebula, 'like a gauze stretched over a hoop' (Sir John Herschel).

Planetary Nebulæ are very curious objects; they derive their name from the fact that, viewed in a good telescope, they appear to have a sharply defined more or less circular disc, immediately suggesting the appearance presented by a planet. These objects are generally of a bluish or greenish hue. Their apparent diameter is small; the largest of them is situated in Ursa Major, and the area it occupies on the heavens is less than one-hundredth part of the area occupied by the full moon. Still the intrinsic dimensions of the object must be great indeed. If it were situated at a distance from us not greater than that of the star 61 Cygni, the diameter of the globe which the planetary nebula occupies would be seven times greater than the diameter of the orbit of the outermost planet of our system.

Among the class of Stellar Nebulæ one of the most superb objects visible in the heavens must be included. The object to which we refer is the great nebula in the Sword-handle of the constellation of Orion. The star θ Orionis consists of four pretty bright stars close together, while in a good telescope at least two others are visible, the whole presenting the almost unique spectacle of a sextuple star. But around this star, and extending to vast distances on all sides of it, is the great nebula in Orion. The most remarkable feature of this nebula is the complexity of detail which it exhibits. The light is of a slightly bluish hue, and under the power of great telescopes portions of it are seen to be thickly strewn with stars. Perhaps it would be more correct to say that portions of it *contain* stars ; for there is good reason to believe that in this nebula, as well as in some others, a part of the light which we receive is due to glowing gas.

The last of the different kinds of nebulæ to which we shall allude is the class of objects known as nebulous stars. By a 'nebulous star' we are to understand a star surrounded by a luminous haze, which is, however, generally so faint as only to be visible in powerful instruments.

§ 143. *Composition of Light.*—We shall now give a brief account of a very remarkable method which has recently been applied with great success to the examination of the heavenly bodies. This method is termed *spectrum analysis.* The peculiar feature of spectrum analysis is, that with the assistance of a telescope it actually gives us information as to the nature of the elementary substances which are present in some of the celestial bodies. To explain how this is accomplished it will be necessary for us to refer again to some properties of light which were already alluded to (§ 8). A ray of ordinary sunlight consists in reality of a number of rays of different colours blended together. The 'white' colour of ordinary sunlight is due to the joint effect of the several different rays. We have, however, the means of separating the constituent rays of a beam of light and examining them individually. This is due to the circumstance that the amount of bending which a ray of light undergoes when it passes through a prism varies with the colour of the light.

§ **144.** *Construction of the Spectroscope.*—Suppose A B C (Fig. 131) to represent a prism of flint glass. If a ray of ordinary white light travelling along the direction P Q falls upon the prism at Q, it is bent by refraction, so that the direction in which it traverses the prism is different from the direction in which it was moving when it first encountered the prism. The amount of the bend-

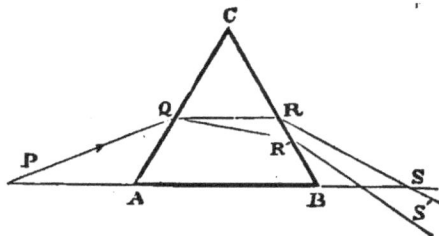

FIG. 131.

ing is, however, dependent upon the colour of the light. In a beam of white light we have blended together the seven well-known prismatic colours, viz. red, orange, yellow, green, blue, indigo, violet. We shall trace the course of the first of these and the last. The red light is the least bent ; it travels along (let us suppose) the direction Q R until it meets the second surface of the prism at R ; it is then again bent at emergence, and finally travels along the direction R S. On the other hand, the violet portion of the incident beam, which originally travelled along the direction P Q, is more bent at each refraction than the red rays. Consequently after the first refraction it assumes the direction Q R', and after the second refraction, the direction R′ s′. The intermediate rays of orange, yellow, green, blue, and indigo, after passing the prism, are found to be more refracted than the red rays, and less refracted than the violet rays ; they are, therefore, found in the interval between the lines R s and R′ s′.

We have therefore, in the prism, a means of decomposing a ray of light and examining the different constituents of which it is made. We shall now show how this is practically applied in the instrument known as the *spectroscope*. The principle of this instrument may be explained by reference to Fig. 132. At s is a narrow slit, which is supposed to be perpendicular to the plane of the paper. Through this slit

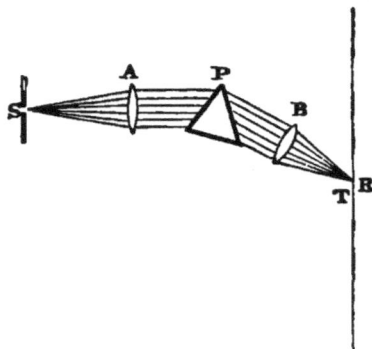

a thin line of light passes, and it is this thin line of light which is to be examined in the spectroscope. After passing through s, the light diverges until it falls upon an achromatic lens placed at A. This lens is to be so placed that the distance A s is equal to the focal length of the lens ; it therefore follows that the beam diverging from s will, after refraction through the lens A, emerge as a beam of which all the constituent rays are parallel. Let us now for a moment fix our attention upon the rays of some particular colour. Suppose, for example, the *red rays*. The parallel beam of red rays will fall upon the prism P. Since each of these rays has the same colour, it will, on passing through the prism, be deflected through the same angle, and, therefore, the beam which consisted of parallel rays before incidence upon the prism will consist of parallel rays after refraction through the prism, the only difference being that the entire system of parallel rays will be bent from the direction which they had before. In this condition the rays will fall upon the achromatic lens B, which will bring them to a focus at a point R, where we shall suppose a suitable screen to be placed. Thus the red rays which pass through the slit at s will form a red image of the slit upon the screen at R.

But what will be the case with the violet constituents of the light which passes through s? The violet rays will fall upon the lens A, and will emerge as a parallel beam (for we have supposed the lenses A and B to be both achromatic), the parallel violet beam will then fall upon P, and it will emerge from P also as a beam of parallel rays. It will, however, be *more deflected* than the beam of red rays, but still not so much so as to prevent it falling upon the lens B, which will make it converge so as to form an image at T near to the red image at R, but somewhat below it.

Let us suppose the slit at s to be exceedingly narrow, and let us suppose that the beam of light which originally passed through s contained rays of *every degree* of refrangibility from the extreme red to the extreme violet. We should then have on the screen an indefinitely great number of images of the slit in different hues, and these images would be so exceedingly close together that the appearance presented would be a band of light equal in width to the length of the image of the slit, and extending from R to T. This band, the colour of which gradually changes from red at R to violet at T, is known as the *prismatic spectrum.* Instead of the screen the eye itself may be employed to receive the light which emerges from the lens B, so that the spectrum is impressed upon the retina. For the more delicate purposes of spectrum analysis this plan is always adopted.

Suppose that the light which was being examined consisted only of rays of certain special refrangibilities, the spectrum which would be produced would then only show images of the slit corresponding to the particular rays which were present in the beam. Consequently, the spectrum

would be 'interrupted,' and the character of the spectrum
would reveal the nature of the light of which the beam was
composed.

This may be made to give us most valuable information
with reference to the nature of the source from which the
·light emanates.   We do not here attempt to enter into the
matter further than is necessary to show how the method can
be applied astronomically.   When the light from some of
the nebulæ (especially those of a bluish hue) was examined
in the spectroscope, it was found by Huggins that by far the
greater portion of the light is concentrated into two or three
bright lines.   This proves that a great portion of the light
from nebulæ of this particular character consists of rays
possessing the special refrangibilities corresponding to the
observed rays.

We have thus an accurate means of comparing the light
which comes from the nebulæ with the light from other
sources.   If a glass tube contain a small quantity of gas,
and if a galvanic current be passed through the tube (we do
not here attempt to enter into details) the gas inside the
tube may be raised to a temperature so exceedingly high
that it will become luminous, and the light which emanates
from it can be examined by means of the spectroscope.   It
is found that each different kind of gas yields light which in
the spectrum forms lines of so marked a character as to
make the spectrum characteristic of the gas.   It has thus
been discovered by Huggins that the light from several of
the nebulæ brings evidence that in some of these distant
objects substances are present with which we are familiar on
the earth.   He has thus found that there is excellent reason
to believe that several of the nebulæ are at least partially
composed of glowing gaseous material, and that among their
constituents are to be found hydrogen and nitrogen, which
are both elements of much importance on the earth.

It has been ascertained, by the aid of spectrum analysis,

that the majority of the fixed stars are probably bodies of the same general character as our sun, but with individual peculiarities, while in the sun and in several of the stars elementary substances are found identical with those on the earth.

CHAPTER XI.

THE STRUCTURE OF THE SUN.

§ **145.** *The Sun's Spots.*—Spots on the surface of the sun generally consist of two well-marked portions; the interior portion, or *nucleus*, which appears black by contrast with the general brilliancy of the sun, and the surrounding portion or *penumbra*, which is of a greyish colour.

It was discovered by Wilson that the solar spots are really cavities actually hollowed out in the brilliant surface of the sun, which is called the *Photosphere.* On November 22, 1769, Wilson observed a spot, the nucleus of which was nearly circular, and the penumbra was also very nearly a circle concentric with the nucleus. Wilson watched this spot as it gradually drew near to the edge of the sun by reason of the rotation of the sun. He observed that the penumbra soon ceased to be symmetrical, and that while the portion of the penumbra nearest the centre of the sun gradually diminished, that on the other side preserved very nearly constant dimensions.

Let A (Fig. 135) denote the appearance of the spot when first observed; then as the spot approached the sun's limb it passed successively through the phases B, C, D, E, in the last of which even the nucleus has disappeared, and only a small trace remains of the penumbra. As the spots do actually change in form with considerable rapidity, it was necessary to ascertain that the observed changes were not due to intrinsic alterations in the spot itself. This was completely established by Wilson, who, after an interval of about fourteen days, watched the reappearance of the spot on the

FIG. 133.

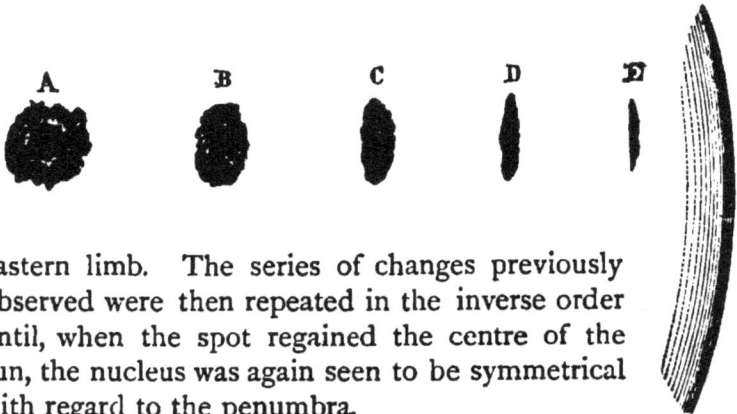

eastern limb. The series of changes previously observed were then repeated in the inverse order until, when the spot regained the centre of the sun, the nucleus was again seen to be symmetrical with regard to the penumbra.

These observations showed beyond doubt that the observed alterations in the form of the spot as it approaches the limb of the sun, can only be attributed to the effects of perspective. This will be made plain by Fig. 134. Let *a b c d* be a truncated cone, *a d* and *b c* being the two diameters of its extremities. When this cavity is viewed from a direction perpendicular to *a d*, it will present the symmetrical appearance shown in A. When the cavity has moved into the position corresponding to B, the penumbra *a b* is foreshortened to a considerable degree; the nucleus is also foreshortened to some extent, while the penumbra *c d* has become slightly elongated. In the last position corresponding to c the portion of the penumbra to the left of the nucleus has become totally invisible, while the nucleus itself is greatly attenuated. It can also readily be understood how, by measurements of the position of the spot at its different phases, the depth of the spot can actually be calculated.

The discovery made by Wilson has been frequently confirmed by observations of other astronomers. Cassini

Fig. 134.

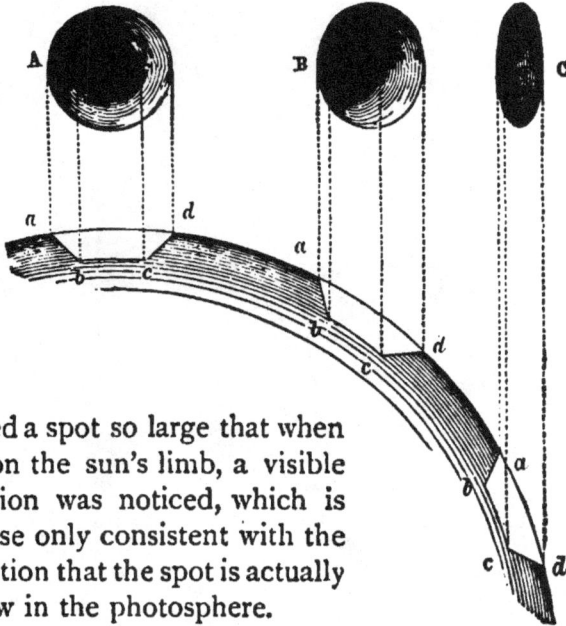

observed a spot so large that when it was on the sun's limb, a visible depression was noticed, which is of course only consistent with the supposition that the spot is actually a hollow in the photosphere.

The group of solar spots shown in Fig. 135 is from a drawing by Nasmyth. This drawing shows also the remarkable objects which have been likened to 'rice grains' or 'willow leaves' on the surface of the sun.

It was long since remarked that the number of spots on the sun's surface was very variable, and that at some epochs few spots, or even none at all, could be detected, while at others they were exhibited in great abundance (see § 69). According to Wolf, the period of the change in the spots is 11·11 years. After having attained a minimum, the abundance of the spots commences to increase rapidly, until in 3·52 years the maximum is reached. The decline to the next minimum takes place more slowly, and a period of 7·55 years is required. During the period of the maximum the area of the sun which is covered by spots is perhaps tenfold the spot area during the minimum.

FIG. 135.

§ 146. *Sun Spots and Terrestrial Magnetism.*—It seems that the phenomena of terrestrial magnetism are in some way connected with the state of the sun's surface. It has been shown that certain magnetic disturbances recur at regular intervals, and that the period of these disturbances is equal to the period between the successive maxima of spots on the sun. The presumption thus raised in favour of a connection between magnetic storms and that form of solar activity which produces sun spots has been confirmed by a remarkable circumstance.

Mr. Carrington (to whose labours on sun spots science

is deeply indebted) was observing a group of spots on September 1, 1859, when suddenly he noticed two patches of intensely bright and white light break out within the area of a remarkably large group of spots. The two spots travelled in five minutes over a space of 35,000 miles, when they vanished as two rapidly fading dots of white light. The instant of the outburst was $11^h 18^m$ Greenwich mean time, and the disappearance at $11^h 23^m$. This observation was independently confirmed by another observer, Mr. Hodgson, who records the occurrence of a bright spot which lasted for some five minutes, and disappeared at about $11^h 25^m$ A.M.

It was proved that at the moment when the sun exhibited this remarkable phenomenon, a magnetic disturbance had broken out. Auroras were observed in both hemispheres, and the electric equilibrium of the earth was so much deranged, that in many places the telegraphs would not work ; at Washington and Philadelphia the telegraph signal men received severe shocks, while at a station in Norway the telegraphic apparatus was set on fire.

§147. *Spectrum of a Spot.*—By bringing the image of the sun formed in the focus of a telescope into the same plane as the slit of a spectroscope, it is possible to bring different parts of the sun on the slit, and thus to examine in detail the different portions of the sun's surface. In this manner, when the image of a spot is crossed by the slit of the spectroscope, the spectrum seen is that of the light which radiates from a spot on the sun. The principal lines in the solar spectrum are invariably seen, whatever be the part of the sun to which the attention is directed ; but in the spots the appearance of the spectrum undergoes great modification. The general brilliancy of the light is diminished, and the dark lines are more or less increased in intensity.

§ 148. *Phenomena witnessed during Eclipses.*—On the somewhat rare occasions when the sun is seen to be totally eclipsed, certain very remarkable appendages to the sun are revealed, which, under ordinary circumstances, are invisible in consequence of the brilliancy of the diffused light from the sun. The appendages witnessed on such occasions are termed the *protuberances* and the *corona*.

The general appearance of these appendages is shown in Fig. 136, which represents the appearance presented during the total eclipse of July 29, 1878. The drawing was made

FIG. 136.

at Capitol Hill, in the city of Denver, Colorado, by Mr. Lewis Swift, an astronomer, who had the great advantage of having been also a witness of the eclipse of 1869. The telescope employed was an achromatic of $4\frac{1}{2}$ inches aperture. Mr. Swift writes : ' By far the most wonderful and unaccountable phenomena attending a total eclipse are the luminous pencils of light extending from the sun, like the radii of a circle, to the distance of several millions of miles, though seldom or never reaching to equal lengths in all directions. The drawing is a very fair representation of it as it appeared to my eye and mind. It will not escape notice that the pencils of light extend to a far greater distance in the direction of the moon's path, though this is not the case in all eclipses. That there should be any variation from an exact circular contour is only another of the many mysteries which environ the subject of solar physics.

'The salient features of the corona must not be omitted, viz. the curvature of the rays at A and the tangential direction of those at B. For about 60° of the moon's circumference the rays were decidedly curved, as represented in the diagram, and for about 30° they were perfectly straight; but if their paths were traced backwards, they would not meet at the sun's centre, as would all the others.

'The *chromosphere* was as beautiful as it was unexpected, for I saw nothing of it during the eclipse of 1869. Its thickness must have been considerable to have been visible with a power of only 25. Immediately on sighting the chromosphere, I saw two prominences—the only ones seen. They were near together and near the centre of the chromospherical crescent. They were of the same colour as the chromosphere, a bright pink. I have represented them in the diagram, together with the visible portion of the chromosphere. One of them was shaped like an awning hook, the other had the form of two stalks of wheat held close together, their drooping heads pointing in opposite directions.'

The light radiated from the prominences differs from the light from the sun. The spectrum of the prominences consists chiefly of bright lines on a dark ground. According to Young, there are nine bright lines in the spectrum of a prominence, of which four are coincident with the known lines of hydrogen. It thus appears that the prominences are vast masses of glowing gas, in which hydrogen is a chief constituent.

It would seem that the prominences are only salient masses of a stratum of a luminous gas entirely surrounding the sun. This luminous gas, which yields a spectrum of bright lines, is the *chromosphere*.

It was discovered independently by Janssen and

Lockyer that the spectrum of the prominences could be viewed by the spectroscope without an eclipse of the sun. Under ordinary circumstances, the diffused light from the sun screens the faint light of the prominences, but if the diffused light could be got rid of by any other means, the prominences would become visible. A spectroscope of a high dispersing power enables this to be accomplished. When the slit of the spectroscope is swept round the limb of the sun, the thin slice of light which is admitted, when a prominence is met with, consists partly of the diffused light, and partly of the light from the prominence. The dispersion of the spectroscope spreads out the diffused light into a long spectrum, whereby its intensity becomes greatly enfeebled. As the light from the prominences produces only a spectrum of bright lines, the intensity of these lines is not impaired by the high dispersion which has weakened the diffused lines, and consequently the lines become visible. By widening the slit considerably it is possible to see the protuberance in full daylight. In the parts of the spectrum where the different lines were seen with a narrow slit, distinct images of the prominence will now appear.

The spectrum of the corona consists of bright lines, and it is not a little remarkable that the spectrum of the corona appears to have some lines in common with that of the Aurora Borealis.

CHAPTER XII.

ASTRONOMICAL CONSTANTS.

§ 149. In the present chapter we record the principal numerical determinations which have been made of the important Constants cf Astronomy. Care has been taken as far as possible to give in each case the most recent results, while, both for their intrinsic value as well as for their historical interest, earlier values are often added. In many cases also bibliographical information will be found, which will enable the reader to refer to the original sources. In the preparation of this chapter the greatest assistance has been afforded by the admirable ' Répertoire des Constantes Astronomiques,' by J. C. Houzeau, Director of the Royal Observatory at Brussels.[1]

The following abbreviations are used to facilitate reference :—

'A. N.' 'Astronomische Nachrichten.' Altona, 1823–1875 ; Kiel, 1873 et seq. 90 vols. in 4to. Each volume contains 24 numbers, and reference is only made to the numbers.

'C. d. T.' 'Connaissance des Temps ou des Mouvements Célestes.' Années 1679–1880. Paris.

'C. R.' 'Comptes Rendus de l'Académie des Sciences de Paris.' Paris, 1835–1878. 86 volumes, 4to.

'M. A. S.' 'Memoirs of the Royal Astronomical Society of London.' London, 1822–1880. Vols. 1 to 44.

'M. N.' 'Monthly Notices of the Royal Astronomical Society of London.' 1828–1878.

'Ph. Tr.' 'Philosophical Transactions of the Royal Society of London.' 1665–1877. 167 vols. in 4to.

[1] *Annales de l'Observatoire Royal de Bruxelles.* Nouvelle série, 'Astronomie,' tome 1, 1878.

A A

SPHERICAL ASTRONOMY.

§ 150. *Obliquity of the Ecliptic.*

B.C.		
1100.	Tcheou-Kong, by the solstitial shadows at Loyang in China (Laplace, 'C. d. T.' 1811, p. 450)	23° 54′ 2′
140.	Hipparchus, with the astrolabe at Alexandria (Ptolemy, 'Math. Comp.' lib. i cap. 11 and 13)	23 51 20
A.D.		
130.	Ptolemy, with the astrolabe (Ptolemy, 'Math. Comp.' lib. i. cap. 11) . . .	23 51 15
629.	Litchou-Foung, by shadows (Laplace, *loc. cit.*)	23 40 4·1
1460.	Regiomontanus at Vienna, with a quadrant (Clavius, 'Opera Math.' t. iii. p. 149) .	23 30 49
1525.	Copernicus ('De Revol. Orbium Cœlest.' lib. ii. cap. 2)	23 28 24
1590.	Tycho Brahe (see Bugge, 'Berliner Jahrbuch,' 1794, p. 100)	23 29 46
1627.	Kepler ('Tab. Rudol.' p. 116) . .	23 30 30
1689·5.	Flamsteed, by solstitial observations at Greenwich ('Hist. Cœl. Proleg.' p. 114) .	23 28 56
1755·0.	Bradley, as deduced by Bessel ('Fund. Astron.' p. 61)	23 28 15·3
1795·0.	Maskelyne, from 20 solstices, observed at Greenwich (see Bessel,' Fund. Astron.' p.60)	23 27 57·6
1800·0.	Bessel, by comparing the sun's declination with Delambre's Tables ('Fund. Astron.' p. 61)	23 27 54·8
1813·0.	Brinkley, from 16 solstices by Oriani, Pond, Arago, Mathieu, and himself. ('Ph. Tr.' 1819, p. 241)	23 27 50·4
1850·0.	Value adopted by Leverrier ('Annal. Obs. Paris,' Mém. t. iv. pp. 51 and 203) .	23 27 31·8
1868·0.	Airy (see Powalky, 'A. N.' 1903) .	23 27 22·3

§ 151. *Secular Diminution of the Obliquity of the Ecliptic*
—In early times the Arabians had noticed that the obliquity of the ecliptic was decreasing. The following are the most recent determinations of the present rate of decrease per century:—

A.D.

1856.   Leverrier, by the theory of attraction ('Annal. Obs. Paris,' Mém. t. ii. p. 174)   .   .   47″·566

1858.   Leverrier, by observations extending over a century (*ibid.* t. iv. p. 51) .   .   .   45″·76

1867.   Lehmann, by the formulæ of attraction ('A. N.' No. 1619) .   .   .   .   .   47″·244

1872.   Powalky, by altering a little the masses used by Leverrier in accordance with recent researches ('A. N.' 1903).   .   .   47″·00

The variations in the obliquity of the ecliptic can only make the obliquity fluctuate within certain limits. The range of these fluctuations has been determined as follows:—

1825.   Laplace ('C. d. T.' 1827, p. 234)   .   .   3° 7′ 30″

1873.   Stockwell, by using more accurate determinations of the masses of the planets ('Smithsonian Contribs. to Knowl.' vol. xviii. p. 12)   2 37 22

According to this author the limits of the obliquity are—

$$21° \ 58′ \ 36″,$$

and

$$24 \quad 35 \quad 58.$$

§ 152. *Precession of the Equinoxes.*—The precession of the equinoxes was discovered by Hipparchus (Ptolemy, 'Math. Comp.' lib. iii. cap. 2). Newton first explained its cause ('Principia,' lib. iv. cap. 6). The first explicit theory of the subject was given by D'Alembert ('Recherches sur la Précession des Equinoxes,' Paris, 1749). The following are a few of the different values which have been assigned to the coefficient of precession (§ 72) :—

B.C.   Per Annum.

127.   Hipparchus (minimum value), Delambre, 'Hist. de l'Astron. Anc.' t. ii. p. 247   .   .   50″

A.D.

138.   Ptolemy, by comparing his observations with those of Hipparchus ('Math. Comp.' lib. vii. cap. 2)   36″

1525.   Copernicus, by comparing longitudes of this epoch with those of Ptolemy ('De Revol. Orbium Cœlest.' lib. iii. cap. 6)   .   .   50″·201

A A 2

A.D.		
1588.	Tycho Brahe, by the alteration of longitudes ('Astron. Instaur. Progymn.' p. 254)	51″
1712.	Flamsteed ('Historia Cœlestis,' lib. ii.)	50″
1802.	Laplace, by the theory of attraction at the epoch 1750 ('Méc. Cél.' t. iii. lib. vi. ch. 16)	50″·099
1817.	Bessel, by comparing his observations with those of Bradley ('Fund. Astron.' p. 285) for 1750	50″·176
1841.	O. Struve, by comparing Bradley's observations with those of W. Struve (1800)	50″·237
1844.	O. Struve from Bradley's Stars, by taking into account the proper motion of the solar system	50″·234
1856.	Leverrier finds as the value for 1850 ('Annal. Obs. Paris,' Mém. t. ii. p. 175)	50″·236
1873.	Stockwell, by the theory of attraction ('Smithson. Contrib. to Knowl.' vol. xviii. p. xii.), mean value	50″·438

According to Stockwell (*loc. cit.*), the limits of the coefficient of precession are—

$$48''·212$$
and
$$52''·664.$$

The period of the rotation of the pole of the equator around that of the ecliptic is—

$$25694·8 \text{ years,}$$

with inequalities which can alter this value to the extent of 201·2 years. The derangement of the equinox from its mean place is confined within a limit of—

$$3° 56' 26''.$$

§ **153.** *Nutation.*—In 1687 Newton had indicated the existence of nutation as a consequence of the theory of attraction ('Principia,' lib. iii. prop. 21), but he considered its effect would be too small to be appreciable. Bradley discovered nutation in 1747 ('Ph. Tr.' vol. xlvi. p. 1) by observations of stars (§ 76). The following are some of the principal determinations of the constant of nutation:—

A.D.

1747. Bradley, by the observations which led to the discovery of nutation ('Ph. Tr.' No. 485) .   . 9″·0

1799. Laplace, by theoretical calculations, using the mass
of the moon as derived from the tides ('Méc. Cél.'
t. ii. lib. v. ch. i.) .   .   .   .   . 10″·056

1820. Laplace, from observations of the Pole Star ('C. d. T.'
1822, p. 292) .   .   .   .   . 9″·30

1821. Brinkley, by 1,618 observations of stars ('Ph. Tr.'
1821, p. 347) .   .   .   .   . 9″·250

1842. Peters, from the R.A. of the Pole Star 1822–38
('Numerus constans nutationis') .   .   . 9″·223

1844. C. A. F. Peters, from the meridian transits of the Pole
Star observed at Dorpat, 1822–38 ('Mém. Ac. Sc.
St.-Pétersbourg,' 7ᵉ sér. t. iii. p. 125) .   . 9″·216

1856. Leverrier, on reviewing the discussion of Peters,
'Annal. Obs. Paris,' Mém. t. ii. p. 174 .   . 9″·23

1869. Stone, from observations of circumpolar stars at
Greenwich ('M. A. S.' vol. xxxvii. p. 249) . 9″·134

1872. Nyrén, from observations made in the prime vertical
at Pulkowa ('Mém. Ac. St.-Pétersbourg,' 7ᵉ sér.
t. xix. No. 2) .   .   .   .   . 9″·236

§ 154. *Aberration.*—Rœmer in 1675 ('Histoire de
l'Académie des Sciences de Paris,' t. i. p. 214) ascertained
that the eclipses of Jupiter's satellites are retarded according
as the earth is more removed from the planet. Bradley
discovered aberration in 1727 ('Ph. Tr.' No. 406, p. 637)
when observing stars (§ 74). The following are the values
which have been assigned to the coefficient of aberration:—

A.D.

1728. Bradley, from the observations which led to the discovery of aberration ('Ph. Tr.' No. 406) . 20″·25

1821. Brinkley, by 2,633 observations of numerous stars
('Ph. Tr.' 1821, p. 350).   .   .   . 20″·37

1840. Henderson, by the observations of Sirius at the Cape
('M. A. S.' vol. xi. p. 248) .   .   . 20″·41

1843. W. Struve, by observations of seven stars in the
prime vertical at Dorpat ('A. N.' No. 484) . 20″·4451

1844. C. A. F. Peters, by the altitudes of the Pole Star,
observed with the vertical circle at Pulkowa
('A. N.' No. 512) .   .   .   . 20″·503

A. D.
1850.   Maclear, by the observations of α Centauri at the
Cape ('M. A. S.' vol. xx. p. 98)   .   .   .   20″·53
1861.   Main, by observations of γ Draconis, 1852–59
('M. A. S.' vol. xxix. p. 190)   .   .   .   20″·335

§ 155. *Refraction.*—Tycho Brahe formed the first table
of astronomical refractions, and it was by him that refraction
was first taken account of in the reduction of observations
('Opera Omnia,' Frankfort, t. i. p. 79). The following are
the principal values which have been assigned to the refrac
tion at the zenith distance of 45°. This quantity is termed
the coefficient of refraction, and the refraction at zenith
distances is generally nearly equal to the product of the co
efficient of refraction and the tangent of the zenith distance
The amount of refraction at the horizon is also in some
cases given :—

A. D.		Refraction at the Zenith Distance 45°	90°
1604.	Kepler ('Ad Vitellonem Parolipomena,' p. 125, cap. iv.)	40″	1980″
1721.	Newton, from the data given by Halley, bar. 28·8 in., temp. 70° F. ('Ph. Tr.' No. 366, p. 118)	54	2025
1755.	Bradley, from observations at Greenwich, at temp. 50° F. and bar. 29·6 in. ('Astron. Obs. at Greenwich,' vol. i. p. xxxv.) . .	56·9	1976·8
1805.	Laplace, from the observations of Delambre at 0·760 mm. pressure, and 0° C. temp. ('Méc. Cél.' t. iv. liv. x. ch. i.) . .	60·50	
1814.	Brinkley, at 29·6 in. pressure and 50° F. temp. ('Trans. R. I. A.' vol. xii. p. 77) .	56·8	
1814.	Groombridge, at 29·6 in. pressure and 50° F. temp. ('Ph. Tr.' 1814, p. 337) . .	57·43	
1817.	Bessel, at 29·6 in. pressure and 48°·75 F., from Bradley's observations ('Fund. Astron.' pp. 45–50)	57·49	2166·8
1823.	Ivory, at 29 in. pressure and 50° F. temp. ('Ph. Tr.' 1823, p. 409) . . .	58·36	2057·5
1868.	Stone, from observations at Greenwich, above and below the Pole, at 29·6 in. pressure and 50° F. temp. ('M. N.' vol. xxviii. p. 29) .	57·38	

| | | Refraction at the Zenith Distance | |
| | | 45° | 90° |

1866. Gyldén, from observations at Pulkowa at 29·6 in. and 7°·44 Réaumur ('Mém. Ac. Sc. St.-Pétersbourg,' 7ᵉ sér. t. x. No. 1) . . 57″·68 2061·7

§ **156.** *Twilight.*—The following figures indicate the number of degrees at which the sun is below the horizon when the stars become visible to the unaided eye:—

A. D.
130. Ptolemy, appearance of stars of the first magnitude . 12°
    End of astronomical twilight . . . 18
1602. Tycho Brahe, end of astronomical twilight (' Astron. Instaur. Progymn.' pp. 95 and 733) . . 17
1618. Kepler (' Epitome Astronomiæ Copernicanæ,' lib. iii. part v. ed. Fritsch, t. vi. p. 285), appearance of Venus 5
    ,,   ,,   Jupiter and Mercury . 10
    ,,   ,,   Saturn . . . 11
    ,,   ,,   Mars 11½
    ,,   ,,   stars of the 1st magnitude . 12
    ,,   ,,   ,,   2nd   ,, . 13
    ,,   ,,   ,,   3rd   ,, . 14
    ,,   ,,   ,,   4th   ,, . 15
    ,,   ,,   ,,   5th   ,, . 16
    ,,   ,,   ,,   6th   ,, . 17
    ,,   ,,   the smallest lucid stars . 18
1865. J. Schmidt, from the mean results of observations (' A. N.' No. 1495) :—
    Appearance of stars of the 1st magnitude . + 0° 40′[1]
    ,,   ,,   2nd   ,, . − 4 18
    ,,   ,,   3rd   ,, . 5 4
    ,,   ,,   4th   ,, . 6 50
    ,,   ,,   5th   ,, . 8 52
    ,,   ,,   6th   ,, . 11 39
    End of astronomical twilight . . . 15 55

[1] The sign + signifies that the sun is above the horizon, and − that the sun is below the horizon.

The Sun.

§ **157.** *Semidiameter of the Sun.*—The following values express the *semidiameter* of the sun when at its mean distance from the earth :—

B.C.		
270.	Aristarchus of Samos (Wallis, ' Opera Mathematica,' tome iii.)	900″
240.	Archimedes, by the dimensions of a screen which covered the disk at the horizon . . .	899

A.D.		
138.	Ptolemy (' Mathematica, Compositio V.' cap. 14) .	969
1543.	Copernicus (' De Revolutionibus Orbium Cœlestium,' lib. iv. cap. 21)	983·3
1602.	Tycho Brahe ('Astron. Instaur. Progymn.' p. 471) .	930
1618.	Kepler (' Epitome Astronomiæ Copernicanæ,' lib. vi. part v. chap. vii.)	916·5
1750.	Bradley. (See Rosa, 'Studii intorno ai Diametri Solari,' Rome, 1874, p. 95) . . .	959·7
1771.	Lalande, by the transit of Venus in 1769 ('Astronomie,' 2ᵉ éd. t. ii. No. 2159)	958·0
1806.	Delambre (' Tables du Soleil,' Paris, 4to. tab. xxix.)	963·3
1824.	W. Struve, by observations at Dorpat up to the end of 1823 (' Berliner Jahrbuch,' 1827, S. 211).	
	Horizontal semidiameter	960·90
	Vertical semidiameter	960·37
1830.	Bessel, by 1,698 transits of the sun at Königsberg ('Tab. Reg.' p. 50)	960·90
1833.	Bessel, by micrometric measurements ('A. N.' No. 228)	950·895
1835.	Encke, by the transit of Venus in 1769 (' Abhand. d. Ak. zu Berlin,' 1835, Venus Durchgang, p. 95) .	958·42
1845.	Leverrier, by the transits of Mercury (' A. N.' No. 531)	960·01
1852.	W. Struve, by 241 meridian transits at Dorpat, 1822-38 (' Positiones Mediæ,' Petersburg).	
	Horizontal semidiameter	961·12
	Vertical semidiameter	960·66
1853.	Hansen and Olufsen, ' Tables of Sun ' (' Tables du Soleil,' Copenhagen, 4to. p. 165) . .	961·19

A.D.

1875. Fugh. By the discussion of 6,827 observations at Greenwich, 1836–1853, with three meridian instruments; the author finds the ellipticity insensible, and for the semidiameter ('A. N.' No. 2040) . 961"·495

§ 158. *The Sun's Parallax.*—The figures here given denote the angle which an equatorial radius of the earth subtends at the mean distance of the sun :—

B.C.

270. Aristarchus of Samos (see Wallis, ' Opera Mathematica,' tome iii.), not exceeding .   .   . 3'

150. Hipparchus, from the dimensions of the earth's shadow observed in eclipses of the moon (Ptolemy, 'Math. Comp.' lib. v. cap. 17) .   .   . 3'

A.D.

138. Ptolemy, by the shadow (lib. v. cap. 17, *loc. cit.*) . 2' 50"

1543. Copernicus, by the shadow ('De Rev. Orb. Cœl.' lib. iv. cap. 24)   .   .   .   .   . 3' 0

1618. Kepler, by the diurnal Parallax of Mars (' Epitome Astronomiæ Copernicanæ,' p. 479)   .   . 1'

1672. Flamsteed, by the diurnal Parallax of Mars ('Ph. Tr.' No. 89) 0 10

1677. Halley, by the diurnal parallax of Mercury ('Cat. Stell. Austral.' London, 1679)   .   . 0 45

1719. Bradley and Pound, by the diurnal parallax of Mars (See Gehler's ' Physikalisches Wörterbuch,' Bd. viii. S. 822) .   .   .   .   . 0 10·5

1751. Lacaille, by comparing observations of Mars made at the Cape of Good Hope, with those made in Europe ('Ephémérides des Mouvements Célestes depuis 1765 jusqu'en 1774,' Paris, Introd. p. 1) . 0 10·38

1761. Pingré, by the transit of Venus in 1761 ('Mém. Ac. de Paris,' 1761, p. 486) .   .   . 0 10·60

1770. Euler, by the transit of Venus in 1769 ('Novi Commentarii Ac. Sc. Petropol.' t. xiv.) .   . 0 8·8

1771. Maskelyne, by the transit of Venus in 1769 ('Ph. Tr.' vol. lxi. p. 574) .   .   .   . 0 8·723

1804. Laplace, from the Parallactic Equation of the Moon ('C. d. T.' an xii. p. 496)   .   . 0 8·6

1814. Delambre, by the transit of Venus in 1769 ('Astron. Théorique et Pratique,' t. iii. p. 506, t. i. p. xliv.) 0 8·5525

A.D.

1835. Encke, by combining the results of the transits of Venus in 1761 and 1769 ('Abhand. der Ak. zu Berlin,' 1835, Math. Kl. S. 309) . . 0′ 8‴·571

1835. Henderson, by comparing the observations of Mars made at the Cape with those made in Europe ('M. A. S.' vol. viii. p. 103) . . . 0 9·028

1858. Leverrier, by the parallactic equation of the moon ('Ann. de l'Obs. de Paris.' Mém. t. iv. p. 101) 0 8·95

1862. Foucault, by experiments on the velocity of light compared with the constant of aberration ('C. R.' t. lv. p. 502) . . . . . 0 8·86

1863. Hansen, by the parallactic equation and the parallax of the moon given in his tables ('M. N.' vol. xxiii. p. 243) . . . . 0 8·97

1865. E. J. Stone, by observations of Mars at the Cape and Williamstown compared with those at Greenwich ('M. A. S.' vol. xxxiii. p. 97) . . 0 8·943

1865. Hall, by micrometric comparisons of Mars in 1862, at Chili, and at Washington and Upsala ('Astron. Obs. at U.S. Naval Observatory,' 1863, append. p. 64) . . . . . . 0 8·8415

1867. Newcomb, by the observations of Mars in 1862 ('Astron. Obs. U.S. Naval Observatory,' 1865, appendix ii. p. 22) . . . . 0 8·855

1867. Newcomb, from the parallactic inequality of the moon (*ibid.* p. 25) . . . . 0 8·838

1867. Newcomb, from the lunar equation of the earth (*ibid.* p. 28) . . . . . 0 8·809

1868. E. J. Stone, by a new discussion of the transit of Venus in 1769 ('M. N.' vol. xxviii. p. 264) . 0 8·91

1872. Leverrier, by the movement of the perihelion of Mars ('C.R.' t. lxxv. p. 165) . . 0 8·866

1872. Leverrier, by the movement of the node of Venus (*ibid.*) . . . . . . 0 8·853

1872. Leverrier, by the secular variations of Venus in 106 years of observation (*ibid.*) . . 0 8·859

1874. Cornu, by a new measurement of the velocity of light combined with the coefficient of aberration of Struve ('M. N.' vol. xxxv. p. 240) . 0 8·834

1875. Galle, by the observations of Flora in both hemispheres at the opposition of 1873 ('M. N.' vol. xxxvi. p. 202, 'A. N.' No. 2,033) . 0 8·873

A.D.

1876. Von Asten, by the mass of the earth furnished by
the perturbation of Encke's Comet ('Wochenschrift
für Astron.' &c., Halle, 1877, S. 42) . . ♂ 9″·009

1877. Cornu, by the velocity of light measured between
Paris and Montlhéry, compared with the aberration of Struve ('C. R.' t. lxxxiv. p. 369) . o 8·80

1877. Airy, by the internal contacts of Venus during the
transit of 1874 (Christie, 'The Observatory,' vol.
i. p. 141) . . . . . o 8·760

1877. Lindsay and Gill, by observations of Juno in 1874
('Dunecht Observatory Publications,' vol. ii. p.
211), mean result . . . . o 8·765

1879. Gill, from observations of Mars at Ascension Island
in 1877 ('M. N.' xxxix. p. 437) . . . o 8·78

§ 159. *Elements of the Sun's Rotation.*

Period of the tropical rotation.	Longitude of the ascending node.	Date of the Equinox to which the longitude is referred.	Inclination.

1814. Delambre ('Astr. Théor. et Prat.' t, iii. p. 54).

25^d 0^h 16^m 36^s $80°$ $7'$ $4''$ $1775·5$ $7°$ $19'$ $23''$

1863. Carrington, 'Observations of the Spots on the Sun,' London,
pp. 225, 244. (The duration refers to the sun's equator.)

24^d 23^h $19·4^m$ $73°$ $28'$ $1854·0$ $7°$ $17'$

1865. Schwabe, Observations from 1843 to 1854 ('A.N.' No. 1521).

25^d 5^h $1848·5$ $7°$ $17'·7$

1868. Spörer ('Publication der Astron. Gesellschaft,' Bd. xiii. S. 5).

25^d 5^h 37^m $74°$ $36'$ $1866·5$ $6°$ $58'$

1871. Hornstein, from the period of the magnetic declination at
Prague and Vienna ('Sitzungs-Berichte der Akad. der Wiss. zu
Wien,' Bd. lxiv. Abth. ii. Juniheft).

25^d 13^h

§ 160. *Rotation of the Sun.*—Scheiner had observed in
1626 that the period of revolution is not equal for all the
spots on the sun. The following formulæ give the angle of
rotation about the axis of the sun in one day, the latitude of
the spot being λ:—

1863. Carrington (*loc. cit.*):

$$865' - 165' \sin^{\frac{7}{4}} \lambda.$$

1865. Faye, from Carrington's observations ('C. R.' t. lx. p. 816):
$$862' - 186'\sin^2\lambda.$$

1867. Spörer ('A. N.' No. 1612):
$$1011' - 202'\cdot 8\sin(\lambda + 41°13').$$

1872. Zöllner, from Carrington's observations ('A. N.' No. 1850):
$$\frac{863'\cdot 8 - k\sin^2\lambda}{\cos\lambda}$$

where $k = \begin{cases} 613'\cdot 2 \text{ in the Northern hemisphere,} \\ 631\cdot 1 \text{ in the Southern hemisphere.} \end{cases}$

For the effect of refraction in the sun's atmosphere in altering the appearance of rotation see C. H. F. Peters, 'A. N.' 1696.

§ 161. *Bodies seen Passing across the Sun.*—A most interesting account of the different observations which have indicated the existence of bodies passing in front of the sun is given by Leverrier, 'Comptes Rendus,' vol. lxxxiii., pp. 583 and 621. See also C. H. F. Peters, 'Astronomische Nachrichten,' Nos. 1754 and 2253.

§ 162. *Heating and Actinic Power of the Sun.*—The following are the principal memoirs on this subject :—

1742. Bon de St. Hilaire. 'Sur la Chaleur des Rayons directs du Soleil comparée à celle que l'on éprouve à l'ombre, Mémoires de l'Académie de Montpellier,' t. ii. hist. p. 18.

1836. Sir John Herschel ('Reports of the British Association,' 1833, p. 379). See also 'Results of Astronomical Observations made at the Cape of Good Hope.' London : 1847, p. 442.

1838. Pouillet. Experiments with the instrument he calls the Pyrhéliomètre. 'Comptes Rendus,' t. vii. p. 64.

1877. Violle. On the mean temperature at the surface of the sun. See 'Annales de Chimie et de Physique,' 5ᵉ sér. t. x. p. 289.

§ 163. *On the Physical Constitution of the Sun.*—The following is a list of the principal works which treat of the physical constitution of the sun :—

1861-1862. Kirchoff. 'Untersuchungen über das Sonnenspectrum und die Spectren der Chemischen Elemente.' Berlin : two parts, 4to.

1863–1864. E. Gautier. De la Constitution du Soleil : Supplément à la Bibliothèque Universelle. 'Archives,' t. xviii. p. 209 ; t. xix. p. 265 ; t. xxiv. p. 21.

1864. Nasmyth. 'M. N.' vol. xxv. p. 2444.

1865. Warren de La Rue, B. Stewart, Loewy. Researches on Solar Physics. 'Proceedings of the Royal Society of London,' vol. xiv. pp. 71, 72.

1866. Angström and Thalen. 'On the Fraunhofer lines, together with a Diagram of the Violet Part of the Solar Spectrum.' Upsala, 4to.

1868. Angström. 'Spectre Normal du Soleil, atlas contenant les longueurs d'onde des raies Fraunhofériennes données en dix-millionièmes de millimètre.' Upsala, 4to.

1868. G. J. Stoney. 'Proceedings of the Royal Society,' vol. xvii. p. 1.

1869. Zöllner. 'Vierteljahrsschrift der Astronomischen Gesellschaft,' Th. iv. S. 172.

1869–1870. Warren de La Rue, B. Stewart, Benj. Loewy. 'Researches on Solar Physics: Heliographic Positions and Areas of Sun-spots observed with the Kew Photoheliograph.' London : 2 parts, 4to.

1869–1870. Young. 'Journal of the Franklin Institute,' December 1869, June, July, October, November 1870.

1870. Secchi. 'Le Soleil.' Paris. 8vo. 2nd edition, Paris, 1877.

1870. Respighi. Osservazioni Spettroscopiche del borde e dene protuberance solari. 'Atti della Accademia dei Lincei,' vol. xxiii.

1871. E. Becquerel. Sur l'action électrique du Soleil. 'C. R.' vol. lxxii. p. 709.

1874. N. Lockyer. 'Contributions to Solar Physics.' London : 8vo.

1876. Proctor. 'The Sun : Ruler, Fire, Light, and Life of the Planetary System.' 3rd edition. London, 8vo.

1877. Draper. 'Nature,' vol. xvi. p. 364. No. 409.

1879. Ranyard. 'Observations made during Total Solar Eclipses,' M. A. S. vol. xli.

§ 164. *Period of the Solar Spots.*

1844. Schwabe, from his observations, 1826–1842 ('A. N.' No. 495) 10

1852. R. Wolf, from observations of the sun since 1611
　　　　('A. N.' No. 839) . . . . . 11·111

MERCURY.

§ 165. *Elements of the Orbit of Mercury.*

Leverrier,'Annales de l'Observatoire de Paris.' Mémoires, t. v. 1859,
　　pp. 107, 109, and 135.

[*t* represents the number of Julian years (365¼ days) which have elapsed
　　　　since the date of the epoch.]

Epoch 1850.　　　*Noon, Jan.* 1, *mean time at Paris.*

Mean longitude .	$327° 15' 20''·43 + 5381066''·5449t + 0''·00011289t^2$	
Longitude of peri-helion . . .	$75\ \ 7\ \ 13·93 +$	$55''·9138t + 0·00011111t^2$
Longitude of ascend-ing node . .	$46\ \ 33\ \ 8·75 +$	$42''·6430t + 0·0000835t^2$
Greatest equation of the centre . .	$23\ \ 40\ \ 38·01 +$	$0''·0851t - 0·0000036t^2$
Inclination . .	$7\ \ 0\ \ 7·71 +$	$0''·06314t - 0·0000056t^2$

Movement in 100 *years.*

Mean longitude . .	$74°\ \ 4'\ \ 14''·4900$
Longitude of perihelion .	$1\ \ 33\ \ 11·380$
Longitude of node . .	$1\ \ 11\ \ 4·300$

§ 166. *Transits of Mercury across the Sun.*—Kepler predicted the transit of Mercury across the face of the sun in
1631, and the transit was partially observed by Gassendi on
November 9 in that year. The first complete observation
of a transit of Mercury was made by Halley at St. Helena
in 1677.

§ 167. *Diameter of Mercury.*—Before the invention of
the telescope, Tycho Brahe estimated the diameter of
Mercury at 130'' and Gassendi at 120.'' The following are
the measures which have been obtained by the aid of telescopes. The numbers represent the diameter of Mercury
as it would be seen when placed at a distance from the

earth which is equal to the mean distance of the sun from the earth.

1723. Bradley, by micrometric measuements on the sun ('Ph. Tr.' No. 386) . . . . 7″·34
1800. Schröter, by calculation from the micrometric measurements made at different transits ('Berliner Jahrbuch,' 1803, S. 167) . . 1786 . 6·145
     ,,   ,,   .   . 1789 . 6·050
     ,,   ,,   .   . 1799 . 5·978
1833. Bessel, from micrometric measurements made during the transit of 1832 ('A. N.' No. 228) . . 6·6974
1855. Main, with the double image micrometer ('M. A. S.' vol. xxv. p. 43) . . . . . 6·68
1865. Jul. Schmidt, by micrometric observations in 1854 ('A. N.' No. 1543) . . . . 6·454

Schröter (1800) considered the ellipticity of Mercury to be perfectly insensible. Sir William Herschel found the planet perfectly circular during its transit across the sun in 1802.

§ 168. *Mass of Mercury.*—As no satellite of Mercury has yet been detected, our present knowledge of its mass has been inferred from the effect of the perturbation which it produces in other celestial bodies belonging to the solar system. It will be seen that there are considerable discrepancies in the different estimates. The mass of the sun is taken as unity.

1841. Encke, by the perturbation of the comet which bears his name ('A. N.' No. 443) . . . $\frac{1}{4865751}$
1861. Leverrier, by the perturbation of Venus ('Ann. de l'Obs. de l'aris,' Mém. vi. p. 92) . . . $\frac{1}{5310000}$
1861. Leverrier, by the perturbation of the Earth (' Ann. de l'Obs. de Paris,' Mém. vi. p. 92) . . . $\frac{1}{4360000}$
1876. Von Asten, by the perturbation of Encke's comet from 1819–1868 (' Wochenschrift für Astronomie,' Halle, 1877, S. 42) . . . . . $\frac{1}{7036440}$

§ 169. *Rotation of Mercury on its Axis.*—The rotation of Mercury was detected by Schröter in 1800 by observing certain peculiarities in the form of the southern horn of the crescent. The following results have been deduced from these observations :—

A.D.

1801. Schröter, from his own observations and those of Harding on the truncated appearance of the Southern horn (Zach's 'Monatliche Correspondenz,' Gotha, Bd. 14, p. 221) . . 24^h 5^m 30^s

1810. Bessel, by five observations of Schröter during a period of 14 months ('Berliner Jahrbuch,' 1813, p. 253) . . . . . 24 0 52·97

1816. Schröter, from his own calculations and those of Bessel . . . . . . 24 0 50·20

§ 170. *Plane of the Equator of Mercury.*—The situation of this plane has not yet been determined. Schröter considered the inclination of the equator to the plane of the planet's orbit to be about 20°. This was deduced from his observations of a band visible for forty-seven consecutive days in May and June 1801.

§ 171. *Physical Conditions of the Surface of Mercury.*— The following papers may be consulted :—

1801 and 1816. Schröter, 'Hermographische Fragmente.' Göttingen, 2 vols. 8vo.

1803. Sir William Herschel, 'Phil. Trans.' vol. xciii. p. 214.

1840. W. Beer and Mädler, 'Fragments sur les Corps célestes du Système solaire.' Paris, 4to. P. 206.

§ 171. *Brightness of Mercury.*—The phases of Mercury were first observed by Hortensius about 1630. For its photometric brightness see Zöllner, 'Photometrische Untersuchungen über die Physikalische Beschaffenheit des Planeten Merkur,' Leipzig.

VENUS.

§ 173. *Elements of the Orbit.*—The following are the

elements of the orbit of Venus, given by Leverrier in 1861, from a new discussion of the observations ('Annales de l'Observatoire de Paris,' t. v. pp. 95–97) :—

Epoch 1850.   *Noon, Jan.* 1, *mean time at Paris.*

Mean longitude	. 245	33	$14\cdot70 + 2106691''\cdot65043t + 0\cdot000112089t^2$
Longitude of perihelion	. 129	27	$14\cdot5 + \qquad 49''\cdot462t \quad -0\cdot000593t^2$
Longitude of ascending node	. 75	19	$52\cdot3 + \qquad 32''\cdot8899t \; +0\cdot00015508t^2$
Inclination	. 3° 23'		$34''\cdot83 + \qquad 0''\cdot04524t -0\cdot0000156t^2$
Greatest equation of centre	. 0	47	$3\cdot08 - \qquad 0\cdot2227t \; +0\cdot00104t^2$

[*t* is the number of years (365¼ days) which have elapsed since the epoch.]

§ **174.** *Transits of Venus across the Sun.*—Kepler in 1629 was the first to announce the occurrence of the transits of Venus across the sun. The first transit observed was by Horrocks in 1639. The theory of the transits was given by Halley in 1691.

§ **175.** *Apparent Diameter of Venus.*—Before the invention of the telescope Tycho Brahe estimated the diameter of Venus at $3\frac{1}{4}'$ and Kepler at 7'. In 1630 Hortensius estimated it at 57'', and in 1665 Riccioli and Grimaldi at 64''·2. The following figures give the diameter of Venus as it would be seen by an observer situated at a distance equal to the mean distance of the earth from the sun :—

A.D.
1640.	Horrocks, by micrometric measurements (see Wurm, 'Berliner Jahrbuch,' 1807, p. 165).	17''·609
1670.	Flamsteed, by micrometric measurements (see Wurm, 'Berliner Jahrbuch,' 1807, p. 165)	24''·633
1789–1794.	Schröter, by micrometric measurements (see Wurm, 'Berliner Jahrbuch,' 1807, pp. 166–7)	16''·835
1791.	Sir William Herschel, by micrometric measurements (see Wurm, 'Berliner Jahrbuch,' 1807, p. 166)	18''·790
1822.	Encke, by the transit of 1761 ('Entfernung der Sonne,' Gotha, 8vo.) . . .	16''·611

B B

A.D.
1840. Airy, by micrometric observations (Cited in 'Ann. de l'Obs. de Paris,' Mém. t. vi. pp. 26 and 201) . . . . 16″·566

1865. Jul. Schmidt, by micrometric observations ('A. N.' 1543) . . . . 17″·18

1865. E. J. Stone, by observations at Greenwich with the mural circle, 1839–1850, and the meridian circle, 1850–1862 ('M. N.' vol. xxv. p. 59) . . . . . 16″·944

1871. Powalky, from the transits of 1761 and 1769 ('A. N.' 1841) . . . . 16″ 918

1875. Tennant, by micrometric measurements during the transit in 1874 ('M. N.' vol. xxxv. p. 347) . . . . 16″·9036

§ 176. *Ellipticity.*—The ellipticity of Venus is generally regarded as insensible. The observations of Vidal, however, during the inferior conjunction in October 1807, gives 60″·4 for the vertical diameter and 61″·5 for the horizontal diameter. ('Cdt.' 1810, p. 375.)

§ 177. *Mass of Venus.*—The mass of Venus has been determined by the perturbations which it produces in the motions of other bodies belonging to the solar system. The following are the principal results. The mass of the sun is taken as unity :—

A.D.
1779. Lagrange, from the precession of the equinoxes, ('Berliner Jahrbuch,' 1782, p. 115) . . . $\frac{1}{315510}$

Lagrange, from irregularities in the motion of the sun ('Berliner Jahrbuch,' 1782, p. 116) . . $\frac{1}{341420}$

1802. Laplace, by the secular diminution of the obliquity of the ecliptic ('Méc. Cél.' t. iii. vol. vi. c. 6) . . $\frac{1}{383130}$

1802. Delambre, by deducing from the observations of the sun, by Bradley and Maskelyne, the coefficients of the inequalities caused by Venus (see Laplace, 'Méc. Cél.' t. iii. liv. vi. ch. 16) . . . . $\frac{1}{356632}$

1828. Airy, by the inequalities deduced from observations of the sun ('Ph. Tr.' 1828, p. 50) . . . $\frac{1}{401211}$

1843. Leverrier, by the secular variation of the node of Mercury ('C. R.' tom. xvi. p. 1062) . . $\frac{1}{390000}$

1858. Leverrier, by the perturbations of the earth ('Ann. de l'Obs. de Paris,' Mém. t. iv. p. 102) $\frac{1}{400248}$

A.D.
1872. Hill, from the movements of the node of Mercury
('Tables of Venus.' Washington, 4to. p. 2) . $\frac{1}{427170}$
1876. Powalky, by comparing the solar tables of Hansen
and Olufsen with the observations at Dorpat from
1823 to 1839 ('A. N.' 2105) . . . $\frac{1}{397000}$

§ 178. *Rotation of Venus.*—The rotation of Venus was
discovered by Cassini in 1665, and he estimated the period
at a little less than one day. The following are the principal
determinations which have since been made :—

Duration of the sidereal rotation	Longitude of the ascending node of the equator of Venus on the ecliptic	Equinox to which the longitude is referred	Inclination of the Equator of Venus to the ecliptic
1789. Schröter, by the observations of the horns ('Ph. Tr.' vol. lxxxv. p. 117)			
23ʰ 21ᵐ 19ˢ	45°	1789	72°
1840-50. De Vico, by his observations with Palomba at Rome ('Memorie della Specola nel Collegio Romano,' 1840-1841, p. 32; 1843, p. 31; 1849, p. 29; 1850, p. 140).			
23ʰ 21ᵐ 21ˢ·9345	57° 19′ 18″	1839	49° 57′ 34″

§ 179. *Brilliancy of Venus.*—The phases of Venus were
discovered by Galileo in 1610. Delambre gives as the conditions under which Venus appears brightest—

Elongation . . . 39° 43′ 26″
Distance from earth . . 0·4304

This is on the supposition that the orbit is circular: the
actual values of the elongation and distance at the time of
greatest brightness are not quite constant on account of the
ellipticity of the orbit. Reboul states that when Venus is
an evening star, the remarkable brilliancy commences about
50 or 60 days before the calculated time of greatest brilliancy,
and lasts twenty days after. When Venus is a morning star,
the remarkable brightness commences about 20 days before
the calculated date of greatest brightness and lasts for about
50 or 60 days afterwards.

G. P. Bond in 1861 estimated the brilliancy of Venus

when brightest as 4·86 times that of Jupiter, and 1 ÷622,600,000 that of the sun. At the distance 1 from the earth, and 0·7233 from the sun, he found the brightness of Venus to be 1÷514 that of the full moon. The albedo of Venus is 0·809 time that of Jupiter.

In 1876 Plummer determined the brilliancy of Venus when brightest to be 1÷799·5 times that of the full moon.

For researches upon the spectrum of Venus, see Vogel, 'Untersuchungen über die Spectra der Planeten,' Leipzig, 1874, 8vo.

An interesting account of the observations on the visibility of the dark side of Venus is given by Schafarik in the 'Reports of the British Association,' 1873, p. 404.

THE EARTH.

§ 180. *Duration of the Year.*—The period of the revolution of the earth around the sun was the earliest element of our planet which was known with accuracy. The following are the principal values expressed in mean solar time :—

B.C.		Duration of the Tropical Year, 365 days 5 hours.	
3101 (?).	Indian Tables (Bailly, 'Traité de l'Astronomie indienne.' Paris : 1787. 4to. p. 124) .	50^m	35^s
140.	Ptolemy. By his observations and those of Hipparchus ('Math. Comp.' lib. iii. ch. 2)	55	14
A.D.			
1543.	Copernicus ('De Revol. Orb. Cœlest.' lib. iii. ch. 14)	49	6
1602.	Tycho Brahe ('Astron. Instaur. Progymn.' p. 53)	48	45·3
1687	Flamsteed (Newton, 'Phil. Nat. Princip. Math.')	48	57·5
1806.	Delambre ('Tables du Soleil.' Paris, 4to.) .	48	51·61
1853.	Hansen and Olufsen, epoch 1850, with the annual variation —0·00539 ('Tables du Soleil')	48	46·15
1858.	Leverrier, epoch 1800, with the annual variation —0·00539 ('Ann. Obs. Paris,' Mém. t. iv. p. 102)	48	46·045

According to Stockwell ('Smith. Contribs. to Knowl.' vol. xviii. p. xii.), the variations of the tropical year from its mean value are

$$\mp 54''\cdot 20.$$

According to the same authority the variations of the length of the tropical year from its present value have the limits

$$-59''\cdot 13 \text{ and } +49''\cdot 27.$$

§ 181. *Perihelion.*—The movement of the axis major of the earth's orbit is very perceptible within historic times: it was, in fact, known to the Hindoos in the seventh century. The following are the principal results for the longitude of the perihelion as well as for its movement in one hundred years:—

	Longitude of Perihelion.	Movement in 100 Years.
B.C.		
127. Hipparchus (Ptolemy, 'Compos. Math.' lib. iii. c. 4) .	65° 30′	
A.D.		
140. Ptolemy (*ibid.*) . .	65 30	
1515. Copernicus (' De Rev. Orb. Cœlest.' lib. 3, c. 22) .	96 40	1° 40′ 50·4
1588. Tycho Brahe (Lalande, ' Astronomie,' 2nd ed. liv. vi. No. 1313, t. ii. p. 94) . .	95 30	
1758. Lacaille. Epoch, 1700 (' Tabulæ Solares ') . . .	97 35 55	1 49 10
1806. Delambre. Epoch, 1800 (' Tables du Soleil ') .	99 30 5	1 43 35
1816. Airy. 1816 (' Phil. Trans.' 1828, p. 28) . . .	99 46 20·3	
1853. Hansen and Olufsen. Epoch, 1850 (' Tables du Soleil') .	100 21 41·02	
1858. Leverrier. Epoch, 1850 (' Ann. Obs. Paris.' Mém. t. iv. pp. 102 and 105) . .	100 21 21·5	1 42 49·95
1876. Powalky. Epoch, 1850 (' A. N.' No. 2105) . . .	100 21 30·1	

§ 182. *The Eccentricity and Equation of the Centre.*—The inequality in the motion of the sun which is called the equation of the centre was known to the astronomers at Alexandria. The following are the principal determinations of its greatest value ·——

B.C.		Value of the Greatest Equation of the Centre.
140.	Ptolemy ('Comp. Math.' lib. iii. c. 4) .	$2° 23'$
A.D.		
1750.	La Caille ('Mém. Ac. Sc. Paris,' 1750, p. 178)	I 55 40·6
1806.	Delambre ('Tables du Soleil'). Epoch, 1810	I 54 41·8
	Annual variation $-0''·1718$	
1828.	Bessel. Epoch 1800. Eccentricity .	$e = 0·0167922585$
	Annual variation $-0·0000004359$ ('A. N.' No. 133)	
1828.	Airy. Epoch, 1816 ('Ph. Tr.' 1828, p. 29)	I 54 39·9
1853.	Hansen and Olufsen ('Tables of the Sun,' p. 1)	
	Eccentricity	$e = 0·01677120$
1858.	Leverrier ('Ann. Obs. Paris,' Mém. t. iv. p. 102) . . .	I 55 18·78
	Annual variation $-0''·17510$.	

§ 183. *Tables of the Sun.*——The recent tables of the sun in which account is taken of the perturbations are——

1853. Hansen and Olufsen, Copenhagen. 4to., with supplement, 1857.

1858. Leverrier, 'Annal. Obser. Paris,' Mém. t. iv. pp. 119-206.

The elements of the apparent path of the sun are, according to Leverrier——

Mean longitude . $100° 46' 43''·51 + 1296027''·6784 t. + 0''·00011073t^2$
Perihelion . $100° 21' 21''·5 + 61''·6995 t. + 0''·0001823 t^2$
Greatest equation
of centre . $3459''·28 - 0''·08755t. - 0''·00000282t^2$

The letter t represents the number of Julian years which have elapsed since 1850·00

§ 184. *Dimensions of the Earth.*—The following are perhaps the most reliable determinations of the dimensions of the earth as ascertained by actual surveying. They are taken from the recent work on Geodesy by Colonel A. R. Clarke, C.B. (8vo. Clarendon Press Series, Oxford, 1880). If it be supposed that the earth is an ellipsoid with three unequal axes, then we find (p. 308)—

Equatorial semiaxis in feet	$a = 20926629$
,, ,, ,, ...	$b = 20925105$
Polar semiaxis	$c = 20854477$

The axis major of the ellipse in which the earth is cut by the plane of the equator is in longitude—

8° 15′ W. from Greenwich.

'On the ellipsoidal theory of the earth's figure, small as is the difference between the two diameters of the equator, the Indian longitudes are much better represented than by a surface of revolution. But it is nevertheless necessary to guard against an impression that the figure of the equator is thus definitely fixed, for the available data are far too slender to warrant such a conclusion' (p. 309).

If the earth be regarded as a spheroid of revolution, the dimensions are (p. 319)—

$$a = 20926202 \text{ feet,}$$
$$c = 20854895 \text{ feet.}$$

and their ratio—

$$c : a = 292.465 : 293.465$$

§ 185. *Ellipticity of the Earth.*—The following are various determinations of this important constant :—

A.D.
1687. Newton, by the theory of attraction, supposing the earth homogeneous ('Principia,' lib. iii. prop. 19) . . $\frac{1}{230}$

1749. D'Alembert, by the precession of the equinoxes (' Recherches sur la Précession des Équinoxes.' Paris, 4to. ch. ix.) . . . . . $\frac{1}{324}$

A.D.

1789. Legendre, by the measurements of arcs in Peru and France ('Mém. Ac. Sc.' Paris, 1789, p. 422) . $\frac{1}{305}$

1799. Laplace, by 15 measurements with the pendulum ('Méc. Cél.' liv. iii. ch. 5, No. 43, t. 11) . $\frac{1}{335\cdot78}$

1819. Von Lindenau, from the lunar inequalities ('Berliner Jahrbuch,' 1820, p. 212) . . . $\frac{1}{315\cdot82}$

1825. Laplace, from the lunar inequalities deduced from observations of the moon at Greenwich ('Méc. Cél.' liv. xi. ch. 1, No. 1, t. v.) . . $\frac{1}{306}$

1825. Sabine, by pendulum experiments ('Account of Experiments on the figure of the Earth,' London) $\frac{1}{314}$

1849. Airy, by pendulum experiments ('Figure of the Earth,' § 35 in 'Encyclopædia Metropolitana') $\frac{1}{282\cdot9}$

1862. Bæyer, by the triangulation in Prussia ('A. N.' No. 1366) . . . . . $\frac{1}{298\cdot9}$

1880. Clarke, mean result of surveys (*loc. cit.* p. 319) . $\frac{1}{292\cdot96\pm1\cdot07}$

Mean result from pendulums (p. 350) . . $\frac{1}{292\cdot2\pm1\cdot5}$

§ 186. *Gravitation.*—The length of the pendulum which beats one second in a vacuum in an indefinitely small arc at the level of the sea is, according to Ufferdinger (Grünert's 'Archiv für Mathematik und Physik,' 1868), who has given a complete discussion of the observations—

$$P = 0^m\cdot9909757 + 0^m\cdot0022475\sin^2\phi + 0^m\cdot00002085\sin^4\phi.$$

The length of the pendulum P is here expressed in metres, while ϕ is the latitude.

At a height h, above the surface of the earth, the length of the pendulum is P', where R, being the radius of the earth

$$P' = P\left(1 - \frac{2h}{R}\right).$$

Gravity is determined from the length of the seconds pendulum by the equation

$$g = \pi^2 P.$$

§ **187.** *Rotation.*—Laplace had concluded in 1799 that the inequalities of the time of rotation of the earth were insensible ('Méc. Cél.' liv. v. ch. 1, Nos. 8 and 9, t. ii.) Serret in 1859, assuming that the fluctuations of latitude do not exceed a single second, inferred that the period of rotation must be practically constant ('Annal. Obs. Paris,' Mém. t. v. 1859, pp. 290, 291.

Laplace considered, by computation of the eclipses recorded by Ptolemy, that the length of the day had not changed in the lapse of twenty-five centuries by so much as 0·000,000,1 part of its value ('Cdt.' 1821, p. 242). More recently, however, Delaunay has pointed out that the value of the secular acceleration of the moon's motion, indicated by theory, fails to explain accurately the ancient eclipses : he is therefore of opinion that the changes in the period of the earth's rotation are sensible in historic times ('C. R.' t. lxi. 1865, p. 1023). Taking the value of the moon's acceleration which was found by Adams and Delaunay, Schjellerup finds that in order to represent the Chinese eclipses of 708, 600, 548, B.C., it is necessary to apply to intervals measured by the present length of the day a correction—

$$-9^{s}\!\cdot\!547 t^{2},$$

where t denotes the number of centuries from 1800. The change in the length of the day in a period of 2,400 years amounts to 0·01252 second.

§ **188.** *Density.*—Taking the density of water as unity, the following are the principal determinations which have been made of the mean density of the earth :—

A.D.		
1687.	Newton, by theoretical considerations ('Principia,' lib. ii.), between	5 and 6
1775.	Maskelyne, by the action of Mount Shehallien on a leaden plummet ('Ph. Tr.' 1775, p. 500) . .	4·56
1798.	Cavendish, by the torsion balance ('Ph. Tr.' 1798, p. 469)	5·48

A.D.
1821. Carlini, by pendulum observations on Mount Cenis compared with those at the margin of the sea ('Effemeridi Astronomiche di Milano,' 1824, p. 28) . 4·39

1824. Laplace, by the theory of attraction, supposing the density of the crust to be equal to 3 ('Méc. Cél.' liv. xi. ch. ii. No. 5, t. v.) . . . 4·761

1842. Baily, by the torsion balance with extreme precautions ('M. A. S.' vol. xiv. p. 247) . . . 5·6604

1854. Airy, by pendulum observations in Harton Coal-pit ('M. A. S.' vol. xxv. p. 170, and 'Ph. Tr.' 1856, p. 342) . . . . . . 6·566

1856. Haughton, by applying a special method to Airy's observations ('Phil. Mag.' July 1856) . . 5·48

1856. James, by the deviation of a leaden plummet at Arthur's Seat ('Ph. Tr.' 1856, p. 591) . . 5·316

1873. Cornu and Baille, by the torsion balance ('C. R.' t. lxxvi. p. 954) . . . . . 5·56

§ 189. *Mass of the Earth.*—The following are the values which have been assigned to the mass of the earth as compared with the mass of the sun. When not otherwise mentioned, the mass of the earth has been concluded from the known values of gravity and the parallax of the sun:—

A.D.
1687. Newton, supposing the parallax of the sun to be 10″·5. ('Principia,' lib. iii. prop. 8) . . $\frac{1}{169282}$

1782. Lagrange ('Mém. Ac. Berlin,' 1782, p. 181) . $\frac{1}{365361}$

1802. Laplace ('Méc. Cél.' lib. vi. ch. vi. No. 21, t. iii.) $\frac{1}{329630}$

1832. Plana, by the parallactic equation of the moon ('Théorie du Mouvement de la Lune,' Turin, 3 vols. in 4to. t. iii. p. 20) . . . . $\frac{1}{352350}$

1864. Hansen, by the parallactic equation of the moon ('M. N.' vol. xxiv. p. 11) . . . $\frac{1}{319455}$

1867. Newcomb ('Astron. Obs. at U. S. Naval Observatory,' Washington, 1865, append. ii.) . . $\frac{1}{322600}$

1876. Von Asten, from the perturbations of Encke's comet, from 1819-1868. ('Wochenschrift für Astronomie,' &c., Halle, 1877, p. 42) . . $\frac{1}{309654}$

1876. Leverrier. Definitive value ('Annal. Obs. Paris,' Mém. t. xii. p. 9). . . . . . $\frac{1}{324439}$

THE MOON.

§ 190. The duration of the synodic revolution of the moon has been measured in very ancient times. The Chinese give the result (J. B. Biot, ' Précis de l'Hist. de l'Astron. Chinoise,' Paris, 1861, 4to.)—

$$29^d\ 12^h\ 44^m\ 25^s{\cdot}56$$

About four centuries before our era, Eudoxus gave the value—

$$29^d\ 12^h\ 43^m\ 28^s$$

The equation of the centre was known to Hipparchus, (Ptolemy, ' Math. Comp.' lib. v. cap. 7, B.C. 140), and the ancients had also detected the displacement of the perigee as well as that of the node (*ibid.*, lib. iv. cap. 6 and 8).

The evection was discovered by Ptolemy about 130 A.D. (*ibid.*, lib. v. cap. 3), but the name was assigned by Bouliaud in 1657.

The variation was pointed out by Tycho Brahe in 1602 (' Astron. Instaur. Progymn.') The annual equation was also known to Tycho Brahe (Bertrand, ' Les Fondateurs de l'Astronomie Moderne,' p. 94), and definitely announced by Kepler in 1625 (' Opera Omnia,' Francfort, 1858–70, eight vols. in 8vo. t. vi. p. 618).

The equation due to the non-sphericity of the earth was first introduced into the tables by Mason in 1793.

The inclination of the orbit was known to the ancients very probably from an exceedingly remote period (Ptolemy, ' Math. Comp.' lib. v. cap. 12).

§ 191. *Secular Variation.*—Halley remarked in 1692 that, in order to harmonise the observations of the moon in different centuries, it was necessary to introduce into the longitude a term depending upon the square of the time (' Ph. Tr.' No. 204). Laplace showed that an acceleration of this kind was due to the diminution of the eccentricity of the earth's orbit (' Mém. Ac. Sc. Paris,' 1786, p. 235). Doubts have arisen as to whether the secular variation can be completely explained by the variation in the eccentricity of the earth's orbit.

		Secular Variations (sidereal)[1]		
A.D.		Of the Longitude.	Of the Perigee.	Of the Node.
1692.	Halley (' Ph. Tr.' No. 204) .	. $+ 10''\cdot2$		
1771.	Lalande, by the eclipses of Ibn-Iounis, 977 and 978 ('Astron.' 2nd ed. liv. vii. No 1484, t. ii. p. 238) . .	9·9		
1787.	Laplace, by theory ('Mém. Ac. Sc. Paris,' 1786, p. 235) . .	11·1		
1802.	Laplace, by theory, combined with ancient eclipses ('Méc. Cél.' liv. vii. ch. i. Nos. 15 and 16, t. iii.) .	10·2	$-30\cdot9$	$+ 7\cdot6$
1841.	Hansen, theoretically ('A. N.' No. 443)	10·6	36·2	6·8
1847.	Hansen, by theory ('A. N.' No. 597) .	11·5	—	—
1849.	Airy, by Greenwich observations ('M. N.' vol. x. p. 160) . .	—	—	8·3
1853.	Adams, theoretically (' Ph. Tr.' 1853, p. 397)	5·8	—	—
1858.	Airy, eclipses of 1030, A.D., and 556, B.C. (' M. A. S.' vol. xxvi. p. 152).	13·0	—	—
1859.	Delaunay, theoretically ('C. R.' t. xlviii. pp. 825, t. xlix. pp. 311, 313) .	6·1	39·499	6·8
1877.	Newcomb, from observations, ancient and modern ('Amer. Journ. of Sc. and Arts,' 3rd ser. vol. xiv. p. 409) .	8·4		

Sir G. Airy remarks ('The Observatory,' edited by W. H. M. Christie, London, No. 37) : ' In the moon's orbit $1'' =$ 6,000 feet very nearly. In a century, therefore, the moon is accelerated 60,000 feet. In the first year the acceleration is 6 feet ; in the second year 18 feet additional, and so on. The moon's distance is changed in a century by 7·14 feet ; in one year it is changed by 0·0714 feet, or less than an inch. This change proceeds uniformly ; for every year before the epoch

---

[1] For the tropical variation, the secular variation of the precession $+ 1''\cdot121$ must be added.

the distance is additionally greater than the computed distance, and for every year after the epoch it is additionally less' (pp. 419-420). On the ancient eclipses see Newcomb, 'Washington Observations,' 1875, appendix ii.

§ 192. *Periods of the Lunar Movements.*—There are five distinct species of month, the durations of which are here expressed in mean solar days.

1. The average *sidereal* month (27·321661 days), or complete circuit of the heavens.

2. The average *lunation* common month or interval between two conjunctions with the sun (29·530589 days). This is also called the synodic month.

3. The average *anomalistic* month, or revolution from perigee to perigee (27·544600 days).

4. The average *tropical* month, or from the vernal equinox to the vernal equinox again, the equinox being in retrograde motion (27·321582 days).

5. The average *nodical* month, or from a node to a node of the same kind (27·212222 days). This is sometimes called the Draconitic month.

The perigee performs—

			Days.
One sidereal revolution in .	.	.	3232·575
,, tropical ,,	.	.	3231·475

The node makes—

			Days.
One sidereal revolution in .	.	.	6798·279
,, tropical ,,	.	.	6793·414

§ 193. *The Parallax of the Moon.*

A.D.		
130.	Ptolemy, mean parallax at Syzygy ('Math. Comp.' lib. v. cap. 13)	58′ 42″·5
1602.	Tycho Brahe ('Astron. Instaur. Progymn.' pp. 115, 119)	60 51
1687.	Newton ('Principia')	57 30
1751.	Lalande, observations at Berlin, compared with those at Cape of Good Hope ('Astronomie,' 2nd ed. No. 1699) , . . . ~	57 5

A.D.
1782.	Lagrange, by calculation ('Mém. Ac. Berlin,' 1782)	57' 10"·3
1838.	Henderson, by observations at the Cape compared with those in Europe ('M. A. S.' vol. x. p. 294)	57 1·8
1849.	Airy, by modern observations at Greenwich ('A. N.' No. 648)	57 3·16
1857.	Hansen (observations) ('Tables de la lune,' 4to.) .	56 59·57
1866.	E. J. Stone, by the observations at the Cape and Greenwich ('M. A. S.' vol. xxxiv. p. 16) .	57 2·707

§ 194. *Semidiameter of Moon.*—The following are the angular values which have been assigned to the semidiameter of the moon when it is situated at its mean distance from the earth :—

A.D.
140.	Ptolemy ('Math. Comp.' lib. vi. cap. 5) . .	16' 40"
880.	Albategnius ('De Scientia Stellarum,' cap. 30) .	16 12·5
1602.	Tycho Brahe ('Astron. Instaur. Progymn.' p. 134) .	15 25
1666.	Huyghens ('Hist. Ac. Sc. Paris, av. Renouv.' t. 1).	15 42·5
1672.	Horrocks ('Astronomia Kepleriana') . .	15 30
1822.	Groombridge, by the eclipse of 1820 ('M. A. S.' vol. i. p. 139)	15 27·2
1871.	Airy, by the eclipses of 1860 and 1870 (see Neison, 'The Moon,' London, 1876, p. 11) .	15 34·0

§ 195. *The Mass of the Moon.*—The following values of the mass of the moon are referred to the sum of the masses of the earth and the moon as unity :—

A.D.
1687.	Newton, by the tides ('Principia,' lib. iii. prop. 37)	$\frac{1}{39·708}$
1738.	D. Bernouilli, by the tides ('Hydrodynamica,' Strasbourg, 4°)	$\frac{1}{70}$
1755.	D'Alembert, from precession and nutation ('Recherche sur diff. Points du Syst. du Monde,' Paris, 1754–1756, t. ii. p. 182)	$\frac{1}{75}$
1795.	Delambre, from solar observations ('C. d. T.' an. xv. p. 366)	$\frac{1}{69·2}$
1795.	Laplace, by nutation (*ibid.*)	$\frac{1}{71}$
1828.	Bessel, from the lunar inequality ('A. N.' No. 133)	$\frac{1}{85}$
1842.	C. A. F. Peters, by his constant of nutation ('Numerus constans nutationis,' Petersburg, 4to.) .	$\frac{1}{81·24}$

A.D.
1844. Hansen, by the perturbations of the earth (Schumacher's 'Sammlung von Hülfstafeln,' Altona, 1845) $\frac{1}{80}$

1849. Airy, with the parallax derived from Greenwich observations, 1750–1830 ('M. A. S.' vol. xvii. p. 51) . . . . . . $\frac{1}{99\cdot5}$

1858. Leverrier, from nutation ('Annal. Obs. Paris,' Mém. t. iv. p. 103) . . . $\frac{1}{81\cdot84}$

1862. Lubbock, from tides ('M. A. S.' vol. xxx. p. 29) $\frac{1}{67\cdot3}$

1866. Adams, from Peters's constant of nutation (see 'M. N.' vol. xxvi. p. 179) . . . $\frac{1}{81\cdot5}$

1867. Newcomb, from the parallactic inequality of the earth ('Astr. Obs. at U. S. Naval Observatory,' 1865, app. ii.) . . . . . $\frac{1}{81\cdot08}$

1868. E. J. Stone, from precession and nutation ('M. N.' vol. xxviii. p. 43) . . . . $\frac{1}{81\cdot36}$

1876. Von Asten, from the perturbations of Encke's comet ('Wochenschrift für Astronomie,' Halle, 1877, p. 42) $\frac{1}{81\cdot44}$

§ 196. *The Lunar Equator.*—The following are the determinations which have been made of the inclination of the moon's equator to the ecliptic :—

A.D.
1700. J. D. Cassini, by his observations ('Mém. Ac. Sc. Paris,' 1721, p. 108) . . . 2° 30′ 0″

1763. Lalande, by the observations of Manilius ('Astron.' 2nd ed. No. 3026, t. iii.) . . 1 43 0

1819. Poisson, by the observations of Manilius and Nicollet ('C. d. T.' 1821, p. 223) . . 1 28 42

1819. Poisson, from the observations of Manilius made by Nicollet ('C. de. T.' 1821, p. 223). . 1 28 42

1821. Nicollet, from his own observations and those of Bouvard and Arago 1 28 45

§ 197. *Libration.*—As the spots on the surface of the moon are visible to the unaided eye, it is not difficult to perceive that the same side of the moon is always turned towards the earth. This was mentioned by Aristotle (B.C. 325±). The telescope enabled Galileo in 1616 to announce the discovery of the moon's libration. Newton explained the cause of the libration ('Phil. Nat. Princip. Math.' 1687, lib. iii. prop. 1). Cassini deduced from his observations, at the commencement of the eighteenth century, the two following laws ('Mém. Ac. Sc. Paris,' 1721, p. 108):— 1. The rotation of the moon takes place about an axis which remains constantly parallel to itself and in a period equal to that of the revolution. 2. The three planes of the ecliptic, the orbit of the moon, and the equator of the moon, intersect in the same straight line. The ecliptic lies between the two other planes.

The theory of these laws has been given by Laplace, ('Méc. Cél.' liv. v. ch. 2, t. ii. 1799).

§ 198. *Figure of the Moon.*—Let a be the axis of the lunar equator directed towards the earth, b the axis of rotation, and c the third rectangular axis in the direction of the movement of translation. Let A, B, C be the moments of inertia relative to these axes   Then taking c as unity, we have for a the following values :—

A.D.
1761.	D'Alembert ('Opuscules Math.' t. ii. p. 313)	$a = 1\cdot00002678$
1780.	Lagrange ('Mém. Ac. Berlin,' 1780)	$a = 1\cdot00002404$
1799.	Laplace ('Méc. Cél.' liv. v. ch. 2, t. ii.) .	$a = 1\cdot00003304$

The elongation in the direction of a is not the same in both hemispheres. If the elongation in the hemisphere directed towards the earth be taken as 1, then the elongation in the opposite hemisphere has the values—

A.D.
1780.	Lagrange, by theory ('Mém. Ac. Berlin,' 1780)	.	0·25
1821.	Nicollet, by various data ('C. d. T.' 1823, p. 340)	.	0·056
1848.	Wichmann, by his own researches ('A. N.' No. 631) .	0·419

From the discussion of perturbations, Hansen was led to the conclusion that the centre of gravity of the moon is more remote from us than would be inferred from its figure. Taking the distance from the earth to the centre of figure as 1, he has deduced that the distance to the centre of gravity is 1·0001544 ('M. A. S.' vol. xxiv. [1856], p. 31).

§ 199. *Moments of Inertia of the Moon.*—From the movements of the node Laplace finds (*loc. cit.*)—

$$\frac{C-A}{C} = 0.00060.$$

From the coefficient of the annual inequality of the libration (*ibid.*)—

$$\frac{B-A}{C} = 0.00056.$$

Whence, taking A as unity, we conclude—

$$A = 1 \; ; \; B = 1.00056, \; C = 1.00060.$$

The hypothesis of an homogeneous moon would give—

$$A = 1 \; ; \; B = 1.000029, \; C = 1.000038.$$

It would seem as if one or both extremities of the diameter *a* were laden with additional mass. Wichmann has found by the connection which theory indicates between the inclination of the equator of the moon and that of the orbit ('A. N.' No. 621) 1848—

$$\frac{C-A}{A} = 0.000599,$$

and by the inequality of the libration in longitude and the inclination of the equator of the moon—

$$\frac{B-A}{C} = 0.000564.$$

These results confirm those of Laplace.

§ 200. *Topography of the Moon.*—The following are the principal maps which give the details of the surface of the moon as seen in a telescope :—

A.D.
1834. Beer and Mädler, 'Mappa Selenographica totam Lunæ hemisphæram visibilem complectens.' Berlin. Diameter, $0^m \cdot 95$, 4 plates.

1842-74. Julius Schmidt, 'Charte der Gebirge des Mondes.' 25 plates. Berlin, 1878.

For identification of lunar objects the map in Webb's 'Celestial Objects for Common Telescopes,' 2nd edition, London, 1873, is very useful.

For the study of the topography and appearance of the moon refer to—

A.D.
1780.   W. Herschel, 'Astronomical Observations relating to the Mountains of the Moon' ('Phil. Trans.' 1780, p. 507).

1791.   Schröter, 'Selenotopographische Fragmente,' 1-2, Göttingen, 1791-1802.

1874.   Nasmyth and Carpenter, 'The Moon considered as a Planet, a World, and a Satellite.' London, 8vo.

1876.   Neison, 'The Moon and the Condition and Configuration of its Surface.' London, 8vo.

Schröter considered that he had detected changes in certain localities of the moon, especially in Hevelius and Mare Crisium. Ward observed a change of colour in Aristarchus. Schmidt noticed changes in Linné in 1865. For a discussion of the question, see Birt, 'On the Extent of Evidence which we possess Elucidatory of Change on the Moon's Surface' ('Report of the British Association,' 1868, p. 11).

For observations of curious phenomena which sometimes occur in occultations of fixed stars by the moon, see Airy, 'Monthly Notices of the Royal Astronomical Society,' vol. xxviii. 1860, p. 173.

§ 201. *Brilliancy of the Moon.*—The following are the determinations which have been made of the luminous intensity of the moon as compared with that of the sun :—

A.D.
1725.   Bouguer, by the intervention of a candle ('Traité d'Optique sur la Gradation de la Lumière,' Paris, 1760, p. 89)   . . . . . $\frac{1}{300000}$

1829.   Wollaston, by the intervention of a candle ('Ph. Tr.' 1829, p. 27)   . . . . . $\frac{1}{801072}$

1861.   G. P. Bond, by the intervention of a Bengal fire ('Memoirs of the American Academy,' new ser. vol. viii. pp. 222-297).   . . . . $\frac{1}{470980}$

A.D.
1861.   G. P. Bond, by the actinometric action on a photo-
        graphic preparation (*ibid.*)   .   .   . $\frac{1}{340000}$
1865.   Zöllner, with his photometer ('Photometrische Unter·
        suchungen,' Leipzig, 8vo.)   .   .   . $\frac{1}{618000}$

The following determinations have been given of the
' Albedo ' of the moon :—

> Wollaston, 0·12.   G. P. Bond, 0·071.   Zöllner, 0·1736.

§ 202. *Heat from the Moon.*—Piazzi Smyth, in 1856,
attempted to detect the heat radiated from the moon by
direct experiment on the Peak of Teneriffe. He found this
to be equal to one-third of that of a candle placed at a distance of 4·75 mètres (' Proceedings of the Royal Institution,'
vol. ii. p. 493). An extensive series of experiments have been
made by Lord Rosse on the heat radiated from the moon
by the aid of the 3-feet Parsonstown telescope. He found
that the moon radiates as if its surface were at a temperature
of 360° F.=182° C. (the Bakerian Lecture, 'Ph. Tr.' vol.
163, pp. 587–627. See also ' C. R.' t. lxix. p. 922). Marie-
Davy found the lunar heat to be equal to that of a disk of
iron of the same apparent size heated to 100° and placed at
a distance of 35 mètres ('C. R.' t. lxix. 1867, pp. 705,
922, 1154).

MARS.

§ 203. *Elements.*—The following are the elements of the
planet Mars given by Leverrier for the epoch 1850, Jan. 1,
mean noon at Paris (' Annal. Obs. Paris,' Mém. t. vi. 1861,
pp. 309 and 310) :—

Mean longitude	83° 40′ 31″·33 +	689101″·05375t + 0″·00011341t^2.
Longitude of perihelion	333° 17′ 53″·67 +	66″·2411t + 0″·00012093t^2.
Longitude of node	48° 23′ 53″·1 +	27″·992t − 0″·000217t^2.
Inclination	1° 51′ 2″·28 −	0″·02431t + 0″·00000945t^2.
Eccentricity	0·09326113 +	0·19679t − 0·00000252t^2.

The time t is computed in Julian years of $365\frac{1}{4}$ days.

The eccentricity of the orbit of Mars is considerable ; the following values have been obtained for the greatest equation of the centre :—

A.D.		
1719.	Halley ('Tabulæ Astronomicæ') .	10° 40′ 2″
1740.	J. Cassini ('Eléments d'Astronomie')	10 39 19
1771.	Lalande ('Astronomie,' 2nd ed. t. iii.)	10 42 47
1811.	Von Lindenau ('Tabulæ Martis,' Eisenberg, epoch 1800) . .	10 41 33·1 + 0″·3735t
1861.	Leverrier ('Annal. Obs. Paris,' Mém. t. vi. p. 344, epoch 1850) .	10 41 51·48 + 0·3947t

§ 204. *Diameter of Mars.*—The following are the values of the apparent diameter of Mars when situated at the mean distance of the earth from the sun. The ellipticities of the planet are also given :—

A.D.		Equatorial Diameter.	Ellipticity.
1672.	Picard, with the micrometer ('Hist. Ac. Sc. Paris,' t. vii. p. 330) . .	11″·4	
1784.	W. Herschel, with the micrometer ('Ph. Tr.. 1784, pp. 271, 273) . .	9″·13	$\frac{1}{16·5}$
1837.	Bessel, by micrometric observations ('A. N.' No. 1135) . . .	9·3278	Doubtful
1852.	Johnson, with the heliometer at Oxford ('Radcliffe Obs.' vol. xi. p. 292) .	8·992	Elongated
1864.	Kaiser, micrometric measures in 1862 ('A. N.' No. 1468) . . .	9·518	$\frac{1}{118}$
1864.	Dawes (see Webb, 'Celestial Objects for Common Telescopes,' 1873, p. 135) .		Insensible
1865.	Jul. Schmidt, by his micrometric measures ('A. N.' No. 1543) in 1845 . .	9·436	
	In 1854 and 1856 . . .	9·734	
1873.	Engelmann, by micrometric measures at the opposition of 1873 ('A. N.' No. 1966)	9·403	

§ 205. *Rotation of Mars.*—The rotation of Mars was first noticed by Huyghens in 1659 : the following are the different determinations of its period :—

A.D.

1666. Jean Dom. Cassini ('Martis circa proprium axem revolubilis observationes,' Bologna) . $24^h 40^m$

1784. W. Herschel, by 6 years of observation ('Ph. Tr.' vol. lxxiv. p. 233) . . . 24 37 27[*]

1859. Secchi, by several years of observation ('C. R.' t. xlix. p. 346). 24 37 35

1866. R. Wolf, by comparing his observations in 1864 with those of Secchi in 1862 'A. N.' No. 1623) . . . . . 24 37 22·9

1867. Proctor, by the observations of Hooke in 1666, W. Herschel in 1783, and Dawes, 1856-1867 ('M. N.' vol. xxviii. p. 39 ; xxix. p. 232) . 24 37 22·735

1873. Kaiser. by comparing Huyghens's observations with modern ones ('Annalen der Sternwarte,' Leiden, vol. iii. p. 80.) . . . 24 37 22·591

1873. Jul. Schmidt, by the comparison of Huyghens's observations with his own in 1845-1873 ('A. N.' No. 1965) . . . . 24 37 22·57

§ 206. *Direction of the Axis of Rotation of Mars.*

1853. Oudemans, from Sir W. Herschel's observations ('A. N.' No. 838)—

Longitude of North Pole   347 43 equinox 1782
Latitude   ,,   + 58 31   ,,

Oudemans, from Bessel's observations ('A. N.' No. 838)—
Longitude of North Pole   349° 1' equinox 1833·5
Latitude   ,,   + 61° 9'   ,,

§ 207. *Mass of Mars.*

A.D.

1802. Delambre, by the perturbations of the earth (see Laplace, 'Méc. Cél.' liv. vi. ch. 16, t. iii.) . $\frac{1}{2546320}$

1807. Piazzi, from a discussion of solar observations ('Osservatorio di Palermo,' liv. vi.) . . $\frac{1}{3588500}$

1828. Airy, in correcting the solar tables of Delambre by the observations at Greenwich ('Ph. Tr.' 1828, p. 30) . . . . . $\frac{1}{3734602}$

1853. Hansen and Olufsen, by the perturbations of the earth ('Tables du Soleil,' p. 1) . . $\frac{1}{3200900}$

1858. Leverrier, by the perturbations of the earth ('Ann. Obs. Paris,' Mém. t. iv. p. 102) . $\frac{1}{2994790}$

A.D.
1876. Leverrier, by the perturbations of Jupiter (*ibid.*
       t. xii. p. 9) .   .   .   .   . $\frac{1}{2812526}$
1876. Powalky, by comparing the observations of the
       sun made at Dorpat, 1823-1829, with the
       tables of Hansen and Olufsen ('A. N.' 2105) . $\frac{1}{2876000}$
1878. Asaph Hall, by observations of the satellites
       ('Observations and Orbits of the Satellites of
       Mars, with Data or Ephemerides in 1879'),
       4to. Washington, 1878, p. 37 .   .   . $\frac{1}{3093500\pm3295}$

§ **208**. *Brilliancy of Mars.*—In 1859 Seidel, with Stein-heil's photometer, determined the brilliancy of Mars to be 2·97 times that of Vega. In 1865 Zöllner, with his photometer, determined the brightness of Mars to be 1÷6,994,000,000 that of the sun. The albedo of Mars is 0·2672. These measures refer to the appearance of Mars at the time of opposition.

§ **209**. *Physical Condition of Mars.*—The white spot at the pole of Mars was first noticed by Maralui in 1716. Sir W. Herschel ('Phil. Trans.' 1784, p. 273) pointed out that the white regions are not centred about the poles. The following values have been given for the eccentricity of the brilliant southern region :—

A D.
1783. Sir W. Herschel   .   .   .   .   8° 8'
1877. Asaph Hall ('A. N.' No. 2174)   .   .   5 11
1877-74. Schiaparelli ('A. N.' No. 2178)   .   .   6 9

The details of the surface of Mars, together with the nomenclature of its different parts, are given in a map by Proctor ('Chart of Mars, from 27 Drawings by Mr. Dawes,' London, 1874). See another map by Green, from observations in 1877 ('M. A. S.' vol. xliv. pp. 123-140). Terby has, in his 'Aréographie,' Brussels, 1874, discussed all drawings made up to that date.

§ **210**. *Satellites of Mars.*—The following are the elements of the satellites of Mars, as assigned by their discoverer, Asaph Hall ('Observations and Orbits of the Satellites of Mars,' Washington, 1878, pp. 24, 25). The periodic time of Phobos here given has been corrected by $-1^s\cdot074$ (A. Hall, ' C. R.' vol. lxxxix. p. 778) :—

Let a denote the semiaxis major of the orbit at distance unity.

,, J ,, the inclination of the plane of the orbit to the plane of the equator.

,, N ,, the right ascension of the ascending node on the equator.

,, ω ,, angular distance of the node from the apse.

,, e ,, eccentricity.

,, u ,, the angular distance of the satellite from the node at the epoch.

Then we have for Deimos :—

> Epoch 1877, August 28·0 Greenwich mean time.
> Period $= 1^d\cdot262429 = 30^h \ 17^m \ 53^s\cdot86 \pm 0^s\cdot985.$
> Log $\omega = 2\cdot4550955$
> $\qquad a = 32''\cdot3541 \pm 0''\cdot0118$
> $\qquad \left. \begin{array}{l} J = 35^\circ \ 38'\cdot7 \\ N = 48^\circ \ \ 5'\cdot7 \end{array} \right\}$ Equator, August 28, 1877.
> $\qquad \omega = 40^\circ \ 53'\cdot6 \pm 5^\circ \ 4'\cdot98$
> $\qquad e = 0\cdot005741 \pm 0\cdot0004898$
> $\qquad u = 357^\circ \ 30'\cdot5$

PHOBOS.

> Epoch 1877, August 28·0 Greenwich mean time.
> Period $= 0^d\cdot3189244 = 7^h \ 39^m \ 13^s\cdot996.$
> Log $\omega = 3\cdot0526147$
> $\qquad a = 12''\cdot9531 \pm 0''\cdot0142$
> $\qquad \left. \begin{array}{l} J = 36^\circ \ 47'\cdot1 \\ N = 47^\circ \ 13'\cdot2 \end{array} \right\}$ Equator, August 28, 1877.
> $\qquad \omega = 45^\circ \ 30'\cdot4 \pm 2^\circ \ 11''\cdot16$
> $\qquad e = 0\cdot032079 \pm 0\cdot001407$
> $\qquad u = 285^\circ \ 20'\cdot2$

THE MINOR PLANETS.

§ 211. *Eccentricities.*—There are now (1880) above 200 of these bodies known, and an ephemeris of all those of which

the orbit nas been computed is published each year in the
'Berliner Astronomisches Jahrbuch.' The following are
remarkable for the great eccentricities of their orbits. The
sine of the angle ϕ is the eccentricity ('Berliner Astron.
Jahrbuch,' 1882), pp. 114–125 :—

No. of Planet.	Name.	ϕ			Authority.
33	Polyhymnia	19°	51′	1″·0	Schubert
36	Atalante	17	20	45·0	Schubert
41	Daphne	15	38	11·0	Maywald
50	Virginia	16	34	35·2	Powalky
75	Eurydike	17	52	50·8	Stockwell
109	Felicitas	17	27	35·5	Rogers
132	Aethra	22	31	39·8	Watson
156	Xanthippe	15	17	23·2	A. Schmidt
164	Eva	20	16	59·5	Tietjen
167	Urda	18	10	30·6	C. H. F. Peters
175	Andromache	20	22	12·9	Watson
183	Istria	20	40	17·7	Donner
193	Ambrosia	16	34	52·0	Leman

The majority of the minor planets have but small
eccentricities.

§ 212. *Inclinations.*—The majority of the minor planets
have orbits not much inclined to the ecliptic ; the following
exceptional cases may be noticed in which the inclination
is considerable (*loc. cit.*) :—

No. of Planet.	Name of Planet.	Inclination.			Authority.
2	Pallas	34°	43′	28″·4	Farley
25	Phocæa	21	35	20·3	Maywald
31	Euphrosyne	26	28	46·6	Hill
71	Niobe	23	15	56·0	Becker
105	Artemis	21	32	4·4	Watson
130	Elektra	22	54	49·5	Powalky
132	Aethra	24	56	56·5	Watson
148	Gallia	25	21	16·6	Bossert
154	Bertha	20	59	20·9	Anton
164	Eva	24	25	18·1	Tietjen

§ 213. *General Features.*—The mean distance of No. 149 Medusa is 2·132, and of No. 153 Hilda is 3·944. Between these limits lie the distances of the other planets, No. 1 to No. 211. Leverrier has found that the total mass of the planets between the mean distances of 2·20 and 3·16 from the sun cannot attain a mass of more than one-fourth that of the earth ('C. R.' t. xxxvii., p. 797).

The minor planets are telescopic, but Heis records that Vesta has been seen with the unaided eye.

Stampfer, Argelander, and Littrow have determined the approximate diameters of the 70 first minor planets by supposing the albedo the same as that of the outer planets (Littrow, 'Wunder des Himmels,' Stuttgart, 1866, p. 1002). The diameters range between 250 and 15 miles.

JUPITER.

§ 214. *Elements.*—The following elements have been assigned by Leverrier in 1876 ('Ann. Obs. Paris,' Mém. t. xii. pp. 26-29).

Epoch 1850, January 1, mean noon Paris.

Mean longitude	.	.	160°	1′	10″·26 + 156°	18′	7″·2130*t*
Longitude of perihelion	.	.	11	54	58 ·41 +	1	36	30 ·321 *t*
Longitude of node	.	.	98	56	17 ·0 +	1	0	36 ·617 *t*
Inclination	.	.	1	18	41 ·37 −	0	0	20 ·520 *t*
Greatest equation of centre	.	5	31	45 ·32 +	0	1	8 ·752 *t*

In this result *t* denotes the number of Julian centuries of 36,525 days.

§ 215. *Great Inequality of Jupiter.*—The great inequality in the motion of Jupiter, which is due to the perturbing influence of Saturn, was first explained by Laplace, 1785 ('Méc. Cél.' liv. vi. ch. xii.) This inequality depends upon the close approximation which exists between five times the mean movement of Saturn and twice that of Jupiter.

The principal coefficient of this inequality in longitude,

according to Leverrier, 1876 ('Annal. Obs. Paris,' Mém. t. xii. p. 33), epoch 1850, is

$$1205''{\cdot}96 - 0''{\cdot}05536t + 0''{\cdot}00009577t^2$$
$$+ 0''{\cdot}00000000501t^3$$

The period is 929 years (Laplace). The coefficient expresses the amplitude of the oscillation of the longitude on either side of its mean value. The inequality was zero in the year 1790.

§ 216. *Dimensions of Jupiter.*—The following measures give the equatorial diameter of Jupiter referred to the mean distance between the sun and Jupiter. The ellipticity is also added :—

A.D.		Equatorial Diameter.	Ellipticity.
1650.	Riccioli and Grimaldi (Riccioli, 'Astronomia Reformata,' p. 356) . . .	49″·77	
1691.	Jean Dom. Cassini ('Hist. Ac. Sc. Paris, avant renouv.' t. ii. p. 130) . .		$\frac{1}{15}$
1719.	Pound, micrometrically (Newton, 'Principia,' 1726, lib. iii. prop. 8) . .	39·5	$\frac{1}{13\cdot37}$
1799.	Laplace, by the movements of the nodes and perijoves of the satellites ('Méc. Cél.' liv. iii. ch. 5, No. 43, t. 11) . .		$\frac{1}{14\cdot39}$
1833.	Bessel, by the heliometer ('Königsb. Beobacht.' Abth. xix. S. 102) . .	37·60	$\frac{1}{15\cdot73}$
1836.	Damoiseau, by the theory of the satellites ('Tables écliptiques des Satellites de Jupiter,' Paris, 4to. introd. p. ii.) . .		$\frac{1}{13\cdot492}$
1848.	Airy, Greenwich observations ('M. N.' vol. xiii. p. 22)		$\frac{1}{17\cdot9}$
1850.	Johnson, micrometrically ('Radcliffe Obs.' vol. xi. p. 293)	37·380	$\frac{1}{15\cdot6}$
1857.	De la Rue, micrometrically ('M. N.' vol. xvii. p. 7)	37·141	$\frac{1}{18\cdot62}$
1865.	Jul. Schmidt, micrometrically ('A. N.' No. 1543)	38·91	$\frac{1}{15\cdot6}$
1875.	Bellamy, double-image micrometer ('Radcliffe Obs.' vol. xxxv. p. 216) . .	37·00	$\frac{1}{17\cdot91}$

§ 217. *Rotation of Jupiter.*—In 1610, Kepler inferred the rotation of Jupiter from the movements of the satellites. The rotation was first discovered by Jean Dominique Cassini in 1664. The following are the determinations of the period of the sidereal rotation of Jupiter :—

A.D.
1685.	Jean Dom. Cassini ('Hist. Ac. Sc. Paris avant le renouv.' t. x. pp. 1, 513, 707)	$9^h\ 55^m\ 50^s$
1781.	W. Herschel ('Phil. Trans.' vol. lxxi.)	9 55 40
1816.	Schröter ('Hermographische Fragmente,' Bd. ii.)	9 55 33
1835.	Bessel ('Königsberg. Beobacht.' Abth. xx. S. 77, 78)	9 55 26·0
1879.	Pratt, by observations of a red spot ('M. N.' xl. p. 156)	9 55 33·91

§ 218. *Situation of the Axis of Rotation of Jupiter.*—Let I represent the inclination of the plane of the equator of Jupiter to the plane of his orbit, and let N be the longitude of the node of this equator on the orbit; E is the epoch and t the number of years since the epoch.

A.D.
1693. Jean Dom. Cassini ('Les Hypothèses et les Tables des Satellites de Jupiter,' Paris, 4to.)

$$I = 2°\ 55' \qquad N = 314\tfrac{1}{2}° \qquad E = 1693$$

1719. Bradley ('Ph. Tr.' No. 394)

$$N = 311\tfrac{1}{3}° \qquad E = 1718$$

1805. Delambre, by the eclipses of the satellites with annual variations by Laplace.[1] (See 'Méc. Cél.' liv. viii. ch. 10, No. 28, t. iv.)

$$I = 3°\ 5'\ 30'' + 0''·02279t$$
$$N = 313°45'\ 33'' - 0''·2676t \qquad E = 1750·0$$

1836. Damoiseau, by the eclipses of the first satellite ('Tables des Satellites de Jupiter,' Paris, 4to. introd. p. 1)

$$I = 3°\ 4'\ 5'' \qquad N = 313°\ 21'\ 55''$$
$$E = 1750·0$$

[1] The motion of the node here indicated is the true precession of Jupiter. It is necessary to add for our method of computing longitudes the precession of the terrestrial equator.

§ 219. *Mass of Jupiter and his System.*—The mass of the sun being taken as unity, the following values have been assigned to the mass of Jupiter :—

A D.

1687. Newton, from the elongation of satellite iv. observed by Pound ('Principia,' lib. iii. prop. 8)   .   . $\frac{1}{1067}$

1802. Laplace, from the same observation ('Méc. Cél.' liv. vi. ch. 6, No. 21, t. iii.)   .   .   .   . $\frac{1}{1067\cdot09}$

1821. Bouvard, by the perturbations of Saturn ('Tables astronomiques,' introd. p. 11)   .   .   . $\frac{1}{1070\cdot5}$

1825. Nicolai, by the perturbations of Juno ('Berliner Jahrbuch,' 1826, p. 226)   .   .   .   . $\frac{1}{1053\cdot924}$

1826. Encke, by the perturbations of Vesta and Ceres ('Abhandlungen der Akad. zu Berlin,' 1826, Math. Kl. S. 267)   .   .   .   .   .   . $\frac{1}{1050\cdot117}$

1837. Airy, by the elongation of satellite iv. ('M. A. S.' vol. x. p. 47)   .   .   .   .   .   . $\frac{1}{1046\cdot77}$

1841. Bessel, by the elongations of the four satellites ('C.R.' t. xiii. p. 59)   .   .   .   .   . $\frac{1}{1047\cdot879}$

1866. Krüger, by the perturbations of Thémis ('A. N.' No. 1941)   .   .   .   .   .   . $\frac{1}{1047\cdot538}$

1871. Möller, by the perturbations of Faye's comet ('Vierteljahrsschrift der Astron. Gesellschaft,' Bd. vii. p. 95) .   .   .   . $\frac{1}{1047\cdot788}$

1872. Von Asten, by the perturbations of Encke's comet, 1819–1868 ('Mém. Ac. St.-Pétersbourg,' 7ᵉ sér. t. xviii. No. 10)   .   .   .   .   . $\frac{1}{1049\cdot632}$

§ 220. *Brilliancy of Jupiter.*—The following determinations give the brightness of Jupiter as reduced to a mean opposition :—

A.D.

1845. Sir John Herschel, Jupiter with respect to the mean full moon ('M. N.' vol. xxi. p. 199)   .   .   . $\frac{1}{8620}$

1859. Seidel, with Steinheil's photometer ('Untersuchungen uber die Lichtstärke der Planeten,' Munich, 4to.):—

Jupiter = 8·24 × Vega.

A.D.

1861. G. P. Bond, comparing Jupiter with the full moon, obtained the ratio . . . . . $\frac{1}{6430}$

and for Jupiter compared with the sun—

$$\text{Jupiter} = \frac{1}{3028350000} \times \text{Sun}$$

('Memoirs of the American Academy,' new ser. vol. viii. p. 222)

1865. Zöllner, with his photometer, found

$$\text{Jupiter} = \frac{1}{5472000000} \times \text{Sun}$$

('Photometrische Untersuchungen,' Leipzig, 4to.)

From this last result the albedo of Jupiter is 0·6328.

§ 221. *Physical Conditions of Jupiter.*—The following works may be consulted on the appearance of the disc of Jupiter :—

A.D.

1798. Schröter ('Beiträge zu astron. Entdeckungen,' Bd. ii. Berlin).

1841. Beer and Mädler ('Beiträge zur physischen Kenntniss der Himmelskörper,' pp. 91-106).

1852. Nasmyth ('Monthly Notices, R. A. S.' vol. xiii. p. 40).

1858. Murray and Lassell ('Monthly Notices, R. A. S.' vol. xix. p. 51).

1862. Huggins ('Monthly Notices,' vol. xxii. p. 294).

1877. Lockyer ('Nature,' vol. xv. p. 282).

On the atmosphere of Jupiter, consult W. Herschel, 'Phil. Trans.' 1791, p. 322, and on the spectrum see Vogel, 'Beobachtungen angestellt auf der Sternwarte in Bothkamp,' Heft i. 1872, and Heft ii. 1873. Leipzig 4to.

§ 222. *Satellites of Jupiter.*—The satellites I., III., IV., were discovered by Galileo on January 7, 1610. The satellite II. was first seen on the following day by Simon Mayer.

As to the brightness, the satellites are arranged in the following order by Sir W. Herschel ('Phil. Trans.' 1797):—

III. much the most brilliant, IV., I., II. Zenger has determined the relative brightness of the four satellites by their appearance and disappearance in twilight. He has found the following values ('M. N.' vol. xxxviii. p. 67) :—

III.=1·000 ; II.=0·070 : I.=0·061 ; IV.=0·820.

III. is generally reckoned as the same brightness as a star of the sixth magnitude.

On January 12, 1610, five days after the discovery of the satellites, Galileo witnessed the first eclipse. In 1643 Fontana first saw the shadow of a satellite on the disc of Jupiter. In 1707 Maraldi first observed the transit of satellites across the disc.

It appears that the satellites sometimes remain visible just after they have passed behind the limb of the planet. See Todd, 'Monthly Notices, R. A. S.' vol. xxxvii. (1877), p. 284. The satellite II. was seen to be eclipsed by III. by Arnold on November 1, 1693. The satellite IV. was similarly seen to be eclipsed by III., by Luthmer of Hanover, October 30, 1822. On August 14, 1820, Flaugergues saw satellite III. occult a small star.

§ **223.** *Elements of the Satellites.*—The mean distances of the satellites are naturally expressed in terms of the equatorial radii of Jupiter. We have thus :—

			Satellite.		
A.D.		I.	II.	III.	IV.
1612.	Galileo ('Intorno alle cose que stanno su l' aqua,' Florence, 4to.)	6	10	16	24
1693.	Jean Dom. Cassini ('Les Hypothèses et les Tables des Satellites de Jupiter,' Paris, 4to.) . .	5·667	9·017	14·384	25·299
1740.	Jacques Cassini ('Élém. d'Astron.' Paris, 4to. p. 633) .	5·67	9·00	14·38	25·30
1805.	Laplace ('Méc. Cél.' liv. viii. ch. vii. No. 20 t. iv.) .	5·698491	9·066548	14·461893	25·43590

- The following values are expressed in arc, referred to the mean distance from Jupiter to the sun :—

Satellite.

	I.	II.	III.	IV.

A.D.
1841. Bessel ('C. R.' t. xiii. p. 58):—

1′ 51″·7423   2′ 57″·7969   4′ 43″·6059   8′ 18″·8657

§ 224. *Synodic Revolutions.*

1612. Galileo (*loc. cit.*):—

d. h. m. s.	d. h. m. s.	d. h. m. s.	d. h. m. s.
1 18 30 0	3 13 30 0	7 4 0 0	16 8 0 0

1614. Simon Mayer ('Mundus Jovialis,' Nuremberg, 4to.):—

| 1 18 28 30 | 3 13 18 0 | 7 3 56 34 | 16 18 9 15 |

1693. Jean Dom. Cassini ('Les Hypothèses et les Tables des Satellites,' Paris, 4to):—

| 1 18 28 37 | 3 13 17 54 | 7 3 59 39 | 16 18 5 8 |

1817. Delambre ('Tables écliptiques des Satellites de Jupiter,' Paris, 4to):—

| 1 18 28 35·95 | 3 13 17 53·73 | 7 3 59 35·83 | 16 18 5 7·02 |

The following expressions for the mean longitude of the satellites have been deduced by Littrow, from the elements of Damoiseau ('Gehler's Physikalisches Wörterbuch neu bearbeitet,' Bd. ix. p. 1033). If t be the number of Julian years since 1850, then we have—

	Mean Longitude.	Diurnal Tropical Movement.
I.	$15°·0128 + (206 \times 360° + 164°·35467)t$	$203°·488993385$
II.	$311°·8404 + (102 \times 360° + 307°·13231)t$	$101°·374762063$
III.	$10°·3541 + (51 \times 360° + 18°·52114)t$	$50°·317646432$
IV.	$72°·5512 + (21 \times 360° + 318°·84714)t$	$21°·57110943$

If n', n'', n''', be the mean movements, either sidereal or synodic, of the satellites I., II., III., and if l', l'', l''', be their absolute longitude, sidereal or synodic, then Laplace has shown that the following relation is always fulfilled ('Méc. Cél.' liv. ii. No. 66, t. i.) :—

$$3n'' = 2n''' + n'$$
$$l' + 2l''' - 3l'' = 180°$$

Bradley discovered in 1726 ('Ph. Tr.' No. 394) that—

247 revolutions of I. require 437 days 3 hours 44 min.
123   ,,   II. ,,   437 ,,   3 ,,   42 ,,
61   ,,   III. ,,   437 ,,   3 ,,   36 ,,

This remarkable relation between the periods does not extend to the fourth satellite, for

26 revolutions of IV. require 435 days 14 hours 16 min.

§ 225. *Eccentricities and Inclinations of the Satellites.*— The undisturbed orbits of the two interior satellites appear to have no appreciable eccentricity. The perturbation produced by the interference of the exterior satellites causes, however, a slight degree of eccentricity. The orbit of the innermost satellite may be considered to be coincident with the plane of Jupiter's equator. The following are the values of the inclinations of the orbits of the remaining satellites to the plane of Jupiter's equator, as well as their longitudes referred to the epoch 1750·0 (Damoiseau, *loc. cit.*):—

Satellite.	Inclination of the Equator of Jupiter.	Longitude of Ascending Node 1750·0.	Movement of Node in one Julian Year.
II.	1' 6"·2	101°·996	− 12°·0918
III.	5' 3"·1	173°·492	− 2°·568
IV.	24' 33"·1	278°·301	− 0°·706

§ 226. *Masses of the Satellites.*—The masses of the different satellites, taking the mass of Jupiter as unity, are determined by the mutual perturbations which these attractions produce. The following values have been assigned:—

A.D.
1766. Bailly ('Essai sur la Théor. des Sat. de Jupiter,' Paris, 1766, 4to.):—

     I.   0·00004247   III.   0·00007624
     II.   0·0000211   IV.   0·00005

1805. Laplace ('Méc. Cél.' liv. viii. ch. ix. t. iv.):—

     I.   0·0000173281   III.   0·00008849
     II.   0·0000232355   IV.   0·000042659

A.D.
1836. De Damoiseau ('Tables écliptiq. des Sat. de Jupiter,' introd. p. ii.):—

I.	0·000016877	III.	0·000088437
II.	0·00002322696	IV.	0·0000424751

§ 227. *Perturbations of the Satellites.*—The principal work on this subject is Laplace, ' Méc. Cél.' liv. viii. ch. 1–16. For an elementary account by the same author see 'Système du Monde,' liv. iv. chap. vi. J. J. Littrow has given a very clear exposition in his 'Elements of Physical Astronomy,' and in the ' Gehler's Phys. Wörterb. neu bearbeitet,' vol. ix. 1838, p. 1022–1075. The tables are :—

A.D.
1817. Delambre (' Tables écliptiques des Satellites de Jupiter,' Paris, 4to).

1836. De Damoiseau (' Tables écliptiques des Satellites de Jupiter,' Paris, 4to).

1876. Todd (' A Continuation of De Damoiseau's Tables of the Satellites of Jupiter to the Year 1900.' Washington, 4to).

§ 228. *Rotation of Jupiter's Satellites.*—Galileo (1610) inferred that the period of rotation of each satellite was equal to that of its revolution, by observations on the comparative brilliancy of these objects. This was confirmed by others, including Sir W. Herschel in 1797 ('Phil. Trans.' 1797, p. 332), who noticed that the maximum brilliancy of each satellite always occurred at the same point of its orbit, thus:—

For   I.   Between east elongation and conjunction.
„   II.   Between east elongation and conjunction.
„   III. At elongations.
„   IV.   A little before and a little after opposition.

Schröter having detected certain dark marks or spots on the surfaces of the satellites, was satisfied that they always turned the same aspect to the planet ('Berliner Astron. Jahrbuch,' 1801, p. 126). Beer and Mädler were convinced of the same circumstances with respect to I., II., and IV. (' Fragm. sur les Corps cél.' 1840, p. 143).

The apparent diameters of the satellites at the mean

distance of Jupiter from the sun are as follows : D represents the diameter of Jupiter, in fractions of which the diameters of the satellites are measured by some observers.

A.D.		Satellite.			
		I.	II.	III.	IV.
1734.	Maraldi, by the time of entering on the face of Jupiter ('Mém. Ac. Sc. Paris,' 1734, p. 364) . . .	0·050D	0·050D	0·056D	0·050D
1797.	Sir W. Herschel, by the time II. required to enter the disc, and by measuring the shadow of III. on the planet ('Ph. Tr.' vol. lxxxvii. p. 334) . .		0″·9	1″·6	
1829.	W. Struve, by the micrometer ('M. A. S.' vol. iii. p. 301) .	1·015	0·911	1·488	1·273
1856.	Secchi, by micrometric measures ('Annali di Scienze,' t. vii. p. 51)	0·985	1·054	1·609	1·496
1871.	Engelmann, by micrometric measures, and by the times of entering ('Ueber die Helligkeitsverhältnisse des Jupiters Strabanten,' Leipzig, 8vo) .	1·081	0·910	1·537	1·282

On the colours, the physical conditions, and the atmosphere of the satellites, see

A.D.
1678. Jean Dom. Cassini ('Mém. Ac. Sc. Paris avant le renouv.' t. i. p. 266).

1797. W. Herschel ('Phil. Trans.' vol. lxxxvii. p. 332).

1840. Beer and Mädler ('Fragm. sur les Corps cél. du Syst. sol.' Paris, 4to, p. 143).

Engelmann, *loc. cit.*, deduced from his measures of the

diameters and photometric evaluations the albedo of the satellites as follows :—

Satellite.

	I.	II.	III.	IV.
Albedo	0·2203	0·2665	0·1376	0·0792

SATURN.

§ 229. *Elements.*—The following are the most recent elements of the planet Saturn (Leverrier, ' Ann. Obs. Paris,' Mém. t. xxii. pp. A45 and A48).

1876.   *Epoch* 1850, *Jan.* 1.—*Mean noon, Paris.*

Mean longitude .	14 52 28·30 +	(143° 30′ 30″·3210)t
Longitude of perihelion .	90 6 56·7 +	(1° 57′ 21″·338)t
Greatest equation of centre	6 25 31·24 —	(0° 2′ 21″·280)t
Longitude of ascending node .	112 20 53· 0 +	(0° 52′ 19″·594)t
Inclination	2 29 39·80 —	(0° 0′ 14″·002)t

The unit of t is 100 Julian years.

§ 230. *Great Inequality of Saturn.*—In 1625 Kepler had observed that the mean movements of Jupiter and Saturn no longer coincided with those given by Ptolemy. The great inequality which depends upon five times the mean motion of Saturn, minus twice that of Jupiter, was first calculated by Laplace (1785). The period of this inequality is 929 years. The following are the numerical values of the coefficient of the great inequality of Saturn in longitude as computed by different authors :—

A.D.
1802.   Laplace, epoch 1750 ('Méc. Cél.' liv. vi. ch. xiii. No. 35, t. iii.):—
$$2939''·616 - 0''·085t + 0''·00008t^2$$

1821.   Alexis Bouvard, epoch 1800 ('Tables Astron. Paris,' introd. p. iii.):—
$$2872''·649 - 0·080t + 0''·00008t^2$$

1834.   Pontécoulant, epoch 1800 ('Théorie Anal. Syst. Monde,' t. iii. p. 512):—
$$2906''·661 - 0''·11411t + 0''·00000052t^2$$

A.D.
1876. Leverrier, epoch 1850 ('Ann. Obs. Paris,' Mém. t. xii. p.
A8):—

$$2988''\!\cdot\!95 - 0''\!\cdot\!13814t + 0''\!\cdot\!0002424t^2$$

t denotes the number of Julian years since epoch.

The tables of Saturn are :—

1789. Delambre ('Tables of Jupiter and Saturn,' Paris, 4to).
1821. Alexis Bouvard ('Tables Astronomiques, contenant Jupiter, Saturn, Uranus,' Paris, 4to).
1876. Leverrier ('Annal. Obs. Paris,' Mém. t. xii. pp. A9–A286.

§ 231. The following are the principal measures of the diameter of Saturn obtained since the introduction of the telescope. The numbers indicate the angle which the equatorial diameter of Saturn subtends when placed at a distance equal to the mean distance of Saturn from the sun. The ellipticity is also given :—

A.D.		Equatorial Diameter.	Ellipticity.
1650.	Riccioli and Grimaldi (Riccioli, "Astron. Reform.' p. 356)	26''·67	
1790.	Sir W. Herschel ('Ph. Tr.' vol. lxxx. p. 1, vol. xcvi. p. 461)	20''·6	$\frac{1}{10\cdot368}$
1829.	W. Struve ('M. A. S.' vol. iii. p. 301)	17''·991	
1835.	Bessel ('A. N.' No. 275)	17''·053	$\frac{1}{10\cdot199}$
1848.	Main ('M. A. S.' vol. xviii. pp. 43, 46)	20''·05	$\frac{1}{9\cdot227}$
1853.	Dawes ('M. N.' vol. xiii. p. 78)		$\frac{1}{11\cdot988}$
1856.	De la Rue ('M. N.' vol. xvi. p. 43)	17''·66	

§ 232. *Rotation of Saturn.*—This element has been determined twice, but the number of revolutions employed on each occasion is very small.

A.D.
1794. Sir W. Herschel, from observations of a quintuple band from November 11, 1793, to January 16, 1794 ('Ph. Tr.' vol. lxxxiv. p. 48) . . . . . . 10ʰ 16ᵐ 0ˢ·4
1877. Asaph Hall, by the observations of a brilliant spot from December 7, 1876, to January 2, 1877 . . . . 10ʰ 14ᵐ 23·8 ± 2''·30

§ 233. *Mass of Saturn.*—The mass of Saturn together with his system has been thus determined :—

A.D.

1719. Newton, by the elongations of Titan observed by Pound ('Principia,' 3rd ed. lib. iii. prop. 8 cor. 1) . $\dfrac{1}{3021}$

1802. Laplace, from Pound's observations and those of Jacq. Cassini ('Méc. Cél.' lib. vi. ch. vi. No. 21, t. iii.) . $\dfrac{1}{3359\cdot40}$

1802. Bouvard, by the perturbations of Jupiter ('C. d. T.' an xiv. p. 437) . . . . . $\dfrac{1}{3521\cdot31}$

1834. Bessel, by the elongation of Titan ('A. N.' No. 242) . $\dfrac{1}{3501\cdot6}$

1876. Leverrier, by the perturbations of Jupiter ('Ann. Obs. Paris,' Mém. t. xii. p. 9) . . . . $\dfrac{1}{3529\cdot6}$

§ 234. *Brilliancy of Saturn.*—In 1859 Seidel, by the aid of Steinheil's photometer, estimated the brilliancy of Saturn at 0·482 time that of Vega ('Untersuchungen über die Lichtstärke der Planeten,' Munich, 4to).

In 1865 Zöllner has found by his polarisation photometer :—

$$\text{Brilliancy of Saturn} = \frac{1}{130980000000} \text{ that of the Sun.}$$

Albedo of Saturn = 0·4981 ('Photometrische Untersuchungen,' Leipzig, 4to ; and 'A. N.' No. 1575).

Jean Dom. Cassini observed equatorial bands on Saturn, and he noticed in 1677 that they are inclined to the plane of the ring. This was confirmed by Schwabe in 1833 ('A.N.' No. 239). The principal works on the physical constitution of the globe of Saturn are :—

A.D.

1790. Sir W. Herschel ('Phil. Trans.' vol. lxxx. p. 1).
1792. Sir W. Herschel ('Phil. Trans.' vol. lxxxii. p. 1).
1794. Sir W. Herschel ('Phil. Trans.' vol. lxxxiv. p. 28).
1853. Lassell ('Monthly Notices,' vol. xiii. p. 15).

§ 235. *Rings of Saturn.*—The rings of Saturn were glimpsed in 1610 by Galileo, who, regarding the two ansæ as appendages to the body of the planet, called Saturn 'tricorpor' ('Opera,' Florence, 3 vol. 4to, vol. ii. p. 39).

It was Huyghens, in 1656, who first really showed that the body of the planet was surrounded by a ring ('Systema Saturnium,' La Haye, 1659, 4to).

The annular appendage which surrounds the globe of Saturn is divided into several concentric portions. The brilliant ring in the first place is subdivided into two portions by a circular line usually called after Cassini. This line has been studied by many astronomers. Sir W. Herschel has shown that it can be seen on both faces of the ring, and that it forms a permanent division ('Phil. Trans.' 1791 and 1792).

Other finer lines have been seen by Short, W. Herscneı, Kater, Encke, De Vico, G. P. Bond ('Annals of the Astronomical Observatory of Harvard College,' vol. ii.) The most conspicuous next to Cassini's line is that which bears the name of Encke.

Inside the bright ring is a fainter ring, often called the *dusky ring*. This was more or less perceived by previous astronomers, but G. P. Bond, on October 10, 1850, first clearly detected its real character ('M. N.' vol. xi. p. 20).

O. Struve denotes the part of the bright ring exterior to the line of Cassini by A; the part inside the line of Cassini is B, and the dusky ring is C. The following measures have been made of the diameters of the rings, it being supposed that Saturn is at its mean distance from the sun :—

Diameter.

A.D.	*Exterior of A.*	*Cassini's Line.*	*Interior of B.*	*C.*
1650.	Riccioli and Grimaldi (Riccioli, 'Astron. Reform.' p. 356):—			
	57"			
1719.	Pound (Newton, 'Principia,' 3rd ed. lib. iii. prop. 8, p. 2):—			
	42		30"	
1719.	Bradley (Rigaud, 'Bradley's Miscellaneous Works,' p. 350):—			
	41·25		28·10	
1811.	Bessel ('Berliner Jahrbuch,' 1814, p. 174):—			
	38·2694			
1851.	W. C. Bond ('Gould's Astron. Journal,' vol. ii. p. 5):—			
	39·35	35·22	26·26	23·57

A.D.	*Exterior of A.*	*Cassini's Line.*	*Interior of B.*		*C.*	

1852. O. Struve ('Mém. des Astron. de Poulkowa,' t. i. pp. 350, 351):—

40·12	35·52	34·53	25·29	23·57	21·22

1856. De la Rue ('M. N.' vol. xvi. p. 43):—

39·83	35·33	33·45	26·91

1857. Kaiser ('A. N.' No. 1070):—

39·515

When two numbers are indicated for the line of Cassini, they refer to the external and internal diameters of the dark line. Two numbers refer to the exterior and interior diameters; where only one is given, the interior is understood.

The thickness of the ring has been estimated as follows:—

A.D.

1790. Sir W. Herschel ('Phil. Trans.' vol. lxxx. p. 6) . 0″·3

1850. Sir J. Herschel ('Outlines of Astronomy,' 3rd ed. No. 514) . . . . . <0″·057

1857. W. C. Bond ('Annals. Astron. Obs. Harvard,' vol. ii. part. i. p. 122). . . . . <0″·01

The eccentricity of the ring was pointed out by Gallet ('Journal des Savans,' 1684, p. 198). According to theory, the centre of gravity of each ring revolves around the centre of gravity of Saturn, in a period equal to that of the rotation of the ring (Laplace, 'Méc. Cél.' liv. iii. ch. vi. No. 46, t. ii. 1799). Schwabe concludes, from De Vico's observations, that the period of the eccentricity is not constant ('A. N.' No. 595).

The following are the determinations of the position of the plane of the ring with respect to the variable ecliptic :—

A.D.	*Longitude of Ascending Node.*	*Inclination.*

1656. Huyghens ('Syst. Sat.'):—

152° 30′ $\quad\quad\quad\quad\quad$ 23½°

1740. Jacques Cassini ('Elém. d'Astron.' p. 643):—

168°

1776. Lambert ('Berliner Jahrbuch,' 1778, S. 151):—

167° 5′ $\quad\quad\quad\quad\quad$ 31° 30′

A.D.   *Longitude of Ascending Node.*   *Inclination.*

1814.   Delambre, epoch 1715 ('Astron. théor. et prat.' ch. xxix. t.
         iii. p. 97):—
                  166° 20'   31° 56 13
1835.   Bessel, epoch 1800 ('A. N.' No. 274):—
                  166° 53' 8"·9 + 46"·462t 28° 10 44·7 − 0"·350t

§ 236. *Mass of Saturn's Ring.*—Taking the mass of
Saturn as unity, the following values have been determined
for the mass of the ring :—

A.D.
1831.   Bessel, by the secular inequalities of Titan, without taking
         account of the ellipticity of Saturn ('A. N.' No. 195) . $\frac{1}{118}$
1877.   Tisserand, by the movement of the Perisaturnium of Mimas,
         and taking account of the ellipticity of the planet
         ('C. R.' t. lxxxv. p. 698) .   .   .   . $\frac{1}{620}$

Sir W. Herschel considered, from the movements of certain 'points of light,' that the ring revolved around Saturn
in 10^h 32^m 15^s·4 ('Phil. Trans.' 1790, p. 480); but Schröter
('Berliner Jahrbuch,' 1806, p. 159), Harding ('Berliner
Jahrbuch,' 1806, p. 294), and Schwabe ('A. N.' No. 384), have
not seen any displacement in the 'points,' and do not admit
the rotation of the ring supposed by Herschel.

On the subject of the ring see—

A.D.
1806.   W. Herschel ('Phil. Trans.' xcvi. p. 455).
1853.   O. Struve ('Mémoires présentés à l'Académie [de Pétersbourg]
         par les Astronomes de Poulkova,' t. i., Pétersbourg, 4to, p.
         349).
1857.   W. C. Bond, observations of the planet Saturn, forming part 1
         of vol. ii. of the 'Annals of the Astronomical Observatory of
         Harvard College.'
1875.   Trouvelot ('Silliman's Journal,' 3rd series, vol. xi. No. 66).

Robison records having seen a star through the line of
Cassini (Smyth, 'Cycle of Celestial Objects,' London,
1844, vol. i. p. 193).

The theory of the equilibrium of a ring like Saturn's has
been treated by Laplace (1799) ('Méc. Cél.' book iii. ch. 6,
t. ii.) Maxwell has discussed the case of a ring composed
of discrete particles ('On the Stability of Saturn's Ring,'

London, 1859. See 'M. N.' vol. xix. p. 297). See also Peirce on the Saturnian system, 'Memoirs of the National Academy at Washington,' vol. i. 1866, and Hirn, 'Mémoire sur les Conditions d'Equilibre et sur la Nature probable des Anneaux de Saturne,' Nancy, 1872, 4to.

§ 237. *Satellites of Saturn.*—Mimas was discovered by Sir W. Herschel on July 18, 1789. Its apparent magnitude as compared with stars is about 16 (?). The inclination of the orbit to the plane of the ring is insensible. According to De Vico, 1839 (' Memorie intorno al alc. Oss. fatte nel 1838,' Rome, 4to), the periodic time is 0 day, 22 hours, 36 minutes, 17·05824 seconds. According to Jacob, 1860 ('M. A. S.' vol. xxviii. p. 105), the motion in 365 days is 387 × 360° + 118°·053. The annual movement of the Perisaturnium is 540°, according to Tisserand (' C. R.' t. lxxxv. p. 698).

The diameter of Mimas is $0''·23$ (Sir W. Herschel, ' Ph. Tr.' 1790).

Enceladus was discovered by Sir W. Herschel on September 8, 1789. It is of the 15th magnitude. De Vico in 1839 (*loc. cit.*), by combining his observations with those of Sir W. Herschel, gives the period of 1 day, 8 hours, 52 minutes, 57·275 seconds. According to Jacob ('M. A. S.' vol. xxviii. pp. 90, 104), the mean diurnal motion is 262°·73207; the semidiameter of the orbit (supposed circular) is $34''·99$. The inclination to the plane of the ring is zero. The annual motion of the Perisaturnium is 194°, according to Tisserand (*loc. cit.*), and the diameter is much less than $1''$ according to Sir W. Herschel (' Ph. Tr.' 1790).

Tethys.—Discovered by Jean Dominique Cassini on March 21, 1684 ('Nouvelle Découverte des deux Satellites de Saturne les plus proches,' Paris, 1686, 4to; ' Hist. Ac. Sc. av. renouv.' t. i. p. 415, t. x. p. 694). Mean magnitude, according to Sir J. Herschel, 12–13.

1686. Jean Dom. Cassini (*loc. cit.* t. x. p. 694):—
    Periodic time . . . . 1^d 21^h 19^m
    Semidiameter of orbit . . . $45''·13$

1860. Jacob ('M. A. S.' vol. xxviii. p. 106):—
Longitude 1858·00. Greenwich mean time . . 46° 58′·5
Mean movement in one mean day . . . 190°·697725
Longitude of Perisaturnium . . . . 158° 47′
Eccentricity . . . . . . 0·009675
Semiaxis major . . . . . . 42″·695

The orbit is assumed to be in the plane of the ring.

The annual movement of the Perisaturnium is 87° according to Tisserand, and 51° 7′ according to Jacob.

Diameter, Schröter (reduced by Hind), 0″·125 ('M. A. S.' vol. ii. p. 517).

Dione.—Discovered on March 21, 1684, by Jean Dom. Cassini (*loc. cit.*) Magnitude according to Sir J. Herschel, 12.

1686. Jean Dom. Cassini (*loc. cit.*):—
   Periodic time 2ᵈ 17ʰ 43ᵐ
   Semidiameter . . . . 60″·93
1860. Jacob (*loc. cit.*):—
Longitude 1858·00. Mean time at Greenwich . . 245° 11·5
Mean movement in one mean day . . . 131·534929
Longitude of Perisaturnium . . . . 330° 10′
Eccentricity . . . . . . 0·003945
Semiaxis major . . . . . . 54′·60

Orbit supposed to be in the plane of the ring.

Annual movement of the Perisaturnium, according to Tisserand, is 35°·6, and according to Jacob 184 °10′. The diameter according to Schröter is 0″·13, and according to W. Struve 0″·75 ('M. A. S.' vol. ii. p. 518).

Rhea.—Discovered on December 23, 1672, by Jean Dom. Cassini. ('Découverte de deux nouvelles Planètes autour de Saturne,' Paris, 1673; 'Hist. Ac. Sc. av. renouv.' t. i. p. 159, t. x. p. 584.)

Magnitude according to Sir J. Herschel, 11.

1686. Jean Dom. Cassini (*loc. cit.*):—
   Periodic time 4ᵈ 12ʰ 25ᵐ
   Semidiameter . . . . 84″·86
1860. Jacob (*loc. cit.*):—
Longitude 1858·00. Mean time at Greenwich . . 215°31′·5
Mean movement in one day . . . . 79·690216

Longitude of Perisaturnium	69° 34′
Eccentricity	0·00160
Semidiameter	76″·125
Longitude of ascending node on the ecliptic . .	167°4′·5
Inclination of orbit to the ecliptic . . .	28°10′·9

The orbit may be considered to coincide with the plane of the ring.

Laplace has shown how the effect of the ellipticity of the planet acts to retain the interior satellites in the plane of its ring and of the equator ('Méc. Cél.' liv. viii. ch. xvii. No. 36, t. iv.)

The motion of the Perisaturnium is about 10°·9 per annum according to Tisserand, and 96°·5 according to Jacob. The diameter, according to Schröter, is 0″·32.

Titan.—Discovered by Huyghens on March 25, 1655, with a 12-feet refractor ('Systema Saturnium,' La Haye, 1659, 4to ; 'Cosmotheoros,' La Haye, 1698, 4to, p. 69); mean magnitude 9.

1659. Huyghens (*loc. cit.*):—

Periodic time	15ᵈ 22ʰ 41ᵐ
Semidiameter	206″·5

1860. Jacob (*loc. cit.*):—

Longitude 1858·00. Mean time at Greenwich . .	260° 28′·0
Mean diurnal movement	22°·577033
Longitude of Perisaturnium . . .	257° 27′
Eccentricity	0·028587
Semidiameter of orbit	176″·755
Longitude of ascending node on the ecliptic . .	167° 54′·5
Inclination of orbit to the ecliptic . .	27 35·2

The annual movement of the Perisaturnium is 0°·35 (Jacob), 0°·508 (Bessel), 0°·565 (Tisserand). The annual motion of the node is −0°·07 (Jacob), −0°·005 (Bessel). The diameter, according to Schröter, is 0″·75. The principal inequalities of Titan have been computed by Bessel ('A. N.' 195). See also Plana, 'M. A. S.' vol. ii. p. 351. On March 20, 1692, Jean Dom. Cassini witnessed the occultation of a small star by Titan.

Hyperion.—Discovered by G. P. Bond on September 16, 1848, and seen independently by Lassell on the 18th of the same month. Magnitude 17–18, G. P. Bond.

1877. Asaph Hall ('A. N.' 2137).
Passage through Perisaturnium, 1875. August 24·0036 mean time at Washington.

$$\pi = 166^\circ\ 9' \qquad\qquad p = 21^{d}\!\cdot\!3113$$
$$e = 0\cdot125 \qquad\qquad a = 214''\!\cdot\!22$$

Japetus.—Discovered by Jean Dom. Cassini, October 25, 1671 ('Découverte de deux nouvelles Planètes autour de Saturne,' Paris, 1673; 'Hist. Ac. Sc. av. renouv.' t. i. p. 150, t. x. p. 584). Magnitude 11–12, J. Herschel; 11–12, G. P. Bond.

1687. J. Dom. Cassini (*loc. cit.*) :—

$$P = 79^{d}\ 7^{h}\ 55^{m} \qquad\qquad a = 568''$$

1860. Jacob (*loc. cit.*) :—

Longitude 1858·00. Mean time at Greenwich .	294° 28·6
Mean diurnal movement . . .	4°·5380365
Longitude of Perisaturnium . . .	353° 30′ + 4° 10′t
Eccentricity	0·028201
Semidiameter	514''·65
Longitude of ascending node . . .	142° 53′·7
Inclination	18° 43′·6

In 1673 J. Dom. Cassini remarked that in the eastern portion of its orbit this satellite remains invisible for about thirty days ('Hist Ac. Sc. av. renouv.' t. i. p. 174). Jacques Cassini was convinced that the satellite always turns the same face to the planet ('Mém. Ac. Sc. Paris,' 1714, p. 370). Sir W. Herschel has confirmed this, and has found that the satellite changes its brightness through a range of three magnitudes in the course of a single revolution ('Phil. Trans.' 1792, p. 1).

If n_1, n_2, n_3, n_4, be the mean movements of the four interior satellites, then Kirkwood has shown that the following relation is fulfilled ('The Observatory,' No. 7, p. 199; 'M. N.' vol. xxxviii. 1877, p. 64):—

$$5n_1 - 10n_2 + n_3 + 4n_4 = 0$$

To render this absolutely correct, it is only necessary to add o"·62 to the period of Mimas. The periods are then—

Mimas	.	.	.	.	.	o^d	22^h 36^m	18^s·32
Enceladus	.	.	.	.	.	1	8 53	2·7
Tethys	.	.	.	.	.	1	21 18	33
Dione	.	.	.	.	.	2	17 44	51·2

For a general historical account of Saturn's rings and satellites, see the valuable and interesting volume, Grant, ' History of Physical Astronomy,' London, 8vo, pp. 251–272.

URANUS.

§ 238. *Elements.*—This planet was discovered by Sir W. Herschel on March 13, 1781 (' Ph. Tr.' vol. lxxi. p. 492). It had been observed as a star by Tob. Mayer in 1756, and also by Le Monnier, Bradley, and Flamsteed. *t* is the number of Julian years since the epoch.

1874. Newcomb (' Smithsonian Contributions to Knowledge,' Washing. 4to, vol. xix. pp. 81, 184, 219), epoch 1850, January 0. mean time at Greenwich.

Mean longitude	.	.	.	$29° 12' 43''·73 + 15424'''·797t$
Longitude of perihelion	.		.	$170\ 38\ 48·7 +\quad 53·168t$
Greatest equation of the centre	.		.	$5\ 22\ 42·28 -\quad 0·1090t$
Longitude of the ascending node	.		.	$73\ 14\quad 37·6 +\quad 18·5682t$
Inclination	.	.	.	$0\ 46\ 20·92 +\quad 0·0247t$
Semiaxis major	.	.	.	$19·19209$
Eccentricity	.	o	.	$0·0463592 - 0·0000002627t$

1877. Leverrier ('Annal. Obs. Paris.' Mém. t. xiv. part i. pp. A67, A69). Epoch 1850·00, mean time at Paris.

Mean longitude	.	.	.	$29° 17' 50''·91 + 15475·11138t$
Longitude of perihelion	.		.	$170\ 50\quad 7·1 +\quad 53·4582t$
Greatest eccentricity	.		.	$5\ 19\ 26·1 -\quad 0·10938t$
Longitude of ascending node	.		.	$73\ 13\ 54·4 +\quad 18·0570t$
Inclination	.	.	.	$0\ 46\ 19·72 +\quad 0·01732t$
Semiaxis major	.	.	.	$19·14169$
Eccentricity	.	.	.	$0·04634096 - 0·0000002651t$

§ 239. *Equatorial Diameter of Uranus at its Mean Distance from the Sun.*

A.D.
1788. W. Herschel (' Phil. Trans.' t. lxxviii. p. 378) . $3''·90G$

1867. Lassell and Marth, calculated by Winnecke (' M.
      A. S.' vol. xxxvi. p. 37) .   .   .   . $3''\cdot568$
1869. Vogel (' A. N.' No. 1750) .   .   .   . $3\ \cdot624$

At the oppositions of 1842 and 1843, Mädler measured the ellipticity of the disk (' A. N.' Nos. 460, 493) :—

Opposition of 1842 $\qquad \dfrac{1}{10\cdot47}$

       „   1843 $\qquad \dfrac{1}{9\cdot32}$

§ 240. *Mass of Uranus.*—The following are the principal determinations :—

1789. Wurm, from observations of the exterior satellite and taking $8''\cdot8$ for the parallax of the Sun (' Berliner Jahrbuch,' 1792, p. 214) .   .   .   . $\frac{1}{16959}$

1802. Laplace, from the same observations (' Méc. Cél.' liv. vi. ch. vi. t. iii.) .   .   . $\frac{1}{19504}$

1821. Al. Bouvard, by the perturbations of Jupiter and Saturn(' Tables Astronomiques,' Paris. 4to, Introd. p. ii.) .   .   .   .   . $\frac{1}{17918}$

1849. Adams, by the observations of Oberon of Sir W. Herschel and Lassell (' M. N.' vol. ix. p. 160) . $\frac{1}{21031}$

1871. Von Asten, by the observations of Lamont, Otto Struve, Lassell, Marth, on the two exterior satellites (' Mém. Ac. Pétersbourg,' 7ᵉ sér. t. xviii. No. 5, p. 21) .   .   .   .   . $\frac{1}{22020}$

1875. Copeland, by the observations of the two exterior satellites made with Lord Rosse's telescope (' M.N.' vol. xxxv. p. 304). Mean .   . $\frac{1}{24000}$

1878. Asaph Hall, by observations in 1875-76 (' A. N.' No. 2186). Mean .   .   . $\frac{1}{22800}$

1878. Holden, by the same (' A. N.' No. 2186). Mean . $\frac{1}{22800}$

§ 241. *Rotation and Brilliancy of Uranus.*—The only direct determination of the rotation of Uranus is that of Buffham, who observed spots on the disk in 1870 and 1872, and concluded that the period is about twelve hours (' Revue Scientifique,' December 27, 1873). The situation of the plane of the equator of Uranus is presumed, from theory, to coincide with that in which the satellites move. Newcomb, from observations at Washington in 1874 and 1875, finds for the position of the plane of the equator on the ecliptic, epoch 1850 :—

Longitude of ascending node . . .	$165°·48' + 1°40\text{T}$
Inclination	$97°·85 - 0·013\text{T}$

T denotes the number of centuries which have elapsed since the epoch ('Astron. Obs. at Naval Observatory,' Washington, 1873, append. i. part j. p. 41).

In 1865 Zöllner has determined the brilliancy with his photometer ('Photometrische Untersuchungen,' Leipzig, 4to, 'A. N.' No. 1575), and finds—

$$\text{Uranus} = \frac{1}{8,486,000,000,000,000} \text{ of the Sun.}$$

The albedo of Uranus is 0·6406.

The spectrum of Uranus has been examined in 1871 by Vogel ('A. N.' No. 1864).

§ **242.** *Satellites of Uranus.—Ariel.—*This satellite was discovered by Lassell on October 24, 1851 ('M.N.' vol. xi. p. 248). Holden thinks that it was one of the satellites seen by Sir W. Herschel on March 27, 1794 ('M. N.' vol. xxxv. p. 20). Otto Struve has perhaps seen it in 1847 ('M. N.' vol. viii. pp. 47 and 137). Mean magnitude 16 (Newcomb). The following elements have been deduced by Newcomb, from the Washington observations in 1874–75 ('M. N.' vol. xxxvi. p. 208, 'A. N.' No. 2186) :—

Periodical revolution, taking account of precession .	$2^\text{d}·520383$
Radius of the orbit (supposed circular) . . .	$13''·78$
Longitude at the epoch	$15°·90$

The epoch is December 31, 1871, mean noon at Washington. The longitude is counted from the ascending intersection of the orbit with the plane parallel to the ecliptic and passing through the centre of the planet.

Umbriel was discovered by Lassell on October 24, 1851 ('M. N.' vol. xi. p. 248). Holden thinks that Umbriel was perhaps seen by Sir W. Herschel on January 18 and 20,

1790, and that it was almost certainly seen by him on April 17, 1801 ('M. N.' vol. xxxv. p. 19). Perhaps seen by O. Struve in 1847 ('M. N.' vol. viii. p. 407). Mean magnitude, according to Newcomb, 16–17.

The following elements have been given by Newcomb, 1875 (*loc. cit.*) :—

Periodic time	$4^d \cdot 144181$
Radius of the orbit (supposed circular) . . .	$19'' \cdot 20$
Longitude at the epoch	$130° \cdot 59$

Titania.—Discovered by Sir W. Herschel on January 11, 1787 ('Phil. Trans.' vol. lxxvii. p. 125). Mean magnitude 13–14 (Vogel).

The following elements have been given by Newcomb (*loc. cit.*), 1876 :—

Periodic time	$8^d \cdot 705897$
Longitude reckoned from the ascending node on the ecliptic, December 31, 1871, mean noon at Washington . .	$224° \cdot 00$
Radius of the orbit (supposed circular) . . .	$31 \cdot 48$
Radius, by observations, 1874–75	$31 \cdot 38$

Oberon.—Discovered by Sir W. Herschel on January 11, 1787. ('Phil. Trans..' vol. lxxvii. p. 125). Mean magnitude 14 (Vogel).

The following elements have been given by Newcomb (*loc. cit.*), 1876 :—

Periodic time	$13^d \cdot 463269$
Longitude reckoned from the ascending node on the ecliptic, December 31, 1871, mean noon at Washington	$148° \cdot 90$
Radius of the orbit (supposed circular) . . .	$42'' \cdot 10$
Radius, from the observations of 1874–5 . . .	$42 \cdot 08$

NEPTUNE.

§ 243. *Elements.*—Alexis Bouvard, in 1821, pointed out that the movements of Uranus indicated the existence of some exterior planet ('Tables Astronomiques, contenant des

Tables de Jupiter, de Saturne, et d'Uranus,' Paris, 4to, introd.
p. xv.) Leverrier in 1846 computed that the planet which
produced the disturbance must have the following elements
('C. d. T.' 1849, addit. pp. 1-254):—

Semiaxis major	36·154
Sidereal revolution	217ʸ·387
Mean longitude, January 1, 1847	318° 47'
Longitude of perihelion	284° 45'
Longitude of ascending node	156° 0'
Inclination	6° 0'
Eccentricity	0·10761
Mass (sun being unity)	$\frac{1}{9300}$

Adams similarly deduced the following elements in 1846
('M.A.S.' vol. xvi. pp. 427–460, 'Nautical Almanac,' 1850):—

Semiaxis major	37·25
Sidereal revolution	227ʸ·323
Mean longitude, October 1, 1846	323° 2'
Longitude of perihelion	299° 11
Eccentricity	0·12062
Mass (sun being unity)	$\frac{1}{6665·5}$

From the indications supplied by Leverrier, Galle detected
the planet on September 23, 1846. It was subsequently
found that the planet had been observed by Challis in
1846, by Lamont in 1845, and by Le François Lalande
in 1795 (Grant, 'History of Physical Astronomy,' London,
1852, 8vo, ch. xii. pp. 193, 204; 'M. N.' vol. xi. p. 11).
The name Neptune was proposed for the new planet by the
'Bureau des Longitudes' of France. On the history of the
discovery of Neptune, see Airy, 'M. N.' vol. vii. pp. 121-152,
and 'M. A. S.' vol. xvi. p. 385.

The observations of 1795 have enabled the actual orbit
to be computed with a high degree of precision. The
following is the most recent result. *t* represents the number
of Julian years (365·25 days) since the epoch 1850·00,
mean time Paris (Leverrier, 1877, 'Annal. Obs. Paris,' Mém.
t. xiv. part ii. pp. 42, 43):—

E E

Mean longitude .	.	. $334°$ $33'$ $28''\cdot89$
Longitude of perihelion .	.	45 59 $43''\cdot1 + 51''\cdot12675t$
Longitude of ascending node	.	130 6 $25\cdot1 + 39\cdot56306t$
Inclination .	.	. 1 47 $2\cdot13 - 0\cdot34570t$
Mean annual movement .	.	$7915''\cdot89825$
Semiaxis major .	.	. $20''\cdot92742$
Eccentricity .	.	. $0\cdot00896425 + 0\cdot00000005672t$

§ 244. *Diameter, Brilliancy, Mass.*—The following measures have been made of the diameter of Neptune when at the mean distance of the planet from the sun:—

A.D.

1846. Arago, with the doubly refracting micrometer ('Annuaire du Bureau des Longitudes,' 1865) . . $2''\cdot60$

1867. Lassell and Marth ('M. A. S.' vol. xxxvi. p. 37) . $2''\cdot239$

The period of the rotation of the planet has not yet been ascertained. Zöllner with his polarisation photometer has found for its brilliancy ('Photometrische Untersuchungen,' Leipzig, 1865, 4to; 'A. N.' No. 1575)—

$$\text{Neptune} = \frac{1}{79,620,000,000,000} \text{ of the sun;}$$

and the albedo is $0\cdot4648$.

Taking the mass of the sun as unity, we have for the mass of Neptune:—

A.D.

1847. O. Struve, from the satellite ('A. N.' No. 629) . $\frac{1}{13866}$

1847. Peirce, by the perturbations of Uranus ('A. N.' No. 637) . . . . $\frac{1}{20000}$

1849. Adams, from observations of the satellite by Lassell, W. C. Bond, G. P. Bond, and O. Struve ('M. N.' vol. ix. p. 203) . . . . $\frac{1}{17900}$

1855. Hind, from the observations of the satellite by Lassell at Malta in 1852 ('M. N.' vol. xv. p. 47; 'A. N.' No. 921) . . . . $\frac{1}{17135}$

1874. Newcomb, by the perturbations of Uranus ('Smiths. Contrib. to Knowl.' vol. xix. p. 173) . . . $\frac{1}{19700}$

1876. Holden and Asaph Hall, by the observations of the satellite at Washington in 1874-1876 ('A. N.' No. 2100) . . . . $\frac{1}{18520}$

§ **245.** *The Satellite of Neptune* was suspected by Lassell on October 10, 1846 ('A. N.' No. 589), and announced by him on July 7, 1847 ('A. N.' No. 611). Magnitude, according to the discoverer, is 14.

The following elements of the satellite referred to the plane of the ecliptic have been deduced by Newcomb from the observations at Washington in 1873–74 ('Astron. Observ. at the Naval Observatory,' 1873, append. i. part ii. p. 62):—

Periodic time (adopted) . . .	. $5^d \cdot 8769$
Semiaxis major	. $16'' \cdot 275$
Eccentricity.	. $0 \cdot 0088$
Longitude of node	. $184° \cdot 50 + 0 \cdot 0140 t$
Inclination	. $145° \cdot 12$
Longitude of Perineptune . . .	. $184°$

Mean argument of the latitude reckoned from the intersection of the orbit with a plane parallel to the ecliptic, and passing through the centre of Neptune at the epoch 1874, January $0 \cdot 00$ mean noon at Washington . . . . . . $101° \cdot 07$

t is the number of years since epoch.

COMETS.

§ **246.** *Comets Determined by a Single Apparition.*—A catalogue of orbits of the comets of which only a single observation has been recorded has frequently been compiled. See, for example, Chambers's 'Astronomy,' Guillemin's 'World of Comets,' &c. The most complete list is by Galle (Olbers, 'Abhandlung über die Methode die Bahn eines Cometen zu berechnen,' 3rd ed. Berlin, 1864). The following are the periodic comets which are recognised as permanently belonging to the solar system.

§ **247.** *Halley's Comet.*—Some account of this celebrated comet has been given in § 117. The following are the elements which have been determined from the observations at the successive appearances recorded since the year B.C. 11:—

Perihelion Passage, Mean Time Paris	Longitude of Perihelion	Longitude of Ascending Node	Inclination	Perihelion Distance	Eccentricity	Calculator
B.C. 11. Oct. 8ᵈ 19ʰ 19ᵐ	280°	28°	10°	0·58	?	Hind
A.D. 66. Jan. 14 4 48	325 0′	32 40′	40 30′	0·4446	?	Hind
141. March 29 2 24	251 55	12 50	17 0	0·72	?	Hind
989. Sept. 12 0	264	84	17	0·5683	?	Burckhardt
1066. April 1	264 55	25 50	17 0	0·72	?	Hind
1301. Oct. 24 0	312	138	13	0·64	?	Laugier
1378. Nov. 8 18 28 48ˢ	299 31	47 17	17 56	0·5835	?	Laugier
1456. June 8 22 10	301 0	48 30	17 56	0·58552	?	Pingré
1531. April 25 19 10	301 12	45 30	17 0	0·57994	0·96739I	Halley
1607. Oct. 27 0 21 0	300 46 59″	48 14 9″	17 6 17″	0·58419	0·9670888	Lehmann
1682. Sept. 14 19 14 14	301 55 37	51 11 18	17 44 45	0·5828943	0·96792019	Rosenberger
1759. March 12 13 23 55	303 10 28	53 50 27	17 36 52	0·5845193	0·96768436	Rosenberger
1835. Nov. 15 22 41 22	304 30 48	55 9 15	17 45 5	0·5865695	0·96739091	Westphalen

On the ancient appearances of Halley's comet, see Halley, ' Phil. Trans.' 1705, Laugier, ' Comptes Rendus,' t. xxiii. p. 183, and Hind, ' M. N.' vol. x. p. 51.

The next perihelion passage will take place about 1910.

§ **248**. *Tuttle's Comet.*—Discovered on January 4, 1858, by Tuttle, at Cambridge near Boston, U. S. A. ('A. N.' No. 1125). The identity with the comet of January 1790 was announced by Pape ('A. N.' No. 1125). The comet returned in 1871, according to prediction ('A. N.' No. 1840). The next return to perihelion will be in July 1885. (See also 'Vierteljahrsschrift der Astronomischen Gesellschaft,' Bd. vi. p. 91.)

§ **249**. *Faye's Comet.*—Discovered by Faye at Paris on November 22, 1843 ('C. R.' t. xvii. p. 1248). Argelander showed that the observations could only be represented by an elliptic orbit. The comet has been seen four times since that epoch. Alex. Möller has made this comet the subject of special researches ('A. N.' 1259, 1295, 1358, 1522). The perihelion passages occurred in 1843, October 17; 1851, April 1; 1858, September 13; 1866, February 14; 1873, July 18. The comet has been observed at all these returns. The next perihelion passage occurs in December 1880.

§ **250**. *Biela's Comet.*—This comet was seen in 1772 and in 1805–6, before its periodicity was recognised. It was again discovered by Biela on February 27, 1826. Gambart pointed out the identity of this comet with those of 1772 and 1806 ('A. N.' No. 95). This comet was perceived to be double by Herrich and Bradley on December 29, 1846, and separate orbits were computed for the two portions. The following elements have been computed for the successive appearances. The first line in the orbits of 1846 and 1852 refers to the preceding portion, and the second line to the following portion :—

Perihelion Passage, Mean Time at Paris	Longitude of Perihelion	Longitude of the Ascending Node	Inclination	Perihelion Distance	Eccentricity	Calculator
A.D. 1772. February 16^d 15^h 53^m 1^s	110° 8′ 35″	257° 15′ 38″	17° 3′ 8″	0·9860o	0·724510	Hubbard
1806. January 1 22 10 31	109 28 25	251 16 19	13 36 34	0·907077	0·7457068	Hubbard
1826. March . 18 10 52 30	109 48 47	251 27 19	13 33 54	0·9024190	0·7466012	Hubbard
1832. November 26 3 2 27	110 0 5	248 14 46	13 13 1	0·8790148	0·7514682	Santini
1846. { February 11 0 19 12	109 2 53	245 54 15	12 34 53	0·8564465	0·7566623 }	Hubbard
{ ,, 10 22 19 43	109 2 54	245 54 17	12 34 55	0·8564649	0·7566060 }	
1852. { September 22 6 41 56	109 8 17	245 51 26	12 33 16	0·8606016	0·7559217 }	Hubbard
{ ,, 22 22 47 46	109 8 16	245 51 28	12 33 19	0·8606220	0·7558650 }	

This comet was expected to return in 1859, in 1866, and in 1872, but it was not seen on any of these occasions. A memorable shower of shooting stars which occurred on November 27, 1872, was apparently connected with this comet. See ante, § 119. It should be stated that Bruhns considers the comet found by Pogson ('A. N.' No. 1917) could not be identical with Biela's Comet ('A. N.' No. 2054). The next return of Biela's Comet to perihelion is about December 1883.

§ 251. *D'Arrest's Comet.*—Discovered by D'Arrest at Leipzig, June 27, 1851 ('A. N.' No. 764). The orbit having been calculated by the same astronomer, its periodicity was detected. The comet reappeared in 1857, in 1870, and in 1877, but was not seen, as expected, in 1864. On this comet see Leveau, 'Ann. Obs. Paris,' Mém. t. xiv. part i. 1877.

§ 252. *Winnecke's Comet.*—Discovered by Winnecke at Bonn, March 8, 1858. The discoverer detected the resemblance of the elements to those of the comet of July 1819, and showed that the orbit is elliptic ('A. N.' No. 1133). This comet was not seen in 1864, but was seen in 1869 and 1875. The next return is in December 1880.

§ 253. *Brorsen's Comet.*—Discovered by Brorsen at Kiel on February 26, 1846 ('A. N.' No. 555). The movement is direct. The comet was not seen at perihelion in 1851 or in 1863 ; but it has been seen in 1857, 1868, 1873, 1879. Allowing for the effects of perturbations, the elements have been given by Bruhns ('A. N.' Nos. 1686, 1692) for the apparition of 1868. (See also, for elements in 1873, Plummer, 'A. N.' No. 1960.)

§ 254. *Encke's Comet.*—This comet had been observed and a parabolic orbit calculated in 1786, 1795, and 1805, before its periodicity was recognised. It had been discovered respectively on these occasions by Méchain, Caroline Herschel, and Alexis Bouvard. In 1818 Pons discovered it at Marseilles on November 26 ; Encke detected its periodicity in 1819 ('Correspondance Astronomique,' &c., F. de

Zach, t. ii. p. 207). Since then the comet has been observed at numerous returns to perihelion, the last occasion being in 1878. In 1831 Encke thought ('A. N.' No. 211), by comparing the calculated positions with the observed positions, that the return to perihelion was continually retarded by some cause which he and Olbers attributed to a resisting medium (see also 'A. N.' No. 305). Von Asten, who, in a re-discussion of the whole question, has shown that the planetary perturbations are incompetent to produce the observed effects, and has thus corroborated the conclusion of Encke, considers that the observed movements may be completely explained by the perburbations of the planets. ('Mém. Ac. Sc. Saint-Pétersbourg,' 7° série, t. xviii. 1872, No. 10.)

§ **255.** *Physical Theory of Comets.*—The following works may be consulted :—

A.D.
1872.   G. P. Bond, 'Account of the Great Comet of 1858,' in the 'Annals of the Harvard College Observatory,' vol. iii.
1872.   Zenker, 'Ueber die physikalischen Verhältnisse und die Entwickelung der Cometen' ('A. N.' Nos. 1890-1893).
1872-1876.   Zöllner, 'Ueber die physische Beschaffenheit der Cometen' ('A. N.' Nos. 2057-2060, 2082-2086), and 'Ueber die Natur der Cometen,' Leipzig, 1872.

With reference to the theory of the production of the tails of comets see—

A.D.
1812.   Olbers, 'Ueber den Schweif des grossen Cometen von 1811' ('Monatl. Correspondenz,' vol. xxv.).
1836.   Bessel, 'Beobachtungen über die physische Beschaffenheit des Halley'schen Kometen' ('A. N.' Nos. 300-302).
1847.   Sir John Herschel ('Results of Astronomical Observations at the Cape of Good Hope,' London, 4to, pp. 393-413).
1869.   Tait ('Proceedings of the Royal Soc. of Edin.,' vol. viii. p. 553).

The first comet examined with the spectroscope was that of 1864, to which Donati for the first time applied that mode of enquiry ('A. N.' No. 1488) ; see also Huggins ('Proceedings of the Royal Society of London,' vol. xv. p. 5). The spectroscopic observations of comets, up to the year 1872, are given in résumé by Vogel ('A. N.' No. 1908).

METEORS.

§ 256. *Showers of Stars.*—The periodicity of certain
showers of shooting stars, especially that of the Leonides on
November 12–14, was pointed out by Arago in 1835. The
periodicity of the August shower, known as the Perseids,
was announced by Quetelet in 1837. On the general theory
of shooting stars see Schiaparelli, 'Entwurf einer astronomischen Theorie der Sternschnuppen,' Stettin, 8vo. This is
a German translation by G. von Boguslawski of several important works by Schiaparelli.

§ 257. *The Leonides.*—The true period of the revolution
of this shower was originally suggested by Newton, in 1864, to
be 33·25 years ('Mem. Nat. Ac. Science,' vol. i. Washington,
4to). This was confirmed by the remarkable calculations
of Adams ('M. N.' vol. xxvii. p. 247), which showed that the
assumption of this value for the periodic time enabled the
change in the position of the node to be explained by the
planetary perturbations. The following great showers of
Leonides have been enumerated by Newton:—

Return of Shower.	Day.						Year recorded, A.D.
1st	—	.	.	.	.	.	902
2nd	October	14	.	.	.	.	934
4th	,,	15	.	.	.	.	1002
7th	. ,,	17	.	.	.	.	1101
10th	—		.	.	.	.	1202
15th	,,	22	.	.	.	.	1366
20th	,,	25	.	.	.	.	1533
22nd	—		.	.	.	.	1602
25th	November	9	.	.	.	.	1698
28th	—		.	.	.	.	1799
29th	—		.	.	.	.	1832
30th	,,	14	.	.	.	.	1866

The following orbit of the Leonides is given by Tupman
('M. N.' vol. xxx. p. 32) :—

Longitude of perihelion . . .	62° 36'
,, ascending node . . .	231 44
Inclination	15 38
Perihelion distance	0·9802
Semiaxis major	10·340
Eccentricity	0·9062
Direction of movement . . .	Retrograde

On the Leonides see also Kirkwood. 'Rep. Brit. Ass.' 1875, pp. 224-226.

§ **258.** *Radiant Points.*—On the identity of the orbits of star showers with the orbits of comets see Alexander Herschel, 'Rep. Brit. Ass.' 1875, p. 232, and 'M. N.' For a general catalogue of the positions of radiant points see Greg, 'Rep. Brit. Ass.' 1874, p. 324; also Gruber and Greg, 'Rep. Brit. Ass.' 1875, pp. 220, 221 ; 1876, p. 156.

The following short list contains some of the principal showers :—

Name of Shower	Date of Appearance	Radiant		Remarks
		Right Asc.	North Dec.	
Quarantids	Jan. 2–3	232°	+49°	Maximum 1863–66
Lyraids	April 19–20	272	35	Comet I. 1861. Meteors yellow and white, slightly trained
Perseids	Aug. 9–11	44	56	Great shower comet III. 1862
Orionids	Oct. 19–25	89	15	A well-marked shower
Leonides	Nov. 12–14	149	23	The greatest shower comet I. 1866. Period 33¼ years
Andromedes	,, 27–29	25	43	Great shower Nov. 27, 1872. Biela's Comet
Geminids	Dec. 9–12	105	32	Trained meteors

THE SOLAR SYSTEM AS A WHOLE.

§ **259.** *The Invariable Plane.*—The existence of the invariable plane was first indicated by Laplace in 1796, 'Système du Monde,' liv. iv. ch. iii. The following determinations of the position of this plane have been made :—

A.D.

1802. Laplace ('Méc. Cél.' liv. vi. ch. xvii. No. 46, t. iii.) on the fixed ecliptic of 1750.

Longitude of ascending node . . 102° 57′ 30″

Inclination 1 35 31

1834. Pontécoulant ('Théorie anal. Syst. Monde,' t. iii. p. 529) on the fixed ecliptic of 1800.

Longitude of ascending node . . 103° 8′ 45″

Inclination 1 34 16

1873. Stockwell ('Smithsonian Contributions to Knowledge,' vol. xviii. p. 166) on the fixed ecliptic of 1850.

Longitude of ascending node . . 106° 14′ 6″·00

Inclination . . . . 1 35 19 ·376

The last calculation is founded on more recent values of the masses of the planets. Account is also taken of the influence of Neptune, which was unknown to Laplace and Pontécoulant.

§ **260.** *Motion of the Solar System through Space.*—The movement of the solar system, discovered by Sir W. Herschel, is directed towards a point of the celestial sphere, the position of which has been variously determined as follows :—

A.D.   Right Ascension.   Declination.

1783. Sir W. Herschel, from 27 proper motions given by Lalande ('Phil. Trans.' vol. lxxiii. p. 247)

260° 34′·5   + 26° 17′

1839. Argelander, from 392 stars ('A. N.' No. 363). Equin. 1792·5.

259 47·6   32 29·5

1844. Otto Struve, from 392 stars of Argelander ('A. N.' No. 485). Equin. 1792·5

261 23·1   37 35·7

A.D.	Right Ascension.	Declination.

1856. Mädler, from 2,163 stars (' Beobach. der Sternw. zu Dorpat,'
       Bd. xiv.) Equin. 1800·0
                      261° 31'·8   + 39° 53'·9
1860. Airy, from 113 stars with large proper motions ('M. A. S.'
       vol. xxviii. p. 161). Equin. 1840·0
                      261 29   24 44
1864. Dunkin, from 1,167 stars (' M. A. S.' vol. xxxii. p. 36). Equin.
       1846·0
                      263 43·9   25   0·5
1877. L. de Ball, from 80 southern stars ('Monatsberichte der preuss.
       Akad. Berlin,' Bd. xliv.) Equin. 1860·0
                      269 33   23   11

FIXED STARS.

§ 261. *Star Catalogues.*—In the following list will be
found the titles of some of the principal catalogues of stars.
The stars are arranged in order of right ascension:—

A.D.
1845. Baily, ' The Catalogue of Stars of the British Association for the
       Advancement of Science,' London, 4to. This contains
       places of 8,377 stars for the epoch 1850·0, compiled from the
       catalogues of original observers.
1847. Baily, 'A Catalogue of Stars in the "Histoire céleste" of
       Lalande' (containing 47,390 stars for 1800).
1846–63. Weisse, 'Positiones Mediæ Stellarum' (62,530 stars observed
       by Bessel between — 15° and + 45°).
1849. Airy, six different catalogues ('Greenwich Observations for
       1842, 1847, 1854, 1862, 1868, 1876,' app.)
1874. Ellery, 'First Melbourne General Catalogue,' 4to. This contains
       1,227 southern stars reduced to 1870·0.

One of the most useful works that can be found in an
observatory is the ' Durchmusterung' of Argelander, forming
vols. iii , iv., and v. of the 'Astronomische Beobachtungen
auf der Sternwarte zu Bonn.' This includes all stars
between the pole and 2° of south declination, down to the
9·5th magnitude. The value of this catalogue is greatly

enhanced by the circumstance that it is accompanied by a series of maps, on which all the stars of the catalogue are depicted.

The stars visible to the naked eye from the pole to −26° are shown in many maps. See, for example, Argelander, ' Uranometria Nova,' Berlin, 17 plates.

§ **262.** *Magnitudes of the Stars.*—The division of stars visible to the naked eye into six magnitudes is as old as Ptolemy. Galileo continued the nomenclature to telescopic stars. The relation between the brilliancy of a star of a certain magnitude, and that of the magnitude immediately preceding, has been determined as follows :—

A.D.

1836. Steinheil, by photometric observations of 30 stars of magnitude 1 to 5 ('Elemente der Helligkeitsmessungen,' Munich, 4to, p. 27)	o ·353
1837. W. Struve, between the magnitudes 1 to 12, by narrowing the aperture of the telescope ('Stellarum Compositarum Mensuræ Micrometricæ,' Petersburg, in fol. p. lxix.)	o ·346
1851. Johnson, by 78 stars of 4th to 10th magnitude, according to the illumination which the stars admit in the field ('Radcliffe Observatory,' vol. xii. app. pp. 9, 25 ; 'M. N.' vol. xiii. p. 281)	o ·424
From data by Sir W. Herschel	o ·464
,, ,, Groombridge.	o ·388
,, ,, W. Struve, magnitude 1–12 .	o ·383
,, ,, Otto Struve .	o ·406
,, ,, Argelander .	o ·431
1865. Zöllner, with his polarisation photometer, from 293 stars of magnitude 1 to 6 ('Photometrische Untersuchungen,' Leipzig, 8vo)	o ·363
1869. Rosen, by 110 stars of magnitude 5·0 to 9·5, by the aid of Zöllner's photometer ('Bulletin de l'Acad. des Sc. St.-Pétersbourg,' t. xiv. p. 95)	o ·398

§ **263.** *Measurement of the Brightness of Stars.*—Wollaston in 1829 sought to determine the relative brilliancy of Sirius and the sun, by using the flame of a candle as a medium of comparison ('Phil. Trans.' vol. cxix. p. 28). He finds—

$$Sirius = \frac{1}{20,000,000,000} \times Sun$$

Steinheil in 1836, using the brightness of the moon as a means of comparison, finds—

$$Sirius = \frac{1}{3,840,000,000} \times Sun$$

$$Arcturus = \frac{1}{6,008,000,000} \times Sun$$

In a similar manner G. P. Bond has found in 1861 (' M. N.' vol. xxi. p. 129)—

$$\alpha\ Centauri = \frac{1}{19,490,000,000} \times Sun$$

On the relative brilliancy of stars see—

Sir J. Herschel ('Results of Astronomical Observations at the Cape of Good Hope,' London, 4to, pp. 367   ).
Zöllner ('Grundzüge einer allgemeiner Photometrie des Himmels,' Berlin, 4to).

On the heat from the stars see Huggins, 'Proceedings of the Royal Society,' 1869, No. 109, p. 309.

§ 264. *Variable Stars.*—The following catalogues may be referred to, in which the known variables are arranged in order of right ascension with a brief description :—

A.D.
1865. Chambers ('M. N.' vol. xxv. p. 209, and 'A. N.' No. 1496).
1875. Schönfeld ('Zweiter Catalog von veränderlichen Sternen : Jahresbericht des Mannheimer Vereins für Naturkunde,' xl.)

§ 265. *Colours of Stars.*—Aratus, in the third century before the present era, had spoken of a red star in his poem 'Phænomena.' Ptolemy mentioned five reddish stars. Blue stars were first mentioned by Marriotte in 1740. Sir W. Herschel found stars of all the elementary colours of the spectrum ('Phil. Trans.' 1782, p. 112 ; 1785, p. 40). The following works may be referred to on the subject of coloured stars :—

A.D.
1847. Sir J. Herschel ('Results of Astronomical Observations at the Cape of Good Hope,' London, 4to, p. 448).

1864. W. H. Smyth ('Sidereal Chromatics,' London, 8vo).

1874. Schjellerup, 'Zweiter Catalog der rothen isolirten Sterne: Vierteljahrsschrift der Astronomischen Gesellschaft,' p. 252, Band ix. 3. und 4. Hefte).

1877. J. Birmingham ('The Red Stars Observations and Catalogue: Transactions of the Royal Irish Academy,' vol. xxvi. No. 7).

§ 266. *Scintillation.*—Robert Hook in 1667 had ascertained that the scintillation of the stars is due to our atmosphere. For the general theory of scintillation see—

Montigny, 'Recherches diverses sur la Scintillation' ('Bullet. Ac. Belgique,' 2ᵉ sér. t. xxv. p. 631 ; t. xxix. pp. 80 and 455 ; t. xxxvii. p. 165 ; t. xxxviii. p. 300 ; t. xlii. p. 255 ; t. xliv. p. 694).

§ 267. *Stellar Spectroscopy.*—The first application of the prism to the observation of the spectrum of a star was made by Fraunhofer (1817–1823), who, by placing a prism before the object glass of a telescope, observed that the spectra of stars are different from the spectrum of the sun, as well as possessing differences among themselves. He described the spectra of several different stars. Amici showed about the same time that the arrangement of the lines is different, even in different white stars.

Many researches have been made on the spectra of stars and nebulæ. The principal are as follows :—

A.D.
1862. Huggins and Miller. Note on the lines in the spectra of some of the fixed stars ('Proceedings of the Royal Society of London,' vol. xii. p. 444).

1864. Huggins and Miller. On the spectra of some of the fixed stars ('Phil. Trans.' vol. cliv. p. 413).

1864. Huggins. On the spectra of some of the nebulæ ('Phil. Trans.' cliv. p. 437).

1866–68 Huggins. Further observations on the spectra of some of the stars and nebulæ ('Phil. Trans.' vol. clvi. p. 381 ; vol. clviii. p. 529).

A.D.
1867. C. Wolf and Rayet. 'Spectroscopie stellaire' ('C. R.' t. lxv. p. 292).

1868. Secchi. A catalogue of spectra of red stars ('M. N.' vol. xxviii. p. 196).

1868. Huggins. On the results of spectrum analysis as applied to the heavenly bodies ('Report of the British Association,' 1868, pp. 140-152). This is a general résumé of the subject.

1872-73. Vogel ('Beobachtungen angestellt auf der Sternwarte zu Bothkamp,' Leipzig, 4to, Hefte i. und ii. Analysis by the author in 'A. N.' No. 1864).

1873-74. Vogel. 'Spectralanalytische Mittheilungen' ('A. N.' Nos. 1963 and 2000).

1874. D'Arrest. 'Auffindung neuer ausgezeichneter Sternspectra, vom iii. und iv. Secchi'schen Typus' ('A. N.' Nos. 2009, 2016, 2032, 2044).

On the method of determining the velocity of a celestial body in the direction of the line of sight by the displacement thereby produced in the lines of the spectra, see Huggins, 'Proceedings of the Royal Society of London,' vol. xxii. (1872), p. 251; Vogel, 'A. N.' No. 1963 (1873); Airy, 'M. N.' vol. xxxvi. (1875), p. 27; vol. xxxvii. (1877), p. 22; and Christie, 'M. N.' vol. xxxvii. (1877), p. 18.

On the photographs of stellar spectra see Huggins, 'Proceedings of the Royal Society,' vol. xxv. p. 445.

On photographs of stars see Gould, 'The Observatory,' edited by Christie, London, 8vo, vol. ii. 1878, p. 13.

§ 268. *Angular Diameters of the Stars.*—The following estimates relate to the *apparent* diameters which, according to Schwerd (1835), vary inversely as the aperture of the telescope :—

A.D.
1782. Sir W. Herschel, 'Phil. Trans.' vol. lxxii. pp. 148, 153, 157) with the lamp micrometer, α Lyræ, $0''\cdot36$; α Tauri, $1''\cdot5$.
1847. W. Struve, α Lyræ, $0''\cdot261$.

§ 269. *Parallaxes of the Stars.*—The following values have been determined for the angle which the radius of the earth's orbit subtends at the distance of the star; α and δ

are the right ascension, and the declination referred to the
epoch 1850. The number in brackets at the end of the line
denotes the magnitude of the star.

A.D.	34, **Groombridge**, $\alpha = 0^h\ 9^m\ 50^s$; $\delta = +43°\ 10'\cdot 3$:	[8]
1867.	Auwers, from comparison of the right ascension of this star with that of the neighbouring stars by the chronograph	$0''\cdot307$
	η **Cassiopeiæ**, $\alpha = 0^h\ 40^m\ 3^s$; $\delta = +57°\ 1\cdot 1$.	[4]
1856.	Otto Struve, by micrometric measurements . . .	$0\cdot 154$
	μ **Cassiopeiæ**, $\alpha = 0^h\ 58^m\ 19^s$; $\delta = +54°\ 10'\cdot 9$.	[5½]
1856.	Otto Struve, by micrometric measurements . . .	$0\cdot 342$
	a **Ursæ minoris** $\alpha = 1^h\ 5^m\ 1^s$; $\delta = +88°\ 30'\cdot 6$	[2]
1817.	Von Lindenau, from 890 right ascensions . .	$0\cdot 144$
1845.	C. A. F. Peters, by vertical circle at Pulkowa .	$0\cdot 078$
1847.	W. Struve, by right ascensions at Dorpat . .	$0\cdot 075$
1847.	W. Struve and Preuss, by right ascensions at Dorpat .	$0\cdot 172$
1847.	C. A. F. Peters, by right ascensions at Pulkowa .	$0\cdot 067$
1847.	Lundhal, from the declinations at Dorpat . .	$0\cdot 147$
	a **Aurigæ** (Capella), $\alpha = 5^h\ 5^m\ 37^s$; $\delta = +45°\ 50'\ 4$.	[1]
1846.	C. A. F. Peters, by meridian altitudes at Pulkowa .	$0\cdot 046$
1856.	Otto Struve, micrometrically	$0\cdot 305$
	a **Canis majoris**, $\alpha = 6^h\ 38^m\ 32^s$; $\delta = -16°\ 30'\cdot 8$.	[1]
1840.	Henderson, by meridian altitudes at Cape . .	$0''\cdot 23$
1864.	Gyldén, by altitudes of Maclear at Cape . .	$0\cdot 193$
1868.	Abbe, by altitudes at the Cape . . .	$0\cdot 273$
	a **Geminorum**, $\alpha = 7^h\ 25^m\ 1^s$; $\delta = +38°\ 12'\cdot 7$.	[1½]
1856.	Johnson, by the heliometer	$0\cdot 198$
	a **Canis minoris**, $\alpha = 7^h\ 31^m\ 27^s$; $\delta = +5°\ 36''\cdot 3$.	[1]
1873.	Auwers, by micrometric measurements . .	$0\cdot 123$
	ι **Ursæ majoris**, $\alpha = 8^h\ 48^m\ 55^s$; $\delta = +48°\ 37'\cdot 6$.	[3½]
1846.	C. A. F. Peters, by observations at Pulkowa . .	$0\cdot 133$
	β **Ursæ majoris**, $\alpha = 10^h\ 52^m\ 45^s$; $\delta = +57°\ 11''\cdot 1$.	[2]
1873.	Klinkerfues, by the displacement found by Huggins in the rays of the spectra	$0\cdot 010$
	Lalande 21185, $\alpha = 10^h\ 55^m\ 10^s$; $\delta = +36°\ 37'\cdot 9$.	[7]
1858.	Winnecke, by micrometric observations . .	$0\cdot 511$
1872.	Winnecke ,, ,, . .	$0\cdot 501$
	ξ **Ursæ majoris**, $\alpha = 11^h\ 10^m\ 10^s$; $\delta = +32°\ 22''\cdot 4$	[4]
1873.	Klinkerfues, by the displacement found by Huggins in the lines of the spectrum	$0\cdot 043$
	Groombridge 1830, $\alpha = 11^h 44^m 19^s$; $\delta = +38°\ 47'\cdot 7$	[6½]

A.D.

1846.	C. A. F. Peters, by micrometric measurements	.	0″·226
1847.	Faye ,, ,, .	.	1 ·058
1848.	Wichmann ,, ,, .	.	0 ·182
1850.	Otto Struve ,, ,, .	.	0 ·034
1854.	C. A. F. Peters, from Schlüter and Wichmann's observations	.	0. 141
1854.	Johnson, by the heliometer . . .	.	0 ·033
1873.	Brünnow, by micrometric measurements .	.	0 ·113
1874.	Auwers, from Johnson's observations . .	.	0 ·009

γ Ursæ majoris, $\alpha = 11^h\ 45^m\ 55^s$; $\delta = +54°\ 31'·7$.   [2]

1873.	Klinkerfues, by the displacement found by Huggins of the lines of the spectrum . . .	.	0 ·016

δ Ursæ majoris, $\alpha = 12^h\ 7^m\ 59^s$; $\delta = +57°\ 52'·0$.   [3]

1873.	Klinkerfues, by the same	.	0 ·024

ε Ursæ majoris, $\alpha = 12^h\ 47^m\ 25^s$; $\delta = +56°\ 46'·5$.   [3]

1873.	Klinkerfues, by the same	.	0. 030

β Centauri, $\alpha = 13^h\ 53^m\ 17^s$; $\delta = -59°\ 38'·7$ [1]

1852.	Maclear, from meridian altitudes . .	.	0 ·470
1868.	Mœsta ,, ,, .	.	0. 213

α Bootis (Arcturus), $\alpha = 14^h\ 8^m\ 49^s$; $\delta = +19°\ 57'·9$ [1]

1846.	C. A. F. Peters, by absolute measurements .	.	0 ·127
1856.	Johnson, by the heliometer . . .	.	0 ·138

α Centauri, $\alpha = 14^h\ 29^m\ 28^s$; $\delta = -60°\ 12'·6$ [1]

1840.	Henderson, by absolute measures at the Cape, 1832-1833 (mean)	.	1 ·16
1842.	Henderson, by absolute measures at the Cape, 1839-1840 . . : . .	.	0 ·9·3
1851.	Maclear, by direct and reflected altitudes .	.	0 ·919
1852.	C. A. F. Peters, from Henderso 's results .	.	0 ·976
1868.	Moesta, measures at Santiago .	.	0 ·88

α Herculis, $\alpha = 17^h\ 7^m\ 49^s$; $\delta = +14°\ 33'·9$.   [3½]

1854.	Jacob, by absolute measures at Madras	.	0 ·60
1858.	Jacob ,, ,, .	.	0 ·061

Œltzen 17415, $\alpha = 17^h\ 37^m\ 18^s$; $\delta = +68°\ 28'·6$.   [9]

1863.	Krüger, by micrometric observations . .	.	0 ·247

γ Draconis, $\alpha = 17^h\ 53^m\ 8^s$; $\delta = +51°\ 30'·5$.   [2]

1817.	Pond, Greenwich observations . . .	.	0 ·051

70 *p* Ophiuchi, $\alpha = 17^h\ 57^m\ 52^s$; $\delta = +2°\ 32'·4$.   [4½]

1859.	Krüger, micrometrically . . .	.	0 ·169
1803.	Krüger, observations in 1859-1862 .	.	0 ·162

α Lyræ, $\alpha = 18^h\ 31^m\ 51^s$; $\delta = +38°\ 38'·8$.   [1]

1817.	Pond, by the mural circle at Greenwich .	.	0 ·007

A.D.		
1840.	W. Struve, micrometrically . . .	0″·262
1846.	C. A. F. Peters, altitudes at Pulkowa. .	0 ·103
1853.	C. A. F. Peters, from O. Struve's observations	0 ·142
1856.	Johnson, by the heliometer . . .	0 ·14
1859.	Otto Struve, micrometrically by distances .	0 ·119
1859.	,, ,, by position angles	0 ·161
1865.	Main, micrometrically	0 ·154
1870.	Brünnow, micrometric measures at Dunsink .	0 212
1873.	Brünnow, by a new series ,, .	0 ·188
	σ **Draconis**, α = 19ʰ 32ᵐ 38ˢ ; δ = + 69° 24′·3 .	[5]
1870.	Brünnow, by micrometric observations at Dunsink	0 ·222
1873.	Brünnow ,, ,, ,,	0 ·246
	α **Aquilæ**, α = 19ʰ 43ᵐ 28ˢ ; δ = + 8° 28″·5	[1½]
1817.	Pond, meridian altitudes at Greenwich .	0 ·160
1834.	Taylor, by meridian altitudes . . .	0 ·978
	61 **Cygni**, α = 21ʰ 0ᵐ 11ˢ ; δ = + 38° 0′·9 .	[5½]
1812.	Arago and Mathieu, by absolute altitudes .	0 ·55
1840.	Bessel, by the heliometer (final result) .	0 ·348
1846.	C. A. F. Peters, absolute measures at Pulkowa .	0 ·349
1849.	,, rediscussion of Bessel's observation .	0 ·360
1853.	Pogson, by the heliometer . . .	0 ·384
1854.	Johnson, by the heliometer . . .	0 ·397
1854.	Woldstedt, from Otto Struve's observations .	0 ·523
1854.	C. A. F. Peters, from Schlüter's observations .	0 ·360
1859.	Otto Struve, micrometric observations of, 1852–1853.	
	,, ,, distances . .	0 ·509
	,, ,, angles of position .	0 ·501
1863.	Auwers, micrometrically . . .	0 ·564
1878.	Ball, by differences of declination at Dunsink .	0 ·465
	3077 **Bradley**, α = 23ʰ 6ᵐ 5ˢ ; δ = + 56° 20′·5	
1873.	Brünnow, by micrometric observations at Dunsink .	0 ·069
	85 **Pegasi**, σ = 23ʰ 54ᵐ 21ˢ ; δ = + 26° 17′·3 .	[6]
1873.	Brünnow, by micrometric observations at Dunsink .	0 ·054

For a list of memoirs and researches on annual parallax
see Knobel, ' M. N.' vol. xxxvi. pp. 384–388.

§ **270.** *Proper Motions of the Stars.*—Ptolemy and the
ancients generally seem to have considered the configuration
of the stars to have been perpetually invariable. It was
only in the 18th century, in 1718, that Halley discovered
displacements in the latitudes of Aldebaran, Sirius, and

Arcturus ('Phil. Trans.' vol. xxxi. p. 736). Jacques Cassini
established the proper motion of Arcturus beyond doubt
by showing that the apparent displacements were not
participated in by the neighbouring star η Bootis. The
following list is selected from that given by Knobel
('M. A. S.' vol. xliii. pp. 62, 63). See also 'M. N.' vol.
xxxvi. pp. 381–384.

A.D.

1775. Tob Mayer, 'Opera inedita,' t. i. Göttingen, 4to, p.
98. Contains the proper motions of 80 stars found by
comparing his observations with those of Römer, which
were made 50 years previously.

1806. Lalande, 'Connaissance des Temps,' 1808, p. 354. 516
proper motions.

1818. Bessel, 'Fundamenta Astronomiæ,' p. 311. In comparing his observations with those of Bradley, Bessel
announced that the proper motions were usually met
with in double stars.

1835. Argelander, 'DLX. stellarum fixarum positiones mediæ,'
Helsingfors, 4to. 560 stars from the observations at
Abo compared with those of Bradley.

1851–1860. Main, in 'M. A. S.' vol xix. p. 136, and vol. xxviii. p. 127.
The proper motions of 1,170 and 270 stars respectively
found by comparing the modern observations at Greenwich with those of Bradley.

1856. Mädler in 'Beobachtungen zu Dorpat,' vol. xiv. (3,222
star s observed by Bradley). This is the most important work on proper motions hitherto published.

1860. Jacob, in 'M. A. S.' vol. xxviii. p. 1. List of 317
stars with large proper motions.

1865–1875. E. J. Stone, in 'M. A. S.' vol. xxxiii. p. 61, and vol.
xlii. p. 129. 460 and 406 stars respectively. The
first deduced from Greenwich observations ; the second,
of southern stars, from observations at the Cape, compared with more ancient ones.

For interesting points relating to the proper motions of
stars, see —

A.D.

1870. Proctor, 'Proceedings of the Royal Society of London,' vol.
xviii. p. 169.

§ 271. *Double Stars.*—The earliest mention of double stars, requiring for their separation the powers of the telescope, is by Jean Dom. Cassini, who in 1678 alluded in this way to β Scorpionis, α Geminorum, and γ Arietis (' Hist. Ac. Sc. Paris av. renouv.' t. i. p. 266). When Bode, in 1781, formed a first list of double stars, he only enumerated 79 (' Berliner Jahrbuch,' 1784, S. 183). The first really systematic labours are those of Sir W. Herschel (' Phil. Trans.' 1782, p. 112). For a list of catalogues of double stars see Knobel (' M. A. S.' vol. xliii. 1877, pp. 21–61), and for an enumeration of works relating to double stars see Knobel, ' M. N. vol. xxxvi. pp. 367–372.

In 1782 Sir W. Herschel commenced the practice of recording the distance and position angle of the two components of a double star. The following are the principal works relating to double stars :—

A.D.
1782–1804. Sir W. Herschel (' Phil. Trans.' lxxii. p. 112, lxxv. p. 40, xciii. p. 339, xciv. p. 353, ' M. A. S.' vol. xxxv.)

1824. Sir J. Herschel and Sir J. South (' Phil. Trans.' cxiv. part iii.)

1837. W. Struve, ' Stellarum duplicium et multiplicium mensuræ micrometricæ.' This is the chief work of the author, containing measures of 3,134 multiple stars. Lord Lindsay has given an index to these measures in ' Dunecht Observatory Publications,' vol. i. (1876). Aberdeen, 4to.

1844–1860. W. H. Smyth, ' A Cycle of Celestial Objects.' London, 1844, 2 vols. 8vo. This is a most interesting work, especially to the amateur, but is now unfortunately very scarce.

1847. Sir J. Herschel, ' Results of Astronomical Observation at the Cape of Good Hope,' London, 4to. Micrometric measures of 2,520 multiple stars in the southern heavens.

1867. Dawes (' M. A. S.' vol. xxxv. p. 164).

1874. Sir J. Herschel, published by Main and Pritchard in ' M. A. S.' vol. xl. p. 1. This contains 10,317 double or multiple stars, forming a general catalogue of all those known up to 1872.

A.D.
1876.   Duner, 'Mesures micrométriques d'étoiles doubles.' Lund, 4to.

1873-1877.   Burnham, 'M. N.' vol. xxxiii. pp. 351, 437, vol. xxxiv. pp. 59, 382, vol. xxxv. p. 31, vol. xxxviii. p. 79. These papers contain measurements of many new double stars.

1879.   Burnham. Double-star observations made in 1877-8 at Chicago with the 18½-inch refractor of the Dearborn Observatory, comprising (1) a catalogue of 251 new double stars with measures, and (2) micrometrical measures of 500 double stars ('M. A. S.' vol. xliv. pp. 141-305).

The following may also be consulted on the appearance and conditions of the more interesting groups :—

A.D.
1835.   E. J. von Littrow, 'Die Doppelsterne,' Vienna, 8vo.
1835.   Sir J. Herschel, 'A List of Test Objects, principally Double Stars, for the Trial of Telescopes' ('M. A. S.' vol. viii. p. 25).
1847-1848.   Mädler, 'Untersuchungen über die Fixstern Systeme,' Milan and Leipzig, 2 parts.
1868.   Webb, 'Celestial Objects for common Telescopes.' 2nd ed. London, 8vo.
1878.   Flammarion, 'Catalogue des Étoiles doubles et multiples en mouvement relatif certain.' Paris, 1878.

Sir W. Herschel, in reobserving double stars after an interval of twenty years, discovered a gradual change in the angle of position in α Geminorum, γ Leonis, ε Arietis, ξ Herculis, δ Serpentis, and γ Virginis ('Phil. Trans.' 1803, pp. 339, 365, 372, 377, 382). Savary computed the first orbit of a binary star, that of ξ Ursæ majoris, in 1828 ('Connaissance des Temps,' 1830).

Orbits of binary stars computed by Doberck and others will be found in Flammarion's work already referred to.

NEBULÆ.

§ 272. *List of Authorities.*—Aratus in the third century before the present era mentions the object in the constellation Cancer, which is now known as the Præsepe Cluster. Simon Mayer (Marius), in 1612, calls attention to the luminous spots in the heavens, which he attributes to collections of stars. Mairan in 1754 first alludes to gaseous materials in such objects. Messier, 1771, distinguished clearly between nebulæ and clusters. The great development of this branch of astronomy is due to Sir W. Herschel, who arranged the clusters in three classes and the nebulæ proper in five others ('Ph. Tr.' 1786, p. 457).

The following are the principal works on nebulæ :—

A.D.

1781-2. Messier, 'Connaissance des Temps,' 1783, p. 225, and 1784, p. 254. Contains lists of 103 of the most remarkable nebulæ in the heavens.

1786. Sir W. Herschel ('Phil. Trans.' vol. lxxvi. p. 471). 1,000 new nebulæ and clusters.

1789. Sir W. Herschel ('Phil. Trans.' vol. lxxix. p. 226). 1,000 new nebulæ and clusters.

1802. Sir W. Herschel ('Phil. Trans. vol. xcii. p. 503). 500 new objects.

1833. Sir J. Herschel, 'Observation of Nebulæ and Clusters' ('Phil. Trans.' 1833).

1847. Sir J. Herschel, 'Results of Astronomical Observations at the Cape of Good Hope.' London, 4to, ch. i. p. 51.

1862. D'Arrest ('A. N.' Nos. 1366 and 1369). List of nebulæ possessing special features, such as variability, duplicity, or proper motion.

1864. Sir J. Herschel. A general catalogue of nebulæ and clusters of stars ('Phil. Trans.' cliv. p. 1). This great work contains the places of 5,079 nebulæ and clusters brought to the epoch 1860·0.

1867. D'Arrest, 'Siderum nebulosorum observationes Havinenses,' Copenhagen, 4to. 1,942 nebulæ.

1867. Marth ('M. A. S.' vol. xxxvi. p. 53). 600 nebulæ which are not represented in Sir J. Herschel's general catalogue.

1867. Vogel, 'Beobachtungen von Nebeosflecken und Sternhaufen,' Leipzig, 1867, 8vo.

1878. Dreyer. Supplement to Sir J. Herschel's 'General Catalogue' ('Trans. R. Irish Acad.' vol. xxvi.) Places of 1,172 nebulæ and notes to the Gen. Cat.

A.D.
1879–80. Lord Rosse, Observations of nebulæ, 1848-78 ('Trans. R. Dublin Soc.' vol. ii.)

On the distribution of nebulæ see—

A.D.
1789. Sir W. Herschel ('Phil Trans.' vol. lxxix. p. 212).
1814. ,, ,, ,, civ. p. 282).

For the descriptions of remarkable nebulæ with drawings see—

A.D.
1844. Lord Rosse ('Phil. Trans.' vol. cxxxiv. p. 321).
1850. ,, ,, ,, cxl. p. 499).
1861. ,, ,, ,, cli. p. 681).
1868. ,, ,, ,, clviii. p. 57).
1867. G. P. Bond ('Annals of Harvard College Observatory,' vol. v.)

For a complete list of drawings of nebulæ up to 1877 see Holden, 'Smithsonian Miscellaneous Collections,' Washington, 1877, 8vo, pp. 78-90.

On the possible changes in the forms of nebulæ see—

A.D.
1877. Holden ('American Journal of Science and Art,' vol. xiv. p. 433).

For the spectra of nebulæ see under the head of Stellar Spectroscopy.

§ 273. *The Milky Way.*—That the Milky Way is an agglomeration of small stars was admitted by the ancients. Galileo adopted the same view when he first saw the Milky Way through the telescope. The constitution of the Milky Way has been carefully studied by Sir W. Herschel ('Phil. Trans.' 1785, p. 258 ; 1814, p. 280). The pole of the Milky Way has been thus determined :—

A.D.
1847. W. Struve, 'Etudes d'Astronomie stellaire,' St.-Pétersbourg, 8vo, p. 62.
$$\alpha = 12^h\ 38^m;\ \delta = +31°\cdot5\ ;\ \text{Equin. } 1825\cdot0.$$
1878. J. C. Houzeau ('Annales astronomiques de l'Observatoire de Bruxelles,' t. i. Uranom. p. 21).
$$\alpha = 12^h\ 49^m\cdot1\ ;\ \delta = +27°\ 30'\ ;\ \text{Equin. } 1880\cdot0.$$

§ 274. *The Zodiacal Light.*—This phenomenon does not appear to have been noticed by the ancients. Humboldt says that the Mexicans had detected it in 1509. Childrey was the first to mention it in Europe (1661). Jean Dominique Cassini discovered the zodiacal light independently, and commenced a scientific study of it in 1683. One of the best general descriptions of the zodiacal light is that of Argelander, 'Schumacher's Jahrbuch,' 1844, p. 148. See a.so Jul. ,F. Schmidt, ' Das Zodiacallicht : Uebersicht der seitherigen Forschungen nebst neuen Beobachtungen über diese Erscheinung in den Jahren 1843–1855,' Brunswick, 8vo.

On the spectrum of the zodiacal light see—

A.D.
1872. Angström, in ' Roscoe Spectral Analysis,' London, 8vo.
1872. Vogel ('A. N.' No. 1893).
1875. Arthur W. Wright ('American Journal of Science and Arts,' 3rd series, vol. viii.)
1877. Piazzi Smyth, ' Edinburgh Astron. Obs.' vol. xiv.

On the various hypotheses which have been advanced to explain the zodiacal light see—

A D.
1799. Laplace, ' Mécanique céleste,' t. ii. p. 169.
1807. Thomas Young, 'Nat. Phil.' London, 2 vols. 4to, vol. i. p. 502.
1835. Arago, 'Astron. popul.' Paris, 1855, 4 vols. 8vo, t. ii. p. 183.
1843. J. C. Houzeau, ' A. N.' No. 492.

INDEX.

Index.

4
5
9# WOL

Wol
f and Rayet, stellar spectroscopy,
432
Woll
aston, albedo of the moon, 387
-- brillian
cy of the moon, 385
— bril
liancy of Si
rius, 430
Wright, A. W., spectrum of z
odiacal
light, 441
Wurm, mass of Uran
us, 414

YEAR, Juli
an, 388
— le
ngth of, 136, 372
— tropi
cal, variation of. 37
3
Young, consti
tution of the sun, 365
— spe
ctrum of protuberanc
es of the sun,
351
Young, Th
omas, theory of the zodiacal
light, 441 .

ZENGER, brilliancy of Jupiter's satellites, 397

ZON

Zen
ith, 69
Zenith dist
ance, 95
Zenit
h distance of sun, means of finding,
113
Zenker, physical theory of com
ets, 424
Zodiac, the, 251
Zod
iacal light, 440
Z
öllner, albedo of the moon, 3
— albedo of Saturn, 405
— bri
lliancy of Jupiter, 397
— bri
lliancy of Mars, 390
— b
rilliancy of Mercury, 368
— brilliancy of the moon, 387
— bri
lliancy of Neptune, 418
— br
illiancy of Saturn, 405
— bril
liancy of stars, 429, 430
— bril
liancy of Uranus, 415
— cons
titution of the sun, 365
— phy
sical theory of comets, 424
— rot
ation of the sun, 364
Zon
es of the earth, 149-152

www.ingramcontent.com/pod-product-compliance
Lightning Source LLC
Chambersburg PA
CBHW020904210326
41598CB00018B/1762